LIFEBUOY MEN,

LUX WOMEN

BODY, COMMODITY, TEXT

Studies of Objectifying Practice

A series edited by

Arjun Appadurai,

Jean Comaroff, and

Judith Farquhar

LIFEBUOY MEN,

LUX WOMEN

Commodification, Consumption,

and Cleanliness in Modern

Zimbabwe

TIMOTHY BURKE

Duke University Press

Durham & London

1996

Fourth printing, 2003
© 1996 Duke University Press
All rights reserved
Printed in the United States of America on acid-free paper ∞
Typeset by Keystone Typesetting, Inc. in Minion
Library of Congress Cataloging-in-Publication Data
appear on the last printed page of this book.

CONTENTS

ACKNOWLEDGMENTS

The writing of this study, as with most scholarly works, was made possible only through the kindness of strangers and the generosity of friends. In particular, this work could not have been completed without the advice and assistance of David William Cohen. His tireless enthusiasm and interest in this study, and in my progress as a graduate student generally, made everything possible. I owe him for the quiet integrity of his support and for his profoundly enlightened approach to academic practice, which always seeks to empower every student to succeed on his or her own terms.

I must also thank my other teachers, fellow students who formed a sort of University of the Witwatersrand extension school for their fortunate American colleagues. Carolyn Hamilton, Catherine Burns, Keith Breckenridge, and Keith Shear have taught me a great deal over the last six years about our common subject matter and about everything else under the sun besides. Their selfless counsel while this work took shape was irreplaceable and precious. I have relied heavily on their good judgment and wise advice in the past, and will doubtless do so in the future. Many other professors and students from Johns Hopkins University also helped develop this study. Foremost among them are Sara Berry, whose support in a time of need was deeply appreciated. Jonathan Sadowsky, Carol Summers, and Garrey Dennie have all provided useful advice and insightful critiques. Norma Kriger, Michel-Rolph Trouillot, and Gillian Feeley-Harnik also provided invaluable readings of this manuscript while serving as members of my dissertation committee.

My research in England was immensely rewarding largely because of the assistance of enthusiastic strangers. In particular, I am grateful to Shula Marks, whose immediate interest in and later promotion of this project filled me with great self-confidence at an uncertain time. I am also indebted to Terence Ranger, whose innumerable contributions and practical counsel are now interwoven everywhere into the fabric of this work, both early and late in its construction. Other important assistance was provided by Megan Vaughan, whom I met all too briefly but whose remarks were substantial and helpful; Margaret Pelling, who filled some gaps in my understanding of hygienic history; Tim Marchant, whose advice and friendship made my work in London possible in more ways than one. I must thank the Unilever Archives for allowing me to examine some useful records and documents.

I must thank also the patient and highly skilled staff of the Zimbabwe National Archives; Professor Victor Machingaidze, whose many struggles with bureaucracy on my behalf are deeply appreciated and whose quiet and substantive advice on the design and conduct of this work was indispensible; the Department of Economic History at the University of Zimbabwe, whose members impressed me with their intellectual vigor and helpful spirit; and fellow researchers Timothy Scarnecchia, Randall Smith, Volker Wild, Lynette Jackson, Martha Lane, Steven Robins, and Patrick Bond. I was impressed by and am thankful for the honesty and intelligence displayed by the manufacturers and advertisers that I had the opportunity to speak with. Numerous individuals in Harare and Murewa who were kind enough to receive pesky researchers into their homes also made this study possible, and I can only hope that something of their own interests and experiences comes through in this work.

Colleagues at Northwestern University, Rutgers University, and Emory University have provided both specific comments on this work and general encouragement. Ivan Karp and Corrine Kratz in particular have provided a great deal of meaningful advice and thoughtful reflections on this work and on academic work in general. Scholars who played some role in my writing and revision of this work have included Misty Bastian, Mark Auslander, Debra Spitulnik, Jean Comaroff, Belinda Davis, Paul Landau, Jane Parpart, Michael O. West, Erica Rappaport, Luise White, Jim Campbell, Jeff Rice, Barbara Moss, Elisha Renne, Victoria de Grazia, Belinda Davis, Aisha Khan, Karen Tranberg Hansen, and Nancy Rose Hunt. My colleagues in the Department of History at Swarthmore College provided a crucial endorsement of this book by hiring me, and their enthusiasm and support has helped me greatly through the final stages of its preparation.

James Ferguson provided me with an astonishingly detailed road map for the revision of this study: he saved me from countless small humiliations, and I am deeply in his debt. I am also grateful for the work of my other readers, who provided invaluable assistance with the revision of my original dissertation manuscript.

Tuso Tapera, who made possible many of the interviews that are invaluable to this work, deserves more than thanks. She has a tangible share in the authorship of this work. By helping to record the testimonies on which a small but indispensible portion of this work relies, she helped to write this study.

Last, but most emphatically not least, I must thank Melissa Mandos for continual support, patient understanding, and an informed reading of this text. She has enriched my character and my goals, my sense of the world and my place within it.

INTRODUCTION

A commodity appears, at first sight, a very trivial thing, and easily understood. Its analysis shows that it is, in reality, a very queer thing, abounding in metaphysical subtleties and theological niceties.—Karl Marx

In Africa, you want more, I think. People get avid. This takes different forms in different people, but it shows up in some form in everybody who stays there any length of time. It can be sudden. I include myself. Obviously I mean whites in Africa and not black Africans. The average black African has the opposite problem: he or she doesn't want enough.—Norman Rush

Dambudzo Marechera, whose disturbing fiction both illuminated and obscured the experience of Africans under white rule in colonial Zimbabwe, once wrote a short story about a "new magazine" looking for articles on "a Modern African Family" who "must be seen to consume the products and manufactures of white civilization."[1] Marechera derides the vision conjured up by the editors of this only barely imaginary magazine, but with his characteristic bitter self-mockery, slowly maps himself into the construct of the "Modern African Family":

What would the ideal modern black home be like? . . . Foam rubber seats and cushions and poufs and couches and folding beds? On a use-now-pay-later basis, of course. A television. A giant radiogram. And on all the walls, portraits of such heroes as Cecil Rhodes and Chief Tugela. I would have a complete wardrobe of clean underwear. Clean socks, for once. And we would polish everything with Mr. Shine. And eat rice and roast chicken until we felt sick of it. My eldest son would be a credit to the family. Three cars in the garage at the back. And house-servants to clean up after us everywhere we went in the house. Advertisement agencies would desperately want us to pose for photographs recommending Ambi

Skin-Lightening Cream, Coca-Cola, Castle and Lion Lagers, Benson and
Hedges, Pure Wool Suits, and, yes, Fanta Orange Tastes So Good . . .
We would consume the Christian religion until our jaws ached. We
would consume chunks of sermons, chunks of earnest prayers, and consume to the hilt the knowledge of our station in the human hierarchy. . . .
We would consume every sugared stick of Family Planning and screw each
other only when there was a sword between her loins and my loins. We
would utter speeches condemning the practice of polygamy, the evils of
lobola, the superstitions of magic and witchcraft, and, in short, cast out—[2]

This study focuses on one southern African nation, Zimbabwe, and takes up
the challenge of understanding the nature and role of commodities, consumption, and needs in its modern history. Marechera's mockery of the equation of
colonialism with modernity and consumption underlines an area of historical
experience that scholars have long understood to be important but that nevertheless has remained largely unexplored. For example, experts on the Atlantic
slave trade, from Walter Rodney to David Eltis, have argued that the specific
nature of the goods imported by Africans are somehow crucial to an overall
understanding of the slave trade.[3]

Similarly, studies of modern colonialism have frequently suggested that
consumption played an important role in the development and maintenance of
colonial domination. For example, canonical analyses of the construction of
Zimbabwe's political economy from 1890 to the 1950s, most notably by Ian
Phimister and Giovanni Arrighi, have identified the appearance of "new needs"
among Africans as one of the key transformative processes connected with colonial rule. Phimister writes: "The advance of commodity relations was quick-
ened and the 'material reproduction of natural economy' further undermined
in almost every case by the interaction of two processes. One was the widespread withdrawal of labour from both use-value production in agriculture and
in craft activity, and from surviving networks of local and regional trade. The
other was the development of new needs."[4] Arrighi similarly argues: "In analyzing the process whereby the sale of labour-time became a necessity for the
African population of Rhodesia, attention must be focused upon . . . the transformation of 'discretionary' cash requirements into 'necessary' requirements."[5]
The study of transnational political economy and the development of the multinational corporations has also been discussed in several disciplines under the
heading of "taste transfer."[6] Those scholars who have asserted the importance
of such experiences in southern Africa or in the modern world-system generally have usually failed to investigate vital questions: How do new needs de-

velop? What makes a luxury into a necessity? What causes "tastes" to "transfer"? What changes the relationships between things and people? How do people acquire deeply felt and expressed desires for things they never had or wanted before?

Zimbabwe is a particularly apt setting for an investigation of these questions. In Zimbabwe, as in Nigeria, South Africa, and Kenya, capitalist manufacturing went through particularly rapid and significant expansion at the end of World War II. This expansion was accompanied by a wave of new activity in advertising, promotion, and marketing directed at the indigenous peoples of the region and a simultaneous growth in the range and type of manufactured goods available to and "needed" by African consumers. Most characterizations of what has variously been called "modern," "monopoly," or, more optimistically, "late" capitalism agree that advertising and cultural hegemony have become increasingly crucial to its functioning. This is no less true in Zimbabwe's historical experience with modern manufacturing capital and advertising. In addition, Zimbabwe's unusually late independence from colonial rule means that its secondary industry developed under conditions of formal racial segregation while the global economy transformed itself after World War II. Alongside the particulars of Zimbabwe's history and social structure, the above factors give this study potential comparative significance while also granting it meaningful historical specificity.

A number of recent studies of advertising—as well as other studies of the relationship between commodity culture and modern capitalism in the history of European and North American societies—have provided some suggestive hints about how an analysis of commodification in colonial Zimbabwe should proceed. In particular, Judith Williamson's work *Decoding Advertisements* contains a crucial insight that has helped determine the basic structure of this book: "the subject drawn into the work of advertising is one who knows. To fill in gaps we must know what to fill in, to decipher and solve problems we must know the rules of the game. Advertisements clearly produce knowledge . . . but this knowledge is always produced from something already known, that acts as a guarantee, in its anteriority, for the 'truth' in the ad itself."[7] This point is fundamental to the structure of this study. The forms of knowledge, subjectivity, identity, and consciousness produced by or through the process of commodification and the transformation of needs in Zimbabwe cannot be understood without a detailed map of "prior meanings"—the cultural and social raw material from which "the social life of things" was shaped. To understand the making of consumption and material culture, it is necessary to investigate the processes by which goods have acquired their significance. To investigate these processes, it is also necessary to examine the colonial political economy and the general functioning of cultural hegemony under colonial rule. As Arjun Appadurai comments, "commodities represent very complex social forms and distributions of knowledge."[8] It is therefore crucial to somehow limit the focus of this work.

While much of this study will be concerned with the general development of commodification, I have chosen to focus specifically in my analysis on goods made for use on the body, toiletries like soap, Vaseline-type body lotions, skin lighteners, cosmetics, shampoos, perfumes, and deodorants. This may at first glance appear to be a trivial, or as Jan Vansina has suggested, "nice" (and thus irrelevant) group of goods to focus on.[9] Part of the justification for such a focus is that toiletry production has historically been one of the most developed sectors of light manufacturing in southern Africa. Additionally, a very significant proportion of advertising and promotional activity aimed at Africans by businesses from the 1950s to the 1970s was focused on toiletries. At the same time, because toiletries work on or through the body, they function in one of the most intensely contentious aspects of modern identity. The material body is often in plain sight, but the social body, as an artifact of the self and a canvas for identity, is both indispensible and invisible.

But in the end, I must stress that such a focus is also basically heuristic. A successful investigation of the general role of commodities and consumption in shaping distinctive forms of consciousness in twentieth-century Zimbabwe necessarily demands close attention to particular commodities, particularly given the absence of a secondary literature on the social history of consumption in Africa, but I do not want to misrepresent this essentially tactical choice. Toiletries are not a coherent and separate group of commodities: the history that has given them distinct meaning manifests itself in other goods as well, such as clothing, and the consciousness and identities produced through toiletries, or any other goods, have been part of many other aspects of social life and practice in modern Zimbabwean history. Moreover, the different toiletries discussed in this work all have their own peculiar history, which I discuss in detail in chapter 6.

As a consequence, this book will follow what may seem to some readers to be a disjointed structure, dealing first with the social history of hygiene, then with the development of merchant capital and manufacturing in the colonial political economy, and finally with advertising, post–World War II commodity culture, and the role of consumption in shaping African identities in Zimbabwe. The final two chapters are the realization of the arguments and materials covered in the previous chapters, and as such may draw the primary attention of many readers. I believe, however, that this book clearly demonstrates that the exploration of postwar commodity culture contained in these culminating chapters necessarily depends upon the exposition offered in the first four chapters.

Commodity Fetishism, "False Needs," and Use-Value

One of my hopes in this book is to bring political economy, cultural studies, and critical theory into productive dialogue with each other. In service to this goal, I will devote the rest of this introduction to an exploration of some of the most important theoretical considerations that affected the writing of this study.

In trying to frame the history of commodity culture in Zimbabwe, I was particularly drawn to the concept of *fetishism*. This term comes with both advantages and drawbacks. Marx recognized that the concept of commodity fetishism was a singular intellectual "discovery" even as he also admitted that it was the single aspect of his theoretical work whose implications he himself did not fully understand.[10] The richness of Marx's exposition of fetishism has attracted generations of radical scholars and political activists, tantalizing them as the possible wellspring of a fully realized marxian theory of ideology, culture, agency, or consciousness. More than a few critics have suggested that some of these efforts have been pursued "at the price of theoretical incoherence."[11]

My own interest in the term does not stem from a crypto-theological desire to render theory from an original source. Rather, I am simply attracted by the rich potential of Marx's argument that commodities are able to assume an independent life, that relations between things—and between people and things—accompany, conceal, or displace the actual state of relations between people. This leaves open the question of how commodities are meaningful in noncapitalist societies, which has invited among other things the problematic long-term distinction between "gift" and "commodity" in anthropological scholarship.[12] Marx felt that fetishism was particular to capitalism, that it was the key engine of the alienation of the producer from his or her labor, which he identified as the unique consequence of the capitalist mode of production. Clearly, however, goods had social meanings and social power in precolonial southern African societies and in other noncapitalist cultures. Fetishism, then, is more than (but includes) the meanings invested in goods; it is also the accumulated power of commodities to actually constitute, organize, and relate to people, institutions, and discourses, to contain within themselves the forms of consciousness through which capitalism manufactures its subjects. The anthropological perspective of Mary Douglas and Baron Isherwood's *The World of Goods* is thus too general, too vague. Douglas and Isherwood see goods as possessing no inherent characteristics: they are instead imbued with characteristics only through being part of culture. "Goods are neutral, their uses are social; they can be used as fences or bridges," they write.[13] As Jean Comaroff and John Comaroff point out in a recent article, "it is just this, the fetishism of things, that is most conspicuously absent from recent anthropological accounts of the nature of objects in the social world."[14]

Yet fetishism in this sense is only a starting point. Marx says little about the mechanisms that make fetishism. This is a key issue, because clearly not all commodities are equally fetishized or equally powerful or constitutive in their social meanings. Here, the "biography" of a commodity, as Igor Kopytoff has termed it[15]—its "prior meanings"—clearly is what gives various goods their rich individuality within a specific place and time. For each commodity there is "a peculiar aptitude for abstracting and congealing wealth, for making and breaking meaningful associations."[16]

This sense of fetishism, to my mind, goes far beyond Marx's original sense of the term. After all, the institutions in modern capitalism and in modern culture that most centrally deal with the meaning of goods—advertising in particular—were unfamiliar to Marx. Wolfgang Fritz Haug's masterly treatment of "commodity aesthetics" has done much to build on the foundation laid in *Capital* by its appreciation of such institutions, their powers, and their limitations,[17] as have a number of other historical and critical studies of advertising.[18] The work of Jean Baudrillard, though often problematic, has also contributed an important concept, "sign value," which adds to an understanding of how the meanings of goods are produced and, more importantly, how the making of meaning has become central to the power of modern capitalism. Baudrillard suggestively argues that the capacity to manipulate goods-as-signifiers has itself become critical to modern capitalism's ability to generate and control surplus value.[19]

It seems to me that Marx's definition of commodity fetishism does not leave sufficient room for the complexity of the relations between things and people, room for the imaginative possibilities and unexpected consequences of commodification, room for the intricate emotional and intellectual investments made by individuals within commodity culture. Marx's analysis in *Capital* has engendered an interpretive tradition that sees fetishism as a process by which "false needs" are made and "real" relations concealed by the conscious agency of ruling classes.[20] This is a tradition I more or less oppose throughout the body of this work. I willingly accept the basic postulate that lurks behind the idea of "false needs," that the identification and pursuit of needs by consumers in a capitalist mode of production necessarily takes place under a condition of domination. However, arguing that all felt needs under capitalism are thus false, merely *appearance,* shadows that would blow away in a just society, seems to me to be both politically naive and intellectually impoverished: politically naive because such an assumption imagines that any new social dispensation *could* sweep away the accumulated historical meaning that each individual commodity represents; intellectually impoverished because it relies on a theory of power that is monolithic and simplistic. If we turn instead to the Foucauldian sense of power—power as both restricting and productive—we begin to see that the uses and understandings inherent within a commodity in a given time and place are likely to be the result of the intersections of macropowers and micropowers, the partial and challenged hegemony of rulers and the episodic creativity of the ruled, the logics and disjunctures of everyday life. As Nikolas Rose puts it, fetishisms have "a certain history, are inscribed in certain practices and are the site of struggles whose nature, effects and outcomes cannot be simply deduced functionally from certain concepts of the relations of production. . . . To think these struggles within the forms of reduction proposed by the figure of fetishism [in Marx] is to abandon the possibility of theorising them in their own specificity, to reduce them to the empty banality of a single cause, to effectively exclude engagement with them at an appropriate political level."[21] There may of course be a danger here, which James Ferguson and Akhil Gupta have identified as "*celebrating* . . . those 'consumers' . . . who fashion something quite different out of products marketed to them, reinterpreting and remaking them, sometimes quite radically, and sometimes in a direction that promotes resistance rather than conformity. The danger here is the temptation to use scattered examples of the cultural flows dribbling from the 'periphery' to the chief centers of the culture industry as a way of dismissing the 'grand narrative' of capitalism . . . and thus of evading the powerful political issues associated with Western global hegemony."[22] As Bill Livant asks, "Is it just possible that such an account of commodity fetishism as the basis of consumer culture is itself a theory which is a fetish of the commodity?"[23] This is precisely why retaining some sense of Marx's use of fetishism is important, to ensure that an analysis of the interpretations and uses of manufactured commodities always returns—though perhaps indirectly—to the *political* problem of colonial rule and capitalist domination in Zimbabwe's modern history.

In a century of marxian and radical scholarship on the problem of human needs, discussion of the idea of commodity fetishism has been closely mingled with an assessment of the concepts of use-value and exchange-value. These are very deep intellectual waters indeed, filled with lurking monsters of scholarly esoterica, and I can only hope to raise a few simple points of importance. If we dismiss the identification of commodity culture under capitalism as "false needs," then what becomes of Marx's distinction between value seated in an object's fixed utility to human beings and value that derives from the architecture of circulation in a given society? Under capitalism, Marx implies, the "real" value of a commodity's utility is increasingly wrapped up within a socially "real" but alienating value accruing to the commodity through its passages in a capitalist system of exchange.

In some ways I share feelings of ambivalence expressed by Megan Vaughan in her recent history of biomedical discourse in African history. As she notes, the "social constructivist" position with regard to medicine has so many obvious explanatory virtues that it is easy to overlook or finesse its most obvious difficulty, namely, do people really get sick?[24] I am drawing the problem crudely, no doubt, but the dilemma remains. I have adopted in this study an account of use-value that questions the absolute status assigned to it by Marx; I offer a history in which the fundamental utility of the object itself is subject to change, invention, or eradication. Yet there is an absurdity hidden in this perspective, one that *Capital* tries to address precisely by making the distinction between exchange-value and use-value. Goods are *not* pure free-floating signifiers; they are not blank slates upon which history and power can write freely.[25] They have concrete material qualities which limit and prescribe their uses and their nature. On some level, food is for eating, soap is for washing, clothes are for wearing. The limits of utility are far broader than contemporary Western "common sense" about objects might seem to permit, but they do exist. However, since tautology (use-value = what is useful) is the only way to systematically define and separate absolute utility from the cultural making of utility, it is better not to make the attempt. Haug writes: "In all commodity production a double reality is produced: first the use-value; second, and more importantly, the appearance of use-value,"[26] to which I would add a corollary: that this appearance of use-value cannot be simplistically revealed or dismissed. Like all such "appearances," it is a social reality. We simply need to accept that an absolute or essential utility for goods *must* implicitly exist, but that it can be profitably discerned only in *specific* historical and cultural situations.

"Common Sense," Hegemony, and the Complexity of Desire

Another crucially useful feature of Marx's idea of commodity fetishism is his description of the production and reproduction of commodity culture as a process that covers its own tracks. The properties ascribed to each thing, and the power of the thing to act upon other things and upon people, rapidly become "natural" properties with no pedigree or fixed location. This account is related to the view of "common sense" found in Antonio Gramsci's work on the nature of hegemony. For Gramsci, "common sense" was a manufactured artifact consisting of dicta lodged in the public memory of a given society that everyone "knows" to be true, dicta that act to reproduce relations of domination as "natural" relations, thus serving the interests of ruling classes.

Locating the work of fetishization in "common sense" has important consequences for the historian of commodification, particularly in Africa and other non-Western societies. Some of these are practical and methodological. It is difficult to investigate how and when a commodity became "needed" when everyone in a society seems to agree that it has always been needed. It is hard to investigate the history of a commodity when its meanings are all hidden in the unspoken parts of everyday life. This is an instance of the intellectual dilemma that David William Cohen and E. S. Atieno Odhiambo have termed "the anthropology of the shadows."[27] In interviews with Zimbabweans, I did not find any particular reluctance to discuss consumer preferences or associations, but only some goods have actually attracted conscious and explicit interest. However, "common sense" also leaves marks of its production everywhere it touches, outlining a zone of eloquent silence that can be studied and used to more completely delineate the biography of a given commodity.

Understanding commodities as a form of "common sense" in the Gramscian sense of the term implicitly characterizes fetishisms as the product of hegemonic power and the reinvention or misrecognition of the social meaning of goods as inevitably a form of resistance. This is, in Zimbabwe at any rate, a useful perspective on the history of modern commodity culture, but it is incomplete. Left in this form, it invites the shibboleth of "false needs" to creep back into the picture by reducing the interpretation and use of manufactured commodities to a purely functional and straightforward mechanism of domination. One major part of the problem here is the "orientalist" model of hegemony, the idea that the production of knowledge by the powerful to aid in their projects of domination works in unified and coherent ways to achieve predictable objectives. In reality, power exerted by and upon commodities is a much more complicated phenomenon, touching not just on the major structural oppositions in a given society but also on the invocation of multiple identities and forms of social difference in the smallest details of daily life.

In analyzing the commodity culture of Belize, for example, Richard Wilk argues that "by condemning third-world consumption as emulation or imitation, we denigrate the creative and expressive capabilities of those people to take and *use* foreign goods for their own purpose. At the same time, it is clear that people are not making completely free choices about goods."[28] In his work *All That Is Solid Melts into Air*, Marshall Berman analyzes Goethe's *Faust* as the emblematic "tragedy of development," an archetypical fable of modernity. Among Berman's insights into the work—and into modernity itself—is that those who interpret the character of Faust as "the primal 'Growthman' who would tear the whole world apart for the sake of insatiable expansion" effectively turn "tragedy into melodrama."[29] For Berman, the tragic accomplishment of Faust is that the world he transforms is desired by the individuals who come to inhabit it, even when the price of their new world is death, madness, or betrayal. Similarly, if I were to render the use and interpretation of commodities in Zimbabwe *simply* as an arena for monolithic struggle between oppressors and oppressed, I would potentially miss the richness and multiplicity of Africans' interest in manufactured things and thus ignore the vivid reality of desire itself. Consumption did not just mechanically reproduce a singular material repertoire—a composite "coat of many commodities"—for each class or social group in Zimbabwean society. The consumption of commodities was also shaped by individual acts of will and imagination, engagement and disinterest. It was also a part of struggles and practices that were tangential or peripheral to the central political and social issues of colonial life.

Bodies, Commodification, and Colonial Subjectivity

Thus, commodities in Zimbabwe were both concrete material expressions of colonial capitalism's resources for domination and testimony to less visible and more uncertain changes in identity. Megan Vaughan's recent reflections on colonial medicine are a useful starting point for considering some of the issues raised by the relationship between commodification and African subjectivity in Zimbabwe. Vaughan considers the question of "Foucault in Africa": namely, does the model of "bio-power," which Foucault derived from his studies of the genealogies of medicine, punishment, and sexuality, make sense in an African context? At least with regard to the efforts of colonial health officials, propagandists, and medical experts, Vaughan concludes that for the most part, "bio-power" in this sense is not an adequate description of the action of colonial medicine on African societies. She points out that "bio-power" in the Foucauldian sense is defined by its ability to generate and infiltrate an individual and produce its effects from within the individual. Restraints, disciplines, and punishments are generated within the sphere of the self. Vaughan concludes that though colonial biomedical discourse might well have devoted lip service to transforming the subjectivity, the inner being of Africans, in order to re-produce the practices characteristic of the "modern" body, in actual practice, health officials, doctors, filmmakers, and other colonial authorities were too dedicated to seeing Africans as an ineradicably mysterious "Other" who could not truly be understood.[30]

Commodification in Zimbabwe, on the other hand, was shepherded along after World War II by a conglomeration of capitalist functionaries and state officials whose interest in penetrating what they called the "African market" necessarily demanded that they try to change the nature of African selfhood. Their dedication to and understanding of this objective was not always consistently maintained or pursued, and their relative "success" was determined by factors and struggles beyond their control or understanding. What I want to consider briefly at this point is whether toiletries in particular, by virtue of their connections to the body, had any special capacity for remaking subjectivity. In other words, is the body in any way a uniquely powerful site in the process of commodification?

Haug's description of the modern apparatus for producing "commodity aethetics" argues that there is an intense and special connection between sen-suality, the body, and the power of modern commercial discourse. In one typical passage, he writes:

What is being thrust on the public is a whole complex of sexual percep-tion, appearance, and experience. Since capital producing underpants is aiming for a niche in a profitable market, underpants are necessarily in the spotlight. Since they must be saleable at monopoly prices, they must be shown in the right light, and thus, once again, the body is emphasized. . . . the advertised underpants are made into a "hit" by underlining the fact that they make the body a "hit." They become the body's snug marketing package, and thus the concerns of capital appear, as it were, in underwear and, by seeming to take an interest in the body, promote commodities by suggesting that they advertise the body itself. The body, on whose behalf all this advertising is happening, adopts the compulsory traits of a brand-named product; in the same way, it is not the body itself but the effective advertising image which is being promoted. Capital's interest in the body even contains certain aspects which are rather more detrimental to it than the Christian aversion to the flesh, which capital propagates as its mission-ary in a different constellation.[31]

The beginnings of the visual culture of modern consumption in the West were certainly replete with bodily images. In nineteenth-century Britain, for exam-ple, a hugely disproportionate share of advertising was aimed at the promotion of patent medicines and toiletries. Critical readings of twentieth-century advertising have pointed out that this trend has actually accelerated. In the twentieth century the pleasures of consumption as represented by manufacturers have been increasingly and explicitly tied to the satisfaction of the body and its hungers.

Given that commodification and advertising in Zimbabwe were very strongly shaped by transnational practices, it is not surprising to discover that this thematic stress on the consuming body reproduced itself in an African context. However, this still leaves the question of whether the body is a distinctive site for the production of identity and selfhood. A number of scholars have suggested that in Foucault, the body surreptitiously creeps back in to establish itself as the fundamental grounding for a new radical politics of identity. Judith Butler, for example, argues that the "boundary of the body," its unstable surface, is a uniquely important locus for social and political activity.[32] However, I feel that if we pursue the history of "the body" too avidly, we risk separating individuals from their bodies, seeing, for example, the bodies of women as separate from the selves of women.[33] This, in fact, is one of the key reasons the first two chapters of this study are devoted to exploring colonial ideals about hygiene, domesticity and manners, and indigenous practices centering on the body, so that I can integrate an understanding of how commodities worked through ideas about the body with an understanding of how consumption defined and refined colonial personhood. The consumption of toiletries in Zimbabwe—and their relationship to "modern" African bodies—cannot be understood without establishing the *historical* weight of these "prior meanings."

Explanatory Notes: Terminology and Interviews

Before I proceed to this discussion of cleanliness and domesticity in colonial Zimbabwe, there are a few brief points about this text that I should note. Like most Africanists, I face a difficulty in deciding how to refer to the places that I study. In the main, I regard the names used since independence as retroactively applicable: Zimbabwe, Harare, Masvingo rather than Southern Rhodesia, Salisbury, Fort Victoria. In some cases, I use the modifier "colonial" to make it clear that my reference is to the period from 1890 to 1979. In my discussions of precolonial society, I also make reference to the "Zimbabwe plateau" as a distinctive region whose approximate boundaries resembled but were not identical to the present-day national borders. In referring to the inhabitants of colonial Zimbabwe, I often use the words "white" and "African," occasionally "settler" or "European" and "black." In addition, when referring specifically to white institutions and culture and to the colonial state, I have sometimes used the term "Rhodesian."

The interviews used in this work were collected from November 1990 to May 1991 and fall into two groups. The first consists of interviews done by me with Zimbabwean and South African manufacturers and advertisers. The second is made up of interviews conducted by me and my assistant, Ms. Tuso Tapera. The testimonies solicited in these interviews are an invaluable part of this study. In referring to information from the first group of interviews, this study uses the pronoun "I"; with the second group, the pronoun "we" is used in order to reflect the joint nature of the "authorship" of such testimonies.

The design of interviews for this project posed some important difficulties. Advertisers and marketers were a natural source of information, but how does one solicit historical testimonies about commodities from consumers themselves? The narrowly utilitarian and "scientific" methodologies used in most market research were completely unacceptable models for my own inquiries. David William Cohen has urged Africanists to pursue "the undefining of oral tradition" by considering all members of a society as valid sources for historical knowledge differentially derived from their individual access to "the intelligence of everyday life."[34] In my own case, I faced a relevant dilemma: *all* Zimbabweans have experienced the history of commodification, but none of them, save certain executives, have any particularly privileged viewpoint to offer. Most Africanist histories, even if they open themselves to a broad range of testimonies, are still able to exercise some selectivity in identifying informants with institutional or historical experiences of special importance to their inquiries. My interests, on the other hand, lay entirely and amorphously within "the intelligence of everyday life."

As a consequence, Ms. Tapera and I solicited interviews simply by walking through the oldest of Harare's townships, Mbare. We selected at random households in the lower-income area of mixed shanties and permanent houses known to residents as "Joberg" and in the working-class area of settled houses known as the "Nationals." At these homes, we had conversations with individuals and with groups varying in size from two to ten people, of all ages, usually women but sometimes including men. Ms. Tapera also conducted a small number of independent interviews at or around her family's home in the middle-class suburb of Westwood in Kambuzuma Township. Both of us also traveled to the village of Murewa, to the northwest of Harare, and conducted a number of interviews with rural families there. I have no desire to make rigorous "scientific" claims for the representativeness of these interviews. We did try to solicit testimony from rural and urban individuals of both genders, of many ages and income groups, and from people with varying backgrounds, as we did so with the aim of listening to as many distinctive individual voices as possible. In this respect, our interviews resembled to a limited extent the use of "focus groups" by marketers, but with an entirely different intent and shape. The purpose of these interviews was to submit this evolving work to the scrutiny of those who have experienced the history that I was studying. Adding these voices to my archival research made me question my assumptions and added an infinitely rich layer to the history this study recounts.

The format of our interviews did have some important consequences. First, because these interviews were conducted in Harare and Murewa, and because my own meager language skills are in Shona, this study as a whole offers a weak understanding of the specificity of experiences with commodification among Ndebele and Tonga peoples in Zimbabwe. I have tried to discuss this process in terms relevant to all regions of Zimbabwe, but this has not been possible consistently throughout this study. Second, because most interviews were with groups, the testimony we solicited was often made richer by debates and discussions between a number of individuals, but single, idiosyncratic views were probably lost as a result.

Finally, the subject matter of our interviews posed some difficulties. I was told by colleagues that I could not possibly ask questions about specific commodities, that I would have to find some sort of indirect language to get at the subjects that interested me. We did try to keep interviews somewhat conversational, following the thread of discussion wherever it led. However, we found that questions about individual commodities and types of commodities, as well as practices of hygiene and bodily presentation, often produced detailed and complicated answers. Questions about advertising, by contrast, were often ignored. However, these responses were rarely consistent for all commodities or practices of interest. Some goods of importance to this study have been completely absorbed into the invisibility of "common sense," while others have been attached to richly specific historical experiences. There were other problems as well. For one, because so much hygienic practice and use of commodities are subtle though vital parts of everyday life, I at first had difficulty understanding the significance of what I was hearing or not hearing. This, of course, was part of the reason for pursuing the interviews in the first place. Also, because so much of this subject matter involves relationships and institutions that have achieved the illusion of being "natural" and in the realm of "common sense," a good portion of our informants made points with reference to the present or to an indefinite past. The historicity of their comments on commodities was not always clear. I have tried to take all of these problems into account wherever relevant.

In any event, these interviews helped to confirm—or correct—impressions I had formed during archival research in England and Zimbabwe. The evolution of modern Zimbabwean commodity culture and the accompanying transformations of "needs" are not merely crucial points for the scholarly, intellectual understanding of colonialism and capitalism in Zimbabwe. Perceptions of the changing world of things—and changes in relations between things and people—have been and still are critical to the ways that Zimbabweans themselves have experienced and explained capitalist transformation and modernity.

1 CLEANLINESS AND "CIVILIZATION": HYGIENE AND COLONIALISM IN SOUTHERN AFRICA

On the Monday following the 1990 repeal of the Separate Amenities Act in South Africa, *The Weekly Mail* sent two reporters to the town of Ermelo. The reporters, Philip Molefe and Phillip Van Niekerk, discovered that access to previously forbidden facilities was still difficult. The most intense confrontation concerned access to the town's swimming pool. After trying to prevent Molefe's entrance, Ermelo's white swimmers were enraged when he entered the pool: "While Molefe splashed around . . . he [the white lifeguard] screamed angrily, 'Swim. De Klerk said you can swim. . . . I'll give you some soap to wash with.'" Later, at another stop, one man muttered "*Die plek gaan nou vuil word* [This place is going to be dirty now]."[1]

These sorts of visceral feelings about the bodies of Africans, visions of them as dirty or diseased, have been among the most intensely expressed aspects of racist sentiment in southern Africa over the course of the twentieth century. This chapter will review the colonial production of cleanliness, health, and racialized bodies, a process loaded with powerful images and archetypes and marked by intricate interactions between Africans and white settlers. Diana Jeater has described debates in colonial Zimbabwe about marriage and gender as the discursive making of a "moral realm."[2] In the next two chapters, I hope to describe the outlines of a similar colonial domain around health, cleanliness, race, beauty, and bodies.

Cleanliness and Modernity in the Nineteenth Century

Contemporary practices of hygiene and related bodily practices that presently seem fundamental and natural are in fact anything but. Historians have made increasingly clear in recent works that in Western Europe and the United States fastidious personal and household cleanliness was relatively uncommon until

the mid-nineteenth century.[3] The premodern European sense of human bodies
was not simply "pre-hygienic," an ignorant precursor to a great age of modern
scientific cleanliness, but instead represented an entirely different world of
bodily subjectivity, with its own historical implications and complexities.[4]
Along the same lines, new forms of fastidiousness in the nineteenth century
sometimes had little to do with inarguably "rational" changes in the understanding of disease. Some of the strongest, most insistent British advocates of
new hygienic practices disdained "germ theory" until well into the twentieth
century and described "dirt" as an alchemical substance, an essence that oozed
immorality and degeneracy and contravened the law of God.[5] Changes in
sanitation, personal cleanliness, and collective hygiene unquestionably had a
positive impact on general health, but these nineteenth-century transformations were also brokered by a much more pervasive and subtle field of ideologies and institutions, all related to what Frank Mort has called "medico-
moral politics."[6]

These developments had two particularly important consequences in relation to the overall subject of this book. First, images of cleanliness, appearance,
and beauty were increasingly used in nineteenth-century Western Europe to
define social hierarchy and difference. Second, personal and social hygiene was
increasingly portrayed as a key attribute of feminine domesticity. Both of these
visions were given social power through a range of new institutions and official
practices throughout Western Europe and in the United States around the turn
of the century. In but one typical example, "lady health visitors" in Birming-
ham, England were directed in 1899 to "visit from house to house in such
localities as the medical officers of health shall direct; to carry with them
disinfectant powder and use it where required; to direct the attentions of those
they visit to the evils of bad smells, want of fresh air, and dirt conditions of all
kinds . . . to urge, on all possible occasions, the importance of cleanliness,
thrift and temperance."[7] Nineteenth-century European visions of difference
and hierarchy were also constructed within such institutions with reference to
the constant circulation and recombination of colonial and metropolitan stereotypes. When the English working classes were hegemonically portrayed to
be "unwashed" or to have deformed bodies, these characterizations were always
dialectically related to depictions of colonial subjects as also having dirty or
undesirable bodies.[8]

The disciplinary changes that affected racial science in the latter half of the
nineteenth and early twentieth century only added to these emphases, as did
colonial exhibitions, travel narratives, and the like.[9] Sander Gilman, for example, has argued that events like the European exhibition of Sarah Bartmann,

"the Hottentot Venus," were constitutive of the general Western imagination of black bodies, within which images of disease and pathology were especially powerful.[10] Travel narratives equally reflected the degree to which many Europeans were expanding on and elaborating already existing preoccupations with the sensual, physical, and bodily nature of subordinate peoples. Philip Setel has analyzed images of "body-ideology" in a number of memoirs by travelers in Africa and found "a conceptual domain in which health, illness and techniques of bodily display linked Africans to preconceived notions of race, moral status, and savagery . . . [a] hidden template against which the African body / person was measured."[11]

In mid-nineteenth-century south-central Africa, Robert Moffat, the prominent missionary representative to the Ndebele court, occasionally commented on his hosts' hygiene or bodily practices, but he did not seem to imagine his own practices to be particularly superior. Indeed, he was shocked and irritated on one occasion when asked for soap by a Khoisan man during his travels: "A likely thing, to carry soap . . . to sell to a —! What wonderful ideas shoot out of the depraved heart!"[12] His contemporary, Thomas Morgan Thomas, described the Ndebele people as "generally clean" and "not wanting in good taste in respect to beauty, cleanliness and dress."[13] A trader, Herbert David Crook, echoed Thomas's complimentary description of Ndebele physical customs, though he found Shona-speaking peoples the opposite, "utterly regardless of personal cleanliness."[14] Another contemporary, Henry Stabb, was repelled by the lack of cleanliness in an Afrikaner household in Zeerust, but found Lobengula, his court, and the Ndebele generally attractive and well-groomed.[15] As Jean Comaroff has noted, the increasingly common recurrence of the image of "greasy" African bodies in works by Moffat, Livingstone, and other mid-century travelers in the region reflected a growing and racially charged antipathy toward the bodies of "savage" Others, bodies infected by the disorder of the natural world, exuding dangerous contagion at all who came into contact with them.[16]

Racism and African Bodies after 1890

The somewhat fitful interest in the bodies of Africans expressed by precolonial European travelers to south-central Africa contrasted noticeably with accounts written after the 1890 invasion of Matabeleland and Mashonaland. White settlers vehemently and repeatedly characterized Africans as filthy, depraved, and ugly. Most whites, regardless of their institutional or professional affiliations, agreed with the first Anglican bishop of Mashonaland, G. W. H. Knight-Bruce, that Africans were a "repulsive degradation of humanity."[17] The Native Commissioner of Gutu District in 1909 argued, "before European influence . . . natives lived more or less like wild animals and in a general state of filth."[18] His contemporary in Hartley District agreed: "They have no idea of cleanliness. . . . They rarely bathe themselves and only during the warm weather and never wash their clothes or blankets . . . their kraals are swarming with vermin of every description."[19] By contrast, European hygienic practices were defined as the essence of "civilization," coinciding with the cementing of intensive personal and social cleanliness as normative in English social life. Cleanliness was described typically by one official in the 1920s as a "primary art of civilised life."[20] For most whites living or traveling in Zimbabwe after 1890, the African world was a world of universal dirt and filth, while their own social world was its opposite, cleansed and pure.

Such sentiments stand out as a distinctive aspect of colonial racism not merely because of their omnipresence and vehemence, but because of their physicality, the manner in which they influenced whites to react with revulsion or avoidance in the presence of Africans. For example, many early Native Commissioners represented their African charges as generically "syphilitic," though their concept of "syphilis" was clearly vague and quasi-metaphorical. Many of them believed this "syphilis" could be spread to whites through shaking hands, physical proximity, or handling hut taxes—leading some officials to routinely disinfect tax money after collecting it.[21] The intensity of similar hygienic fears is suggested by Nathan Shamuyarira's recollection that at Waddilove mission, "the wife of one missionary was known to have sprayed a chair on which an African teacher had been sitting, and broken a cup from which another had drunk tea."[22] As the example of the Weekly Mail reporters' trip to Ermelo suggests, the physicality of racism has retained its grip over time, even as other expressions of official and unofficial racism have been called into question, in both South Africa and Zimbabwe. C. Frantz and Cyril Rogers's study of Rhodesian racial thought, done in the 1960s, cited many whites continuing to offer hygienic explanations for segregation, such as: "This is not really a color bar, it is a hygiene bar, as most Africans do not have the facilities for proper cleanliness" and "The answer therefore to the question of sharing lots of things would be no. Simply because of the ordinary African's ideas of hygiene."[23] A similar argument appeared in The Citizen in 1956: "the native African is not clean enough, physically, to make social intercourse with him pleasant or even hygienically safe."[24]

The intimate materiality of these readings of race through images of hygiene and appearance were reflected in a number of particular obsessions of whites in colonial society. Odor, for example, was a regular feature of white representations of their interactions with Africans, not only in Zimbabwe but throughout southern Africa. One South African ethnographer commented, "no description of the outward appearance of the Kafirs would be complete if we failed to refer to the omnipresent odour which streams from these people."[25] A Rhodesian farmer wrote, "For it's oh! ye odorous Rhodesian natives . . . Mr. Rhodes spoke of you as 'Africa's greatest asset,' but so far as house-boys were concerned, you were Africa's greatest and smelliest ass."[26] Another lamented, "the unforgettable scent of a bush store—a mingled aroma of coarse native tobacco, 'Manchester goods,' and unwashed nigger."[27] David Caute quoted one Rhodesian professor teaching French at the university in the 1970s: "I cannot tolerate ze smell of zem! Zumtimes I geeve zum black students a leeft in my car and I tell you I can hardly bear ze smell of zem!"[28]

Domestic service formed another critical nexus for the intimate inter-penetration of race, bodies, and hygiene. As Jacklyn Cock has so ably demonstrated, the interlocking of domestic space, privacy, racial identity, labor, and power that pertains in domestic service in southern Africa has been immensely potent.[29] Among Rhodesian texts, the books of Jeannie Boggie on farm life and domestic servants provide the clearest testament to the white perception of these relationships. Boggie endlessly catalogued the supposed errors of her servants, which with critical hindsight look more like deliberate sabotage—allowing meat to rot or putting dung into milk. She also described her own deep-rooted antipathy toward the bodily habits of her servants. In one case she wrote: "I rang the bell three times for Mike. . . . I noticed his hands were dirty, and pictured them foul with—maybe tobacco, or snuff, or maybe nose pickings or beer, or nasty microbes from handling dirty cards at card playing. So I said quite calmly: 'Mike. Go wash hands first.' He continued to place the parts together where the cream goes through. 'Ouch! Go wash hands. Go WASHEE hands.' No result. I said it a fourth time, accompanied by a smack on his back with a home-made brush made from the stalks of some weed, and used for swilling the dairy floor with water."[30] In another case, Boggie described servants as "breathing out beer . . . breathing out perspiration . . . breathing out roast meat . . . take sixteen strips of unwashed tripe hanging from the roof . . . put the total breathings into a 'scent bottle' to get breakfast ready and make toast. . . . Imagination seems to picture that native-handled toast as too abso-lutely smelly."[31] Similarly, in Frantz and Rogers's study, numerous apprehen-sions about the hygiene of domestic workers cropped up.[32]

As with all aspects of racist thought and behavior, such physical revulsion was at least partly performative, part of the public theater of segregation. The pervasiveness of sexual relationships between white men and black women simultaneous with racist visions of "dirty Africans" underlined the flexible and relative nature of such attitudes.[33] Moreover, some colonial texts also drew upon tropes of "noble savagery" to describe African bodies. "Probably the first thing," wrote one visitor, "which strikes a new-comer in Rhodesia is the unexpected beauty of the black race—the beauty, that is, of their bodies."[34] In such descriptions, African bodies emerge as aesthetically pleasing examples of "primitive physiques." Some practices that particularly revolted Europeans at one moment, like smearing oil mixtures on the skin, were then inverted and seen as "splendid."[35] Felix Bryk, in his ethnographic study of African bodies and sexuality, wrote:

> The Negro is not satisfied with the beauty that nature has bestowed upon his skin. . . . He adds to his chocolate brown, black skin new erotic values by painting it, oiling it and branding it. Painting with various earths and vegetable color is only used for temporary purposes. . . . Sometimes they rub themselves, from head to foot, wearing all their clothes and ornaments, with an animal fat mixed with red earth, so that they shine like red copper in the sun. This is a pleasant sight to behold, which does not discount the natural beauty of their unpainted bodies. Others satisfy themselves with just oiling their bodies. This blends the various copper hues of the skin into full force like a freshly varnished oil painting by Rembrandt.[36]

Though seemingly at odds with beliefs that Africans were inherently dirty and diseased, such visions were often placed without any evident sense of contradiction right alongside expressions of disgust. African bodies, in these depictions, were beautiful as long as they remained pure and untouched by "civilization" or change. Africans "kept in their place" were more beautiful than those who accepted civilization. In this light, clothing in particular was seen as something that spoiled the image of the "primitive": "there are many Mashonas on the road, for the most part undisfigured by European dress. A leopard skin is the ideal costume for these bronze figures, with their fine free carriage and movement."[37] The spectacle provided by wild nature in Africa was increasingly prominent in travelers' tales by the late 1800s; Africans clothed or washed like Europeans violated hegemonic presumptions held by such travelers about the unity between "savage" peoples and the "natural world." As one hunter commented in his diary, it was "an outrage to the beautiful wild bush veldt, a native clad in his shabby unbecoming European clothes, and a hat that he lifted. Ugh."[38]

Both tropes, the dirty, unredeemed African body and the noble, pure, "savage" body, were pervasive parts of colonial sentiment and of white subjectivity. As such, they had a powerful and fairly obvious influence on both colonial institutions and colonial culture. Precolonial African bodily practices, while far less visibly influential on the production of racialized bodies in Zimbabwe, also played an important role.

Cleanliness and Bodies in Precolonial Southern Africa

An older man in Murewa, a large rural town northeast of Harare, remarked to us during an interview that some local villagers had ignored the promotion of "modern" domestic practices in the 1950s by Helen Mangwende, the wife of the local chief. We asked why these people had objected to these campaigns. Surprised, he considered for a minute, and then said that he did not know. After all, he said, why would one defend ignorance?[39] From the vantage of the present, resistance to cleanliness and domesticity seems to many Zimbabweans to be a fantastic and surprising concept. European-inspired practices of cleanliness are now a generally accepted part of education and have infiltrated and commingled with other types of "common sense" about the body, health, and manners. The memory of racist statements about African habits seems to have dimmed in some quarters, though not all. Indeed, some Africans never paid much attention to the beliefs of whites about the appearances of their bodies. As I noted in my introduction, "common sense" has a habit of silencing the memory of its own manufacture, thus veiling its complexity and heterodoxy.

The description of precolonial Zimbabwean practices of hygiene, manners, and appearance is therefore quite difficult. These practices have become intangible; they exist both prior to and invisibly within contemporary culture. Furthermore, there is no reason to suppose that practices common in one area in 1850 were common there in 1550, nor should we suppose that practices common in 1850 were not inspiring and affecting practices in 1950—quite the contrary, in fact. The meanings of cleanliness, the practice of manners, the uses and images of the body that were characteristic of colonial Zimbabwean culture by the 1930s were formed with reference to both African and Western views. "Western" personal and communitarian habits helped to formalize and codify previously indistinct, informal, and undefined "African" ways common in the nineteenth century.

For the most part, soap was not used in interior southern Africa in the nineteenth century, though it was used extensively within areas of European settlement. Soap making did take place along the coasts of present-day Mozambique and Angola, possibly a legacy of Portuguese colonialism, but also, in the case of Angola, due to the presence of palm oil. As European settlement spread outward from Cape Town, a perilocal trade in soap also usually appeared. By the middle of the nineteenth century, most southern Africans had seen soap and perhaps used it on rare occasions. Virtually anywhere else in Africa that palm oil was available, small-scape soap manufacture had been going on for centuries.[40] Southern Africans used some substances similar to soap while bathing. Around the Zimbabwean plateau, plants with the Shona names *ruredzo* (also sometimes known as *feso*) and *chitupatupa* (*dicerocaryum zanguebarium* and *urginea sanguinea,* respectively) were often used during washing. *Ruredzo* is a ground-running vine, with leaves coated in a powdery substance that can provide a slick soaplike effect. *Chitupatupa* is a root plant that can be pounded and molded into a soaplike cake, usually referred to as *mvuka.*[41] The root also has been used as a fish poison. Elsewhere in the region, similar plants were used, such as the "soap dogwood" (*noltia africana*) in the Transkei.

For most twentieth-century Westerners, soap has defined hygiene. Where it and similar commodities have been absent, Westerners have perceived a complete absence of cleanliness. However, all southern African peoples prior to colonialism had their own clearly defined hygienic rules and codes as well as ideas about what constituted proper physical appearance and personal manners. The most basic and regionally universal of these bodily regimens was the regular use of a mixture of soil and some kind of oil or fat to coat most or all of the body.[42] The specific soils used tended to vary depending on availability and preference; most typically, thick red or yellow clay was used. (Red soil used as a toiletry is sometimes called *chivomvu* by Shona speakers; a similar term is used in other regional Bantu-family languages.) The oil used was different from place to place: castor bean oil, miscellaneous other plant oils, and animal fat were all common. (Oil or fat in Shona is *mafuta,* unmelted animal fat is *futa,* and the castor bean plant is *muptfuta.*) Some kind of paste from fats and soil, however, was a basic feature of a regular toilette, along with regular washing with water.

Nearly every nineteenth-century work dealing with southern Africa mentions the practice of "smearing" (*chikichidza* in Shona, varies in dialects). One typical account noted:

> No description of first impressions would be true to life that excluded the all-important question of red clay. A native's dress is never complete without this cosmetic. Red ochre and oil are rubbed into the skin, and frequently into the blanket. . . . The effect of the anointing with oil and clay is to form a splendid protection against the sun and the rain.[43]

A trader residing in Bulawayo in 1878 commented:

Like all other tribes of Kafirs the Matabele are great at smearing the body with fat, and it is wonderful to note the difference it makes in the expression of the face. Soloman, I believe, remarked something about "oil giving a man a cheerful countenance" and by that he must have meant the operation of smearing.[44]

The ethnographers Edwin Smith and Andrew Dale wrote about the practices of Ila-speaking peoples:

A substitute for water is butter or castor oil (prepared from the seeds of the plant, which grows plentifully in some districts) rubbed into the skin for the double purpose of cleansing and softening it . . . the skin cracks on exposure to the sun unless an emollient is used.[45]

The best-known scholarly account of these practices comes from the work of Philip and Iona Mayer on Xhosa communities. The Mayers described "traditional" Xhosa, "the people known as *abantu ababomvu*, 'Red people,' or less politely as *amaqaba*, 'smeared ones' (from the smearing of their clothes and bodies with red ochre)."[46] These "Red" Xhosa practices had very specific local meanings, but they were similar in form to practices throughout southern Africa.

There do seem to have been several regional justifications for smearing. First, the fundamental hygienic concept underlying smearing for most peoples seems to have been a belief that the best way to make the body safe from dirt and other menaces in the environment was to coat it with a protective layer. Second, smearing kept the skin from cracking or drying, a serious problem given the climate of the region. Third, the glossy sheen of the body and the sensuous feelings of smearing were widely regarded as fundamental components of an aesthetically pleasing appearance. It might also be noted that the prevalence of waterborne parasites, as well as predators like crocodiles, made bathing in some water sources dangerous by any standards.

In various parts of southern Africa, different oils, colors, and substances as well as various routines of bodily decoration and cleansing were applied to produce a wide range of effects. Here, my discussion necessarily must focus more tightly, though not exclusively, on nineteenth- and twentieth-century cultures on and around the Zimbabwean plateau—namely, those of Ndebele, Shona, and Tonga peoples. Nineteenth-century practices of hygiene, decoration, and fashion on the Zimbabwe plateau were clearly a vital part of the cultural vocabulary of identity. For example, in the nineteenth century, Shona speakers sometimes referred to Ndebele as *mapswiti,* "small stinging ants"; but the Ndebele and their allies returned the favor by using the Shona word *tsvina* (dirt) to describe their antagonists as *chiTsvina*, "dirty people." This was picked up as "*chiSwina*" by nineteenth-century missionaries working at the Ndebele capital and remained a common colonial name for Shona speakers until the 1940s.[47] Similarly, cleanliness was a part of the vocabulary of ethnic competition during the colonial era. Recent work on the "inner Harare" in the first half of the twentieth century describes the evolution of a community of "northern" migrants from present-day Malawi.[48] The complex, interwoven relationships between these migrants and other local communities have not yet been described fully in scholarly publications. There has been, however, a long history of subtle antagonisms and shifting affiliations between these communities. Some Shona speakers, for example, still use the derogatory terms *maBhuran-daya* (people from Blantyre) and *mabwidi*, and on several occasions both Shona speakers and Malawian migrants told me that their opposite numbers had a long history of being regarded as "dirty."[49]

Other markings and preparations of the body also served to identify forms of social difference. One Ndebele source has referred to such markings as *uphawu*, using the examples of pierced ears as well as the *isidhlodhlo*, "hair ring," commenting, "They were a sign to show to which tribe you belonged."[50] *Uphawu* or *nyora* (Shona for scar, tattoo, cicatrix) often had an intensely local character; many markings also were influenced by personal preference and innovation. Scarification was one of the most common methods of marking the body among Shona and Tonga speakers: facial *nyora* were specific to local groups, while other marks on the abdomen were the same over larger geographical areas.[51] Other similar permanent alterations were found in various parts of southern Africa: the removal or filing of teeth, the elongation or alteration of body parts, and circumcision. Body piercing was also common throughout south-central Africa—for example, some Shona speakers in the northeast and east pierced their lips and added plugs, nails, or wire (holes in the lip or elsewhere were referred to as *phuri* or *ringindi*); elsewhere, Tonga speakers, among others, pierced the septum of the nose and added a plug or other object. Less permanent markers like cosmetics were sometimes also used in preparations of the body and especially the hair, most notably what Tswana peoples called *sebilo* (lead hematite), a sparkling black substance mined and traded by Tswana speakers.[52]

Hairstyling was another common way of marking one's social affiliations; such styling included particular ways of shaping and shaving hair, dyeing with local plant inks, and adorning with beads, wire, or other objects. The Nguni *isidhlodhlo* was only one example. One traveler in the Zambezi Valley in 1900 observed as she visited new areas in succcession, "The natives take great interest and pride in their hair, and their ways of dressing it are many and wonderful. It was always an intense amusement to me, when I came in contact with a fresh set of men, to study the new fashions."[53] Two English nurses described particular local hair fashions around colonial Mutare in 1893: "Many of them plait their wool into hundreds of little tails, which stick up all over their heads. Beads, buttons and bits of brass wire are often woven into these tails. Some weave their hair into a sort of bird's nest, others into a castellated structure."[54] In the northwestern Zambezi Valley, Tonga-speaking peoples were especially noted for using red ochre and fat to sculpt their hair into bumps or short thick braids.[55]

Hair was also an important part of other bodily vocabularies dealing with other kinds of social difference, namely those of class, gender, and generation. Nineteenth-century Shona and Ndebele elites, for example, had access to superior hygienic and decorative materials: ostrich fat, imported scents, and high-quality or rare beads among other items. Elite men and women often wore particularly elaborate hairdos, some of which were reserved for them. In the Ndebele court, the bathing and toilettes of elites were often done only in the company of peers and servants.[56] The Shona *svikiro* (medium) of Nehanda in the 1920s avoided coming into contact with those considered "unclean," particularly women who were not virgins.[57] However, clothing and jewelry were frequently more important than hygiene or bodily appearance in defining elite style and authority. The nature of the relatively decentralized polities among Shona and Tonga peoples in the nineteenth century also meant that such ostentatious differentiation was considered less important, though not invariably. Some accounts of the Shona religious and political leadership in the region south of present-day Harare describe a great deal of pomp and circumstance associated with their bodies.[58] A Shona oral tradition collected in the 1950s concerning marriage and treachery described the sumptuous clothing and toilette of an elite woman in precolonial times: "Her beadwork apron, necklaces, armlets and anklets were the best . . . her body was cleansed with oil and her hair elaborately dressed."[59] Elite ceremony was often attended by particular bodily practices: for example, a newly instituted *svikiro* would have his or her hair cut in a special ritual.[60]

Hygiene and decoration were even more a part of gender and generational repertoires. For example, most of the ethnic marks mentioned above were applied simultaneously as a way of signifying and emphasizing gender and age. The *isidhlodhlo* was explicitly reserved for Ndebele men who were judged by contemporaries to have "reached full manhood." The wearer of an *isidhlodhlo* needed help to create it, so the affirmation of other men was doubly necessary.[61]

Likewise, married Ndebele women were "allowed to wear upon the highest part of the back of the head, a peculiar ornament . . . made of oxhair, vermillion clay and fat . . . worked into spindle-shaped lengths, much resembling long thin oats, but being bulbous at one end and pointed at the other."[62] Herbert Crook described the different male and female hairstyles in central Mashonaland during his trip there in the 1870s: "It requires some years to bring this toilette to a state of perfection. The hair on the temples and back of the head is shaved off and that on the top allowed to grow long. The long hair is plaited into innumerable strings, upon which large white and red beads are alternatively threaded. When treated in this way it looks exactly like a cup of beads, and is the style adopted by women. The men plait their hair in divers ways with grass, and I have seen some with plaits at least eighteen inches long . . . each man has his own idea of arranging his hair."[63] Likewise, in many Shona-speaking and Tonga-speaking areas *nyora* and body piercing were particular to one gender or the other. Cicatrices in particular were often feminized. This is especially true for what Gelfand called "private" *nyora,* marks intended only to be seen by a husband or lover, said to enhance sexual play. In any case, if *nyora* were made, their making often was a defining benchmark of puberty, though most Shona-speaking cultures did not have the elaborate forms of initiation schooling found typically to the south. Piercing, elongation, and circumcision were also richly implicated with gender and age. For example, in the late nineteenth and early twentieth centuries in the Fungwe area north of Murewa, local Shona women, but not the men, often sported *phuri* with brass cuffing, brass wire, or nails placed through the hole in the lip.[64]

Gender and age were also a vital part of other hygienic practices. Among Shona speakers, men and women bathed in separate groups, and children bathed separately from the adults. Tsitsi Dangarembga's novel *Nervous Conditions* is rich in evocative memories of these customs: "As children we were not restricted. We could play where we pleased. But the women had their own spot for bathing and the men their own, too. Where the women washed the river was shallow, seldom reaching above my knees, and the rocks were lower and flatter there than in other places, covering most of the riverbed. The women liked their spot because it was sensibly architectured for doing the laundry."[65] Not all peoples around the plateau were able to wash in a river everyday, and even those close to water usually refrained during the winter. However, even where river water was at some distance or when the weather was too cold, women would usually fetch water in the early morning. This water would be used for the washing of the face, hands, and other body parts. Again, washing was made into a ritual affirmation of relationships between men and women: in some places, wives were expected to attend to the washing of a husband's face and hands in the morning. Some households also had a *chibekeswa*, a small hut or enclosure for washing, with partitioned sections used by adult men and women in the household. (In interviews, some people expressed the opinion that *zvibekeswa* [the plural] were largely a late colonial innovation, built at the urging of white administrators.) In general, washing and the providing of water was governed by the same rules of deference and service that wives were supposed to observe in serving food to their husbands.[66]

However, as with all "customs," negotiation, contestation, and ambiguity also have been the rule. One example, albeit with an indeterminate history, is what Gelfand refers to as *kudfadza murume*, "pleasing the man [husband]," the washing of male genitals after sex. This was described by Gelfand in 1971 straightforwardly as a "duty" taught to girls before their marriage by a *tete* or *ambuya* (paternal aunt or grandmother, *vatete* plural) and then carried out faithfully by many wives. Instead, this kind of washing turns out to be a subject of considerable debate and ambiguity among women and men today, debates that some older women insisted were simply a continuation of disagreements from *kare kare* (long ago). In a number of our discussions, some women defined *kufadza murume* as a task husbands had always demanded from their wives, a duty that wives had always been obligated to perform. Many women in their sixties and seventies recalled being taught this duty as young women by *vatete*, though a few claimed with equal certainty that there was no such practice: "when sex is over, men and women simply sleep; all this talk of any formal practices involving wiping is silly, there's no such thing."[67] Many claimed they found the custom repulsive but obligatory, while other women of the same age claimed that such washing had always been voluntary, or that it had been increasingly stigmatized as perverse. A few older women even proudly proclaimed that they had always enjoyed the practice, and many younger women agreed, defining such washing as mutual and part of new ideals of *rudo* (love) in marriage. When men were part of our discussions, they universally and enthusiastically supported the practice. However, they also found themselves under attack by women of all viewpoints. In one instance, younger women complained that men always expect to be washed but never return the favor. The men present countered that this would be a "blasphemy."[68]

Childhood was also marked in other ways by very particular practices of hygiene, manners, and personal appearance. As noted above, children were invariably expected to bathe separately from adults. Infants were bathed carefully by their mothers, who sometimes used water held in their mouths for such washing, particularly among the Ndebele and Karanga.[69] Cleanliness was also a vital part of the expectations that surrounded birth among Shona speakers. In some areas, unmarried girls were only allowed to see a newborn after the baby had been washed and the placenta carefully disposed of by the older women in attendance. After the umbilical cord's remnants fell off several days after birth, the infant was usually bathed carefully, smeared with oil, shaved, and subsequently named.[70] Children also have been consistently "taught [personal] cleanliness (*kushambidzika*) . . . by the age of four, the child should have learnt the use of the right hand, the use of good language, clapping hands and cleanliness."[71] Rules about washing remained important as children aged: upon their first menstruation, for example, young women in some Shona areas were expected to go immediately to the nearest river or stream and wash, with their closest friends keeping a careful watch against *varoyi* (witches).[72] In the region at large, initiation ceremonies have frequently featured rules regarding washing and the presentation of the body.[73]

Concerns about *varoyi* played themselves out elsewhere in matters of cleanliness and sanitation. Spiritual relationships and needs were important determinants of cleanliness and manners in late nineteenth- and early twentieth-century Zimbabwean culture. Many sanitary and hygienic practices were aimed at protecting people from malign influences. It was important to hide feces, urine, and other secretions as they were useful to *varoyi*; thus, sanitation was a private or family affair that could not safely be collectivized. (When garbage and waste built up to the point of being problematic, many Shona villages simply moved their village site.) Likewise, washing and smearing both could acquire special sanctifying powers. The *muti* (medicine) of the *n'anga* (herbalist) included many pastes, unguents, and cleansers to purify, heal, and protect clients. Important occasions and events often had hygienic rules or beliefs connected to them. Among the Ndebele, for example, those who remained behind when an *impi* (battle regiment) was dispatched into battle were supposed to refrain from washing until the regiment returned.[74] Likewise, in a few Shona-speaking areas, the relatives of a dead person did not wash until the body was safely buried.[75] Among most Shona speakers, the corpse was (and still often is) washed carefully by *vasahwira* (close friends or occasionally senior relatives with ritual responsibilities); after the funeral, *vasahwira* and other mourners were careful to wash themselves thoroughly. In some villages, at the *nhaka* (inheritance) ceremony some time after a death, the son of the deceased was ritually washed by the *dunzvi* (a senior paternal relative who led the ceremony).[76]

None of the practices discussed above were ever fixed or rigid or invariably part of all precolonial Shona, Ndebele, or Tonga cultures. Many of the descriptions available of such practices come from colonial ethnography and narrative, so it is frequently impossible to fix precisely the historicity of a given practice. Significantly, none of these repertoires of hair preparation, piercing, scarification, or any other cosmetic marking of the body was ever simply or narrowly functional to identity. Practitioners had rich, complex, and multifaceted explanations of their body decorations. Gelfand pointed out that his interviewees stressed that many *nyora* were individualistic, creative, and "playful."[77] Regardless of the cultural epoch one is describing, not all members of a given African society understand various rules and codes in the same way, nor did all individuals mechanically reproduce all "normal" practices equally in their own lives. Moreover, ideas about the body, cleanliness, appearance, and manners were also clearly influenced by more complicated, pervasive, and contradictory cosmological principles about contamination, purity, and harmony.[78] In any event, repertoires of the body, of appearance, manners, and hygiene in precolonial society, clearly weighed in important, subtle, and sometimes unexpected ways.

The Domain of Colonial Hygiene

One of the general deficiencies of some treatments of "colonial discourse" is that sentiments expressed in texts become, through the mere fact of their expression, a "discourse." Yet a discourse is much more than that. The concept of "discourse" as derived from Foucault and similar theorists is powerful only as long as scholars insist that it means more than a particular form of talk or a particular assemblage of tropes, narratives, and texts. A discourse is talk that works through particular institutional arrangements of power and knowledge, that circulates and organizes conversations between specific producers and audiences, that relies on specific conventions for establishing its authority. And in so doing, a discourse produces the effects of power within the self, as a form of *discipline*.

In chapter 2, I will argue that hygiene and related issues can be considered fully discursive in this rigorous sense. However, the cultural circulation of hygiene and other topics dealing with bodies were also shaped by much more local, subtle, and incidental processes of mutual interpretation and invention, as dominant tropes in colonial culture. The general cultural domain of race and bodies that emerged in pre-1940s colonial Zimbabwe drew legitimacy from many sources, some of them indigenous.

For example, bodies and hygiene formed a major part of ethnographic typologies used by the colonial state. Cleanliness was habitually cited as a useful marker of a "tribe's" relative cultural "elevation" and character. Knight-Bruce, for example, argued that Shona-speaking peoples were inferior to the Zulu and Ndebele: "I am afraid the Mashona are a very dirty race. In this they differ entirely from the Zulus and their cognate tribes . . . the Mashona have no such ideas [of cleanliness]."[79] Knight-Bruce's representation of the Nguni speakers as physically superior was found widely among the early settlers. One Rhodesian agricultural expert noted in 1910 that he felt the Ndebele were "the most robust and useful . . . the Mashonas . . . [are] of inferior physique."[80] This relative valorization of different colonial ethnicities was part of the characteristic "invention of tribalism" so fundamental to colonial rule in southern Africa, but it was also based on the transmission of local indigenous vocabularies for ethnic differentiation into colonial thought—the Ndebele construction of Shona speakers as *chiSwina*, "dirty people." As Stanley Portal Hyatt noted in 1914, the notion was only "the white man's corruption of a Matabele term of contempt."[81]

As another example, envisioning hygiene as a "primary art of civilization" implicitly invoked whites as living examples of true cleanliness, to be observed, scrutinized, and imitated. And indeed, whites frequently suggested that their own bodies were as much a spectacle as Maxim guns, trains, or hypodermic needles. Stories about avid African audiences carefully observing whites while they were bathing or excreting have been a fundamental trope in colonial (and postcolonial) narratives.[82] The habitual reproduction of such stories was conditioned partly by the quasi-plagiarism endemic to much colonial writing, but such stories also surely are witnesses to an African ethnographic gaze turned at whites, an attempt to classify and investigate whiteness through bodies. Ultimately, the fetishizing of white bodies in the "hygienic domain" drew its impetus from several sources.

For example, in precolonial and early colonial Zimbabwe, one of the predominant descriptive phrases used for whites by Africans was "men without knees," occasioned by the habitual wearing of long pants. The veiling of white bodies was often seen as one with the veiling of white intentions, desires, and interests. The bodies of travelers and settlers in precolonial and colonial Zimbabwe were often investigated, discussed, and surveilled. One soldier traveling with the 1890 settlers described being surrounded by African women while bathing: "Steadily they advanced, their faces expressing mingled awe and curiosity. . . . Out onto the pool they marched until they touched us. Then slowly, gravely, they rubbed our white skins to see if the colour were fast. Obviously they believed we had smeared ourselves with some pigment. . . ."[83] After Herbert Crook and his party arrived at Bulawayo in 1878, they drank beer with Lobengula's wife and her companions and found themselves the objects of intense scrutiny: "They took a great interest in the shape of our hands and at their urgent requests we took off our boots and exhibited our feet, and great was the surprise they expressed at the softness and whiteness of the skin."[84] Henry Stabb lamented while staying at Bulawayo in 1875: "One cannot dress or wash without being stared at, by women as well as by men and no matter what portion of your toilet you are performing, you are sure to have a crowd of curious faces of both sexes gazing at you."[85]

In general, smell, appearance, or cleanliness acquired a certain mobility in colonial culture based on their mutual credibility as markers of difference. Maurice Nyagumbo's accounts of his travels in South Africa during this era provide a particularly striking example of this mobility. Nyagumbo recounted his youthful obsession with "smart" clothing and style while working in South Africa and described with horror, doubtless in imitation of his English-speaking missionary teachers, the habits of Afrikaners in the Orange Free State in the 1930s: "As far as cleanliness was concerned, the Boers of this area were completely oblivious to it and did not understand why a 'kaffir' was so sensitive to anything that was dirty. The whole building was infested with swarms of flies. Women came into the dining-room with hair falling everywhere. People blew their noses and spat everywhere in the dining-room."[86] Furthermore, in public discussions during the 1930s and 1940s, cleanliness was appropriated to representations of "authentic" African practices, by both white authorities and African subjects. One letter writer to *The Bantu Mirror* used cleanliness to explain the desire for "fat girls": "We want to marry fat girls because hygiene tells us that they are healthier than thin ones; few fat people get consumption."[87] Another claimed, "Native women and girls should not follow some white women in the fashion of wearing long hair, which . . . is unhygienic."[88] The prohibitions against bodily contact with strangers, so important to protecting oneself from *uroyi*, resurfaced through the language of hygiene: "Africans are far behind in observing the hygienic rules in their daily lives. Why are Africans now very fond of shaking hands with unknown people? Is this habit according to the rules of hygiene? It is a way of inviting infectious diseases to your bodies such as syphilis, gonorrhea and yaws."[89] One attempt by local leaders to prevent a mission school from being established in one reserve called up images of cleanliness originating from conflicts between Christian denominations.[90]

These "African" uses of cleanliness, and their application to interpretations of difference, acquired authority among whites as well, based in part on fragmentary but nevertheless accurate understandings of some precolonial practices. Some white professionals in southern Africa offered support for the superiority of the "natural" practices of Africans. One South African doctor argued that in "their native state," African physique was that of the "finer specimens of humanity." He added that in the case of clothing, "the uncivilised has an advantage over the civilised; the latter is compelled to wear garments which, even when in good order, have but little to recommend them . . . [while] the red blanket Kafir possesses in his only article of attire a costume adapted to heat or cold, to rain or drought, impervious even to rain through its dressing of fat and red ochre."[91] A peculiar inversion took place in such arguments. Where most settlers saw the "primitive" existence as defined by its lack of hygiene and, thus, endemically diseased and grotesque, some doctors and missionaries argued the opposite: "the uncivilised black bathed frequently and then rubbed grease and red clay over his body. His semi-civilised brother obtained a suit of European clothing which he seldom washed. His shirt was not once removed from his body till it was worn out."[92] A column in the *Bantu Mirror* described "good old habits of the Bantu," including "the washing of the body" and "the giving of an enema." The doctor writing the column went on: "your grandfathers still take more care of their teeth than the young people . . . your ancestors did not wear clothes and were healthier . . . the healthiest, fattest people in the country districts to-day are the unmarried girls who wear the least clothing!"[93]

It is important to underline these sorts of overlapping or interlocking uses of bodies to mark or interpret race and difference in colonial society. The power of bodily discipline to construct both individual and collective practice in Zimbabwe was not merely or purely a product of the projection of power onto Africans through colonial institutions, not merely one more mechanistic and inevitable outcome of white domination. Nevertheless, such intermingled definitions and understandings of cleanliness, filth, and contagion would be of relatively little significance were it not for the systematic use of hygiene as the underlying logic of certain colonial institutions concerned with ruling and disciplining Africans.

2 EDUCATION, DOMESTICITY, AND BODILY DISCIPLINE

Neville Jones, the principal of Hope Fountain School, said in 1935: "Cleanliness and due regard to the maintenance of conditions conducive to health and well-being are the concomitants of civilisation. If they are to be inculcated into Native life, it must be by our own initiative . . . until we have at least given him the opportunity of learning . . . hygienic conditions of life, we have left undone the things that we ought to have done."[1] In this chapter, I review the *discursive* making of colonial bodies in Zimbabwe, the institutionalization of hygiene, manners, and appearance. I briefly review the role played by hygiene in segregationist policy, but the bulk of the chapter is devoted to chronologically overlapping historical developments: the evolution of "native education" and its curricula from 1900 to the 1970s; "Jeanes teaching" and domestic training from the 1920s to the 1940s; state propaganda and administrative action from the 1930s to the 1970s; and the "homecraft" movement and women's clubs from the 1940s to the early 1970s. I examine the characteristic strategies pursued by these institutions for the surveillance, control, and representation of African cleanliness. I also examine how these strategies created new modes of social conduct and self-presentation, new forms of disciplining the self.

Disease, Hygiene, and Segregation

Social segregation, especially the segregation of urban space, was frequently justified in colonial Zimbabwe by reference to images of disease, dirt, and pollution. The leading specialists of the new discipline of tropical medicine at the end of the nineteenth century often argued that sanitation and hygiene were crucial for making the tropics safe for Europeans.[2] In so doing, they frequently contended that "dirty" races and their practices were one of the primary sources of contagion. Donald Denoon has described this process as the forging of "an important link between tropical *disease* and tropical *people*."[3]

This tendency to regard the proximity of Africans as a primary source of contagion has been referred to by Maynard Swanson as the "sanitation syndrome." Swanson writes: "Overcrowding, slums, public health and safety . . . were in the colonial context perceived largely in terms of colour differences. Conversely, urban race relations came to be widely conceived and dealt with in the imagery of infection and epidemic disease. This 'sanitation syndrome' can be traced as a major strand in the creation of urban apartheid."[4] The syndrome played a powerful role in colonial urban policy in Zimbabwe. For example, the sanitary boards of colonial Bulawayo and Harare were among the most important arenas for the design of segregated African "locations."[5] Tseuneo Yoshikune's comprehensive study of colonial Harare has revealed the related importance of the "latrine question" in generating concern about the interpenetration of white and black social life in the area around the Kopje west of the city center.[6] The pass system that was established subsequently to control the movement of Africans was also frequently linked to hygiene. Salisbury's municipal leaders typically fretted in 1917. "Many natives suffering from loathsome and contagious diseases are allowed to go about towns possibly spreading the germs among European families . . . all natives should . . . be obliged to go to a Central Depot, where they would be vaccinated, obliged to have baths, wash their clothes . . . no native should be supplied by the Native Department with a pass to seek work without a medical certificate."[7] The building of townships and the relocation of rural communities denied Africans sources of water and moved them away from the material resources necessary for any sort of hygienic or sanitary practices. As a consequence, segregation often actually produced disease and a lack of hygiene, though white officials used racist ideology to locate the source of this squalor in the cultural character of Africans themselves.

As Swanson points out with reference to the Cape Colony, the sanitation syndrome "served to define the issues and set the pattern of response . . . the sanitation syndrome penetrated into all perceptions and prescriptions of the day: urbanization was seen as a pathology in African life and therefore a menace to 'civilized' (i.e. white) society."[8] Segregation, as a spatial inscription of racial categories, played a crucial role in reinforcing white antipathy toward black bodies. Spatial segregation in southern Africa, as generations of scholars have demonstrated, was "about" social control, but the image of the diseased African provided such segregation with much of its intimate and pervasive power in white consciousness, as well as much of its repressive power for urban Africans.

Schools, Christianity, and the Hygienic Curriculum

Segregation was often constructed through images of hygiene and disease, but education and mission Christianity were the critical sites for colonial discourse about racial bodies, cleanliness, and manners. Many missionaries emphasized the production of what they considered "civilized" converts, and such teachings often focused on the personal appearance and habits of their African pupils. Such emphases were echoed and redoubled by state administrators and educators. As the project of "native education" unfolded in the first half of the twentieth century, hygiene and manners invariably played central roles in its conception and development.

The first mission schools in Zimbabwe were controversial for supposedly encouraging Africans to challenge the shaky social and vocational preeminence of the white population, especially through an emphasis on literacy. One critic claimed that the early schools laid "too much emphasis on the 'man-and-brother' theory and overlooked the necessity for starting at the bottom and gradually inculcating ideas of discipline, hygiene and thrift, with the result that the black man was led to regard himself as the equal of the white and become uppish and troublesome."[9] In 1903, the British South Africa Company (BSAC) administrators took some preliminary steps to legally restrict the nature of any education offered to Africans. One of the five basic provisions of this 1903 ordinance was a requirement that African pupils should first and foremost be taught "habits of discipline and cleanliness." The earliest efforts to produce an official knowledge of the "native problem" confirmed that one of the first and foremost duties of mission education was to teach "habits of cleanliness and discipline and . . . simple hygiene and sanitation."[10] The prioritizing of cleanliness in education for indigenous peoples was also part of general developments throughout the British Empire. In 1903, the Colonial Office asked administrators around the globe to write reports on progress in the teaching of hygiene to colonial subjects, a request prompted by philanthropic pressures, the influence of eugenicists, and the interests of tropical medicine. In the first such colonial circular, Rhodesian authorities testified that "a beginning has been made in hygienic instruction, though it has not yet been adopted generally."[11] By 1911, Rhodesian authorities claimed that education had shifted dramatically toward hygiene and sanitation.[12]

Official approval of hygienic training led missionaries to increasingly underline its role as an uncontroversial yet potent component in the making of Christian, "civilized" African communities.[13] By 1932, the Southern Rhodesian Missionary Conference had completed the process of aggressively embracing "the need of instructing Natives in Hygiene," arguing that "The general alertness of natives was increased by cleanliness of person and home. . . . Simple handbooks on hygiene should be supplied in the Native tongue. Hygiene should be a matter of first importance in schools. Dirt should be prohibited by superintendents and teachers, either on the person or around the school. . . . That cleanliness is next to Godliness should be taught."[14] The importance of cleanliness in mission schools was underlined still further when the new Department of Native Education, headed by Harold Jowitt, formalized the already existing school inspectorate in the 1930s. The new inspectorate made "cleanliness" into a central category in their examination of schools and teachers. Around the same time, local and metropolitan authors also updated earlier, cruder guidebooks on hygiene for Africans.

These textbooks and other pedagogical documents from this era give some sense of the slant and tone of appeals made to African pupils.[15] First of all, pupils were exhorted to rigorously pursue habits of personal cleanliness. In many cases, teachers enforced this dictum by inspecting students closely and turning away those who failed to keep the class standards. Teachers who failed to carry out such supervision were excoriated by their mission supervisors and by state inspectors.[16] By contrast, at Hope Fountain, the London Missionary Society's flagship school, inspectors in the 1920s and 1930s gave the school glowing reports, often in substantial measure because of principal Neville Jones's emphasis on teaching personal cleanliness. One inspector rhapsodized, "The art coming next to that of living happily in community is cleanliness. This begins with cleanliness of person, a matter in which I regard Hope Fountain as having set a higher standard and achieved a higher result than any other Rhodesian mission."[17]

A "Catechism of Health" recited by students at a number of Zimbabwean mission stations in the mid-1920s strongly reflected this focus on personal cleanliness: "I shall keep my body clean by washing all over every day in the summer, in the cold weather I shall wash all over as often as is possible, but I shall always wash all over once a week. . . . I must always wash my head and face and neck and chest and back and arms and legs and my teeth."[18] Another textbook used in colonial Zimbabwe written in the 1930s by missionary G. E. P. Broderick told students: "Many kinds of enemies which may make you sick get into the body-house when there is anything wrong with the skin-wall. . . . The first way in which you can look after the skin is by keeping it clearn. You must wash your body every day, for in your work and in your play you are constantly getting dirty . . . you can see, then, how important it is to wash your body every day."[19] All similar materials stressed that personal cleanliness should be regarded as the "keystone" of hygiene.

A second common emphasis in these courses concerned the care of clothing. As one missionary commented in 1900, "Civilization produces funny puzzles. Having taught the natives to wear clothes (personally I wish clothes were unnecessary) we must teach them to wash them."[20] Broderick's textbook typically instructed: "Your clothes must be washed too. . . . You must learn to look after your clothes, for they are a second wall for your body-house. . . . If they become dirty, or full of holes, or if they are wet, then they can do no good to your body."[21] The teaching of cleanliness also focused heavily on social and communal hygiene. This emphasis gradually became more important after the state hired community demonstrators and later began to publish various forms of propaganda aimed at health issues. Some of the practices emphasized in textbooks and course outlines included the regular sweeping of the home and village, protecting and preparing food, constructing and using trench latrines, and combating insect or rodent infestation. Additionally, lessons on hygiene were often contained within general attempts to promote "civilized" manners and discipline in the comportment of the self and the practice of everyday life.[22]

When mission schools shaped hygienic training for African pupils, they assumed that the teaching of hygiene had to proceed from a solid grounding in Christian principles. Therefore, most textbooks and course outlines began with an explicit linkage between cleanliness and godliness. The 1926 "Catechism of Health" had African pupils reciting "I ought to take care of my body . . . because my body is the house of God and therefore ought to be well kept."[23] Another text added, "Your body is the gift of God. . . . No real Christian . . . will allow his body to be dirty, nor will he let his home and village become dirty places. . . . This will spoil the wonderful body he has been given."[24] Missionaries and their allies often preferred to define dirt as an elemental component of immorality, a distilled, vague, and generalized essence of profane living. The "Catechism" warned students against both "dirt of the mind" and "dirt of the body," giving the following definitions of each: "dirt of the mind" was "thinking bad thoughts or not thinking at all" and "dirt of the body" was "to live in a dirty way and to have dirty habits."[25] In another text, pupils were told that "dirt . . . is anything that makes something unclean" and that disease, especially venereal disease, "comes as a punishment for doing things that are prohibited."[26] Broderick warned students vaguely, "Dirty things are always your enemies."[27]

This emphasis was superficial evidence of deeper, more complicated fissures between missionaries, educational administrators, and medical experts in their approach toward bodies. Missionaries were often subtly predisposed to avoid offering "scientific" explanations of the need for hygiene. Such antipathy was the result of a complex combination of surviving subsurface antagonisms between "germ theory" and the moral, religious promotion of cleanliness plus colonial debates over what forms of knowledge were "appropriate" to the "native mind."[28] Educational authorities in both South Africa and colonial Zimbabwe, such as C. T. Loram and Harold Jowitt, stressed that the teaching of hygiene and health and other matters to Africans should remain focused on moral and not technical forms of knowledge,[29] though some commentators, particularly metropolitan ones, felt that carefully limited "scientific" explanations of disease and hygiene should be given to African pupils.[30] These tensions between "scientific" and "spiritual" explanations have been noted in other aspects of mission education and propaganda.[31]

One choice example of these tensions with regard to hygienic education was the 1956 textbook *Doctor Kalulu,* used throughout British Central Africa. This book used indigenously styled folktales and proverbs to teach lessons about cleanliness and disease. In the stories, the protagonist Kalulu, a hare, is clearly at times meant to represent a wise white doctor or official, and at other times an "advanced" African who has adopted modern ways. Recalcitrant Africans who refused to learn their lessons and remained "superstitious" and "primitive" are represented as warthogs, baboons, or most especially, hyenas. In each story, Hyena and his family harass the wise Kalulu and impede his attempt to teach good habits. In the first story, "Doctor Kalulu," Kalulu washes Warthog's children, which perplexes Mrs. Warthog. Enduring Hyena's taunts, Kalulu patiently explains that "clean children are healthier than dirty children," adding, "my children play with Warthog's children. So by teaching Warthog's wife to wash her children I am helping to keep my own children healthy."[32] In the final story, Hyena is stricken with illness and finally learns from the wise Kalulu that "the greatest enemy of all" is his own superstitious fear and ignorance. Having learned "rationality," Hyena is on the road to recovery.

Most of the African students attending mission and state schools from the late 1920s onward found themselves in a world where they were encouraged to think of themselves as the favored clean among the "great unwashed" of the uneducated and unconverted. Many Zimbabwean activists and writers prominent after World War II were young children attending school during these years and their writings have provided rich autobiographical testimony regarding the ways in which educational discourse about hygiene and manners became integral to a sense of self for many elite Africans, about how hygiene as a form of discipline shaped racial identity in colonial society.

For example, Maurice Nyagumbo recalled in his memoirs that when he arrived at St. Faith's Mission for Standard I lessons, he and the other new pupils were marched to the stream while singing:

Tiri, manyukama,
Tiri, manyukama,
Tine tsvina
Tine tsvina
Hatigoni, kuyeza,
Hatigoni, kudzidza.

We are the newcomers,
We are the newcomers,
We are dirty,
We are dirty,
We do not know how to wash ourselves.
We have not acquired education.[33]

Nathan Shamuyarira similarly noted that "A big change came . . . in 1933 when two boys agreed to go to Waddilove Institution. When they returned they were totally different persons. They were clean, well-fed, respectful and above all able to utter a few words in English."[34] Though he and other students dreaded the gaze of state inspectors and teachers, Shamuyarira also has acknowledged that "These Methodist missionaries made a deeper and more lasting impression on me and my life than any other group of Europeans, although they may not share my political views and conclusions. Besides being at a most impressionable age, I was zealous to learn and copy as much as I could of European ways of living."[35] Lawrence Vambe recounted that many pupils were influenced by the demands and instructions of their teachers, and many came to question or reject those other practices increasingly singled out as "African." For example, Vambe noted: "Some boys had filed gaps in their front teeth, some had had these teeth removed altogether, while others bore scars signifying that they belonged to a certain tribe . . . at Kutama, which represented a gateway to a new world, these marks lost their tribal values and became a source of embarrassment and, inevitably, they felt a sense of shame and grudge against their families for disfiguring them."[36] Some pupils at the time were so committed to adopting these new habits and discourses of the body that an official of the Department of Native Education felt obligated to try to restrain them by warning, "We must not look back on our primitive parents and say with a broken spirit 'Ah look at the dirt and old dilapidated huts in which they live.' It is really wrong for us to say such things."[37]

Still, "conversion" to colonial ideals of hygiene and domestic behavior in Zimbabwe by Africans, students and otherwise, was influenced by many subtle dynamics. For one, colonial schooling, including hygienic and domestic training, was not entirely alien to indigenous institutions for socializing the young. In local African culture, boys and girls were carefully instructed in manners and proper behavior by elder kin of their own gender, though Shona-speaking peoples did not usually make use of the kinds of formal initiation schools common among Ndebele. Shona girls and boys learned domestic skills and proper household behavior around the time of puberty through *mahumbwe* or *mabindikito,* "playing house." In *mahumbwe,* children of roughly similar ages were gathered together and paired up in mock "marriages." Over a period of days, the couples acted out lessons both had received on keeping a household and proper forms of married behavior, explicitly excluding sexual intercourse.[38] In Shona *mahumbwe* or Ndebele initiation schools, domesticity and hygiene were very much a standard emphasis in the education of children. Therefore, colonial schools, though they were promoting practices with different emphases, origins, and meaning, were not educating the young in an entirely unfamiliar manner. In some cases, like *mahumbwe* and the teaching of domestic skills in mission schools, the correspondences between settler and indigenous institutions were even easier to imagine.

Furthermore, the relationships between the young and organized groups of senior men and women within kin networks, like the *vatete,* were often characterized by struggle. Senior relatives exerted authority through diverse institutions and practices of social control, labor, and ownership. Colonial schools may have seemed to be a new resource for the marginal and young in their struggles against the power of senior kin, though the nature of colonial domination often meant that education could not easily or unproblematically be used to such ends, a fact that often did not dawn on ambitious and assertive students until later in life. Discourses and practices of colonial cleanliness and bodily manners allowed for new idioms of differentiation between students and their kin, and displaced the authority of *vatete* and other elders.

Still more significant in this light was the flight of many African women in the 1920s and 1930s to the cities and mission stations. This was not always or simply an emancipatory act; increasing demands for wage labor and forced removals to infertile and peripheral rural areas often placed African women in a crippling position that left flight as the only way to escape slow starvation. However, the four-sided struggle between fleeing women, state administrators, African patriarchs (e.g., spouses and fathers), and missionaries so prevalent in documents from this period suggests that many women who moved and found themselves first in domestic classes (or several decades later, in clubs or other civic organizations) may have felt that new ideals of domesticity and appearance had many positive features. In the end, they may well have "exchanged one form of patriarchal authority for another," as Elizabeth Schmidt points out, but the strategic measurement of gains and losses that accompanied this exchange was calculated differently by each individual woman.[39]

Colonial norms of hygiene, appearance, and manners touched on other forms of social positioning, definition, and struggle in African communities. Where mission stations like Epworth or Chishawasha were established, the missionaries became the new landlords for converts and unbelievers alike. Satisfying the requirements of the missionaries in terms of personal appearance and behavior was often a prerequisite for continued residence near the station, or more importantly, obtaining access to new fertile land owned by the mission. Vambe's Chishawasha kin, for example, confronted these problems in the 1920s and 1930s: "In Mashonganyika village many people, especially those of grandmother's generation, had gone about scantily dressed in their *nhembe* (short leather aprons covering only the genitals) and no one had seen any shame or offence in bared breasts. But now this traditional state of innocence was unacceptable. By now most traditionalists had left the Mission, and the remainder conformed to the demand that people should be fully clothed."[40]

Thus, Western ideals of cleanliness, appearance, and bodily behavior became increasingly powerful within African communities, even among non-elites, during the 1930s. New African elites and whites both publicly explained the growing power of these new behaviors as signs of the struggle between "traditional" and "modern" life, "African" and "European" ways, "heathenism" and "Christianity" and similar binary oppositions. Hygienic language became an important part of these public representations for these men and women. One student wrote to the *Native Mirror* in 1932 fretting, "we let our children go almost naked. We call ourselves civilized, yet we do not show the fruits of that civilization."[41] Another complained: "One day I went to the Native Beer Hall and found all the ladies and gentlemen dressed very nicely. I knew one of these gentlemen very well and so I visited him the next day and saw that his room was very dirty and untidy. What is the good of keeping your body clean when your room and bed are very dirty?"[42] A young mission teacher complained, "We have very fine villages but untidy and neglected. . . . Just see how Europeans keep things properly in their houses. We should not say it is because they have money."[43] Memoirs, autobiographies, testimonies, letters, and other texts generated by men and women of Vambe, Shamuyarira, and Nyagumbo's generation were often marked by ambiguous and contradictory strategies of representing "tribal society" as a limiting and conservative influence on African society, while also defending its value as a cultural resource.

By the mid-1930s, a pedagogy of cleanliness and manners had become a standard part of education for Africans, and had begun to affect conventions of self-presentation in colonial Zimbabwe. This pedagogy was joined in the 1930s by other socializing institutions aimed at Africans. Hygienic training in particular intersected with "home demonstration" and gender-centered forms of schooling that were aimed at generating new practices of domesticity and female behavior in African communities.

"Cleanse Herself and Through Her the Race":
Domesticity and Hygiene

In 1929, Harold Jowitt, the head of the new Department of Native Education, declared, "The greatest need in which this Department is called to assist . . . [is] dynamic activities which shall enrich the life of Native womanhood by reducing human wastage, replacing ignorance and fear by knowledge and self-respect, cleansing the home and through it the race. . . . "[44] Interest in the training and education of African women had been increasing steadily for a decade when Jowitt made his statement. A number of factors motivated this growing interest in domesticity, which had its earliest roots in the first mission schools in Zimbabwe. For one, "black peril" panics among settlers had been a recurrent feature of colonial Zimbabwean life ever since 1890.[45] Such panics invariably focused on the presumed sexual threat posed by male Africans present within the private sphere of the home, such as cooks, servants, and other domestic laborers. In the aftermath of alleged "black peril" incidents, settlers frequently called for the training of African women to replace male domestic workers.

Another related factor was changing European notions about domesticity, the private sphere, and the role of women in the household. Domestic tasks like cleaning, laundry, cooking, and child care were increasingly seen as the sole and inherent province of women. Consequently, male domestics were more likely to be viewed as transgressive, though many of them continued to work in households. In colonial Zambia, men remained the primary servants.[46] As a result, African women were often represented in gendered terms as possessing an innate inclination for reproducing households and maintaining hygiene and health among families. As the 1910–1911 Native Affairs Commission put it, "The desire is so strongly developed in every Native Woman to become a wife and mother . . . that it is irrepressible."[47]

The intersection of the perennial concern with the supply of male laborers with growing anxieties provoked by the increasing migration of African women to urban areas also proved important to the development of domesticity. Authorities, concerned that the appeals of domestic wage labor in the cities were diverting male workers from primary industry, often advocated a more extensive reliance on female servants. For example, the 1910–1911 Native Affairs Commission concluded that "the employment of girls would . . . release a large number of able-bodied men."[48] Likewise, officials were concerned about controlling the women coming to the cities. Though many township administrators tolerated or even encouraged (and sometimes patronized) African prostitutes from the very beginning of female migrancy, they were concerned not to let women escape into the unmonitored interstices of urban life, especially since urban "shebeen queens" appear to have been one of the few groups in the pre–World War II townships capable of substantial capital accumulation.

A related aspect of the growing emphasis on domestic training was the growing number of African women who sought refuge from male relatives or husbands in mission communities. This was a persistent source of tension between missionaries eager for new converts and those Native Commissioners engaged in the construction of "customary law" in collaboration with local male African elites.[49] When the bsac solicited views from local administrators during the first attempt by the state to officially encourage domestic training in 1909, most testified that "customary law" would forbid women entering service without male approval.[50] One commented, "the social conditions of Natives require drastic measures in regard to the control of women; Native law is very clear in regard to that point."[51] Women who ran away to mission stations consistently eroded this understanding between administrators and African patriarchs. Files of complaints made by African men to the Native Department from 1900 to the 1940s are replete with pleas for officials to enforce customary law and return fleeing women to their husbands and fathers.[52] The flight of some women to mission stations provided an extra impetus for teachers to develop "appropriate" forms of education and vocational training for this expanding clientele. Such converts were not necessarily averse to such training, which in some respects resembled indigenous forms of socialization and teaching about the household. Such training was defended by the missions as a contribution to settler society, and it also helped to provide inexpensive domestic workers for the mission itself. Classes in laundry and other domestic subjects were often aimed at accomplishing the station's normal domestic workload. Additionally, as Elizabeth Schmidt has pointed out, missionaries had a strong interest in providing their male preachers and converts with "proper" Christian wives. By acquiring the power to supply brides and contract marriages, the missions appropriated one of the most fundamental forms of power held by elder kin in Shona and Ndebele society alike.[53]

Finally, by the 1930s, officials of the Native Department were also increasingly aware of a growing crisis in the reproduction of labor in both rural and urban areas. Though a minority of white critics like Arthur Cripps pointed out that the problem lay with inherently poor conditions in areas designated as reserves and in the gross inadequacy of urban infrastructure and wages, most officials pursued other explanations. Such explanations often consisted of racist information on African "culture": for example, a supposed African preference for sandveld farming, or an endemic and inbred filthiness and lack of health. The growth of domestic education was connected to such explanations. If the maintenance of the family and home was "naturally" the concern of women, officials reasoned, then intensive education in "proper" domesticity would perhaps address escalating social problems in African communities or at least draw attention away from the naked instrumentality of settler policies.

By 1916, six mission schools had courses in laundry and other domestic subjects. A little over a decade later, the number had at least trebled. Most of these courses offered training in skills to be used by domestic workers: one 1920 syllabus, for example, included lessons on washing "fine linen";[54] another school divided domestic lessons into "theoretical" and "practical" subjects, with the former intended to make the pupils into knowledgeable household servants.[55] The officials of the Department of Native Education offered enthusiastic support in the 1920s for a program of domestic training. They appointed an "Organising Instructress of Domestic Science" to inspect and guide schools offering domesticity training, who wrote upon her appointment: "The study of Home Economics develops women and girls on natural lines, and trains them to be capable and useful women, good wives and mothers. It is of especial value to the natives in Southern Rhodesia, for if it is well taught it should: 1. Raise the Native to a higher state of civilization and mental development. 2. Improve the physical and hygienic conditions of native life."[56] Domestic courses heavily emphasized both personal and social hygiene and stressed the responsibility of the female pupils for maintaining the cleanliness of their children, their adult relatives, their homes, and their villages. One inspector complimented the program at Nengubo Mission, writing, "By example and precept the principles of hygienic living are so thoroughly instilled on all occasions that there is a very reasonable hope that they will be carried out in the kraals when the girls return to home life."[57] Cleanliness was considered possible only through what one author called the "woman factor" and through what another termed the "rightful bringing up by a mission" of African women.[58]

The idea that women should be completely retrained and then sent back out to transform African households found its most powerful expression in the 1930s with the development of a cadre of "home demonstrators," initially referred to as "Jeanes teachers." The Jeanes program, named for an American philanthropist, funded teacher training and "community education" programs in southern and eastern Africa and in the United States. The first Jeanes teachers in colonial Zimbabwe were introduced in 1929. Fifteen men were trained at Domboshawa School, and their wives were given auxiliary training to provide a "good home background" for their husbands.[59] At the same time, the state sponsored the training of twelve Jeanes women, or "home demonstrators," at Hope Fountain. Both Jeanes men and women were expected to train people in new methods and practices of cleanliness and appearance. However, the male instructors were largely expected to handle communal hygiene, especially the building of trench latrines, insect control, and the construction of new types of homes and villages.

An examination of the careers of these Jeanes women in the 1930s gives an insight into the production of African domesticity by the state and missions during this period. Life at Hope Fountain, the school inspectorate's exemplar of cleanliness, was described by one inspector: "Dr. Neville Jones has built an excellent large bath . . . and every girl is required to bathe once a day. The same principle is extended to their clothing. . . . It appears again in their simple clean dormitories, (the blankets of which are hung in the sun daily), and it reappears in their simple scrupulously clean kitchen, and equally clean dining room. The girls are thus learning a second primary art of civilised life, and learning it with spartan and yet cheerful thoroughness."[60] The Jeanes women learned: "home nursing and first aid, maternity work, personal and community hygiene, simple physiology, child welfare and mothercraft, simple cookery with special reference to infants and invalid diet, vegetable gardening relating to the cookery, dressmaking and housecraft, simple household arithmetic, letter reading and writing, simple civics and scripture."[61] Additionally, the demonstrators were expected to "provide an example" by leading lives of exemplary cleanliness and behavior. Both during and after their training, their homes and persons were constantly examined by their local supervisors.[62] After their training, the home demonstrators were assigned to various communities around Zimbabwe, mostly communities near missions.[63] A significant portion of their duties consisted of dispensing medical care of various kinds, particularly midwifery. At their postings, the Jeanes women also taught classes on various domestic subjects. Variable amounts of time were also taken up with visits to villages and homes, where they tried to hold classes or order villagers to clean up. Some demonstrators also scheduled "visiting hours" when local women could come to them and consult them for advice on domestic topics.

In all aspects of their work, the Jeanes women stressed cleanliness, attack-

Jeanes Teachers at Hope Fountain Mission, late 1920s.
(Zimbabwe National Archives)

ing aspects of African life that they now viewed as relentlessly unhygienic. Mary Mutukudzi, assigned to Epworth Mission, described her weekly schedule in 1933: "[On Monday] About 8:30 am I have to go back to the school to teach hygiene to different classes; after that I have to visit the patients in the village . . . if I find the patients in a dirty house, I sometimes clean the house. . . . Tuesday, Wednesday and Thursday, I go to different places. The first thing that I do in every school that I visit is to inspect the school children to see if they come to school clean or dirty. If I find some dirty I will make them clean by washing them. After the inspection I teach hygiene in school. . . . "[64] Another teacher, Elizabeth Mhombochota, described her work in the Mutambara Reserve: "These girls are learning to sew and cook. They can make nice scones and cook nice sadza [thick maize porridge]. . . . We are also trying to lead them to Christ. We teach them to keep the body clean outside and inside."[65]

The Jeanes teachers, as the most directly involved group of aspirant students and teachers, also powerfully internalized colonial regimens of cleanliness and discipline. In their letters and statements, many of the teachers showed an evident pride and confidence in their new habits; they spoke of the filthiness of "heathen" Africans with what at first seems to be a voice slavishly identical to that of their missionary supervisors. Priscilla Moyo saw among her community "many dirty customs, such as eating out of the same dish, sleeping under the same blankets, and spitting on the ground." But she also saw a new community growing: "the kraals are clean and the huts are well-built. Inside the huts everything is tidy and clean because the people have learned how to look after themselves. They keep their garments clean and some of them sit at table to have their meals and have their own plates. Round the yards are beautiful flower gardens."[66] For Moyo, this new vision was emancipatory, empowering. Some of her peers looked still more carefully and saw beyond the terms of vision offered by their supervisors. Rachel Hlazo mused, "The Jeanes woman has to make herself one of the people. Native people, though filthy in some habits, have a great respect for cleanliness. It will be found that in certain departments of domestic life, a certain amount of cleanliness is demanded."[67] Similarly, another teacher, James Rutsate, saw in the Jeanes program an opening for teaching a new conception of African society, through what he called "race pride," which would "make themselves people . . . proud of their own things rather than others'. "[68]

However, new forms of domesticity also provoked opposition and critique. Jeanes teachers, particularly the men, almost immediately drew fire from competing authorities. The Department of Native Agriculture had its own cadre of African agricultural demonstrators who resented interference. When the male Jeanes instructors, with the backing of their missionary supervisors, tried to exercise juridical and political power over the communities in which they were stationed, they drew the wrath of the Native Department. Aggrieved Native Commissioners were eventually responsible for limiting and ultimately cancelling the authority of the male teachers.

In these efforts the administrators often worked in unintentional alliance with African communities who also resented the Jeanes teachers' authority. Both the male teachers and the home demonstrators frequently encountered hostility from other Africans. Native Commissioners' reports and missionary documents make it clear that for the first three decades of colonial rule, many elders forbade their children and grandchildren to attend mission schools; when possible, they sought to prevent such schools from operating. For example, government official "Wiri" Edwards wrote to his superiors in 1911 that "older people" were constantly interfering with mission education in Murewa District.[69] Though some missionaries were wary of state authority, few were inhibited from seeking the assistance of administrators in their struggle with recalcitrant "traditionalists." In 1921, for example, a priest at Kutama Mission appealed to Salisbury to assist the missions: "There is a difficulty of another sort with which Missioner and Instructor cannot cope effectively . . . the opposition of older Natives . . . now sullen and passive, now active and malignant . . . the unredeemed villains . . . should know clearly, and feel plainly . . . the full weight of Government authority."[70]

Still, many other senior family members in African communities aggressively encouraged one or more of their male children to attend mission schools during this same era. Even where strategic, resigned, or even enthusiastic acceptance of mission schooling existed, colonial projects concerning hygiene and domesticity sometimes provoked considerable resistance and anger when demonstrators like the Jeanes teachers entered the scene. The personal journals and correspondence of Jeanes teachers and their supervisors regularly lamented the lack of cooperation, outright opposition, and open ridicule they encountered from many of their appointed targets.[71] Demonstrator Rebecca Ncube reported in 1933: "Sometimes we meet in the kraals of the Christian people, but we often go to the heathen kraals as well. We sometimes have strange experiences. Sometimes the people run away and hide, or lock themselves in their huts when they know I am coming. They often speak roughly to me and refuse to let me go inside their huts."[72] Her colleague Priscilla Moyo reflected, "When a kraal has been cleaned up, the owner of the kraal is sometimes offended, and does not like what has been done."[73] The Native Commissioner at Enkeldoorn reported in 1934, "The Jeanes teachers . . . find disinterested opposition. They are no man's friend . . . [local people] consider that with the reforms propounded further means have been devised to worry them in their quiet lives."[74]

Such opposition stemmed from a number of sources. Many rural people located their sense of self and personal freedom in increasingly circumscribed traditional folkways and resented the demonstrators' open submission to the habits of the white invaders. One frustrated Jeanes teacher who encountered such sentiments wrote, "There is a custom which the old mothers do of bathing the babies' faces. . . . When I came to stop them from doing that, some of them said 'our fathers and mothers were doing this, and we are well and strong.' . . . How shall I answer them when they say this?"[75] In another case, a sympathetic bureaucrat found himself confronted with the complaints of his assistant about a pamphlet on hygiene. The man complained, "None of us were brought up this way and it didn't do us any harm. . . . I was a filthy child myself."[76] Another state official complained: "they ["progressive" Africans] are held back by the . . . older Natives, the 'die-hard' conservatives, whose life is ruled by superstition. . . . The real harm they do lies in their philosophy of life. Any instruction that he may receive in manual work, in agriculture, building, or hygiene is usually wasted, for he is quite convinced that his ancestral methods are good enough."[77]

The demonstrators also violated deeply held notions about propriety, privacy, and the autonomy of households by willfully, regularly, and aggressively cleaning villages, homes, and persons without permission from villagers, homeowners, or those to be cleaned. Besides the straightforward annoyance that such interventions constituted, there were also deep and subtle concerns about permitting strangers to interfere with hygiene, manners, and household property. In particular, the delicate web of spiritual defenses against disease and *uroyi* (witchcraft) was profoundly threatened by the Jeanes teachers' demands for access to the body and its by-products. One male teacher working near Chikore Mission was stymied in his efforts to dig trench latrines (like most of his compatriots) and reported: "The greatest difficulty in community work was in connection with E.C.s (earth closets). When I held meetings with the Chief and headmen, and their people, and talked of these matters, they considered that it is an European way. They said, 'If we Black People do it, we will catch sickness at once, because everybody has to go to the same place.' "[78] The demonstrators were also seen as undeserving of the authority they were attempting to wield; they were outsiders and often considered far too young to be giving orders to men and women twice their age. Mary Mutukudzi, working at Epworth, complained that she was often mockingly dismissed as a "girl" by older women in her area.[79]

The Jeanes teachers were also singled out as the most politically vulnerable of an entire class of authorities working in African communities after the 1920s—the opposition of native commissioners to Jeanes teachers helped encourage communities to target them. Many village leaders and residents strategically hoped to stave off most of the increasingly authoritarian interventions of colonial rule and publicly challenged the motives behind campaigns for personal and social cleanliness and similar transformative projects. For example, Kundhlande found that "cleaning roads coming to the kraals was another difficulty. They did not enjoy it. They said 'The roads lead everything to come into the homes. Native policemen, visitors, motor-cars and witch doctors at night.' "[80] Opposition of this sort affected all kinds of demonstrators and other colonial "civilizing" projects during the 1930s and 1940s. As one African spokesman noted in 1944, "We think when a white man says anything there is a catch in it."[81] Later, nationalist organizations of the 1950s achieved some of their most dramatic and rousing mobilizations by building on this earlier resistance and skepticism with campaigns of public defiance aimed at the Department of Native Affairs.[82]

These kinds of oppositional strategies and moves did not wholly disappear as European-inspired ideals of cleanliness began to ostensibly constitute part of bodily "common sense." After the end of Jeanes training, the institution of "home demonstration" and the promotion of domesticity and hygiene continued to develop, while explicit opposition based on interpretations of African "tradition" began to transmogrify into more subtle, hidden, and fractured forms of critical distance from colonial discourse about bodies and manners.

"If the Home Life is Rotten": The Widening Circle of Official Activity

The Jeanes teachers and other domestic education programs were only the most formal part of a larger attempt by the state, missions, and civic groups to transform the domestic practices of African women and thus reproduce what they regarded as clean and proper-looking African bodies. The wider promulgation of hygiene and health among African audiences, especially the small but growing elite, was staged in particular through newspapers for Africans, but also in a variety of other media and circumstances. Throughout most of the 1930s, the first of these papers, *The Bantu Mirror,* known originally as *The Native Mirror,* was more or less a direct outlet for the Department of Native Affairs. Even in the 1950s, most of these newspapers remained highly permeable to state propaganda. Throughout their history, these papers were read closely by the growing urban elite, but their circulation was also many times larger than official figures, as each copy was passed around to many readers.

The description so common in classrooms of the myriad evils of "dirt" was reproduced regularly at the behest of the Department of Native Affairs in the *Native Mirror* in the 1930s. One article in the second issue informed readers: "the thing to be feared above all others is *dirt.* In dirt is the seed of death."[83] The cover photograph of the July 1932 issue featured a young child being washed, with the caption "Cleanliness Is Next to Godliness." A regular column published in 1939–1940, entitled "Let's Ask Questions" and written by an unnamed "South African Doctor," also featured similar passages on the evils of "dirt" as well as attempts to explain hygiene in "appropriate" idioms.

Articles in African newspapers also celebrated the virtues of domesticity and promoted the vision of womanly behavior then being taught in mission schools. "No nation can rise higher than its womankind," the paper proclaimed. It continued: "It is gratifying to find at present a number of kraals where the marks of improvement are evident such as neatly plastered huts with an orderly arrangement inside; a nicely swept court with the front trash-pile missing; children tidily clothed with garments of the mother's own making; food prepared in a cleanly manner; as well as other indications of a higher plan of living."[84] An article by a missionary warned, "Home life is the very centre of strength for a nation. If the home life is rotten, then the nation is rotten. Girls must see that the nation depends on them."[85] Jeanes home demonstrators

Photograph Published in the *Native Mirror,* 1932, captioned
"Cleanliness Is Next to Godliness." (Zimbabwe National Archives)

and educators began to publish articles offering advice and instruction to women about particular domestic practices and ideas. Women were increasingly warned that they could not succeed as wives in the changing colonial world if they did not harken to the message of domesticity.

The state also accelerated its attempts in the 1930s and 1940s to bring the "gospel of cleanliness and good health" to African audiences through other media. It sponsored lectures in townships by missionaries, doctors, and state officials on hygiene, health, and "development."[86] One typical township lecture, entitled "The Road to Civilisation," given in 1944 and reprinted in the *Bantu Mirror,* compared Africans to ancient Celts in Britain and the settlers to civilizing Romans, noting that "The Romans were cleaner. They had a bath every day."[87] The state also published a variety of pamphlets, leaflets, posters, and other materials promoting collective and personal cleanliness as well as a general program of domesticity and circulated them in townships during the 1950s and 1960s.[88]

One medium that became increasingly popular with authorities was the cinema.[89] In the 1940s, film screenings in townships and mining compounds became increasingly common. Initially, many of the films used were produced in South Africa. A typical offering was *Personal Hygiene for Natives*, described as "an excellent film to encourage cleanliness and a sense of responsibility in natives . . . showing the hygienic handling of food, clean habits of the natives and general healthy living." It was made by the South African Red Cross and shown throughout southern Africa in the mid-1940s.[90] In the late 1940s, the Central African Film Unit was created and began to produce its own "educational" films, many of which contained messages about hygiene and sported titles like *Man of Two Worlds; We Were Primitive; From Fear to Faith;* and *The Wives of Nendi*. These films were shown to larger audiences after the state began, also in the mid-1940s, to operate traveling cinema vans in rural areas. Radio was also of crucial importance. Propaganda campaigns in the newspapers were increasingly echoed over the radio, and the state began to officially promote the installation of radios in African workplaces. In the eyes of the state and civic organizations, radio offered the possibility that "in every hut in every village there may yet be a shiny instrument, a source of entertainment and a new kind of 'fifth column' in the battle against ignorance and apathy."[91]

Officials of the Native Department and township authorities also began in the 1930s to intervene more aggressively and consistently with their growing juridical and administrative powers over everyday life in both urban and rural communities. Though many of these authorities had long railed against "unhygienic" and "primitive" conditions, many had only sporadically *demanded* changes. The increasing presence of various kinds of invasive demonstrators—agricultural, domestic, and otherwise—helped spur these officials to action, partly because demonstrators defined a new field of power and partly because these officials saw the demonstrators as dangerous competitors. These interventions were particularly intense with regard to collective practices of hygiene and sanitation. For example, white authorities began to aggressively compel communities to build trench latrines.

Increasing pressure was also brought to bear on the growing number of Africans working in proximity to whites to conform to colonial ideals of hygiene and manners. African teachers had long had to satisfy missionary supervisors in this regard, but by the 1930s, most Africans dealing officially with settler society were held similarly liable for their appearance by settlers or their agents. For example, those attending meetings of communities in Chiweshe Reserve in the 1930s were required by Matthew Magorimbo, the local Jeanes teacher, to be clean, "even the headmen."[92] Police, soldiers, clerks, and messengers were required to look "smart." It is also important to recall that African education continued to emphasize cleanliness. African students, their numbers growing slowly but steadily after 1945, were always compelled to conform to settler-established norms of bodily appearance and behavior. The government crowed in 1962, "Thousands of African children have been attending schools throughout Southern Rhodesia for three or four decades. One of the subjects that is never left out of their curriculum is hygiene. . . . These children spread the gospel in their homes. The gospel has caught on!"[93]

After World War II, the tone of teachings about cleanliness, African bodies, domesticity, and "civilization" began to change, though often subtly. This shift was due to a number of factors. For one, professionalized advertising and the growth of toiletry manufacture began to have an increasing influence on propaganda about bodily appearance and health.[94] At the same time, the weight and persistence of discourse about hygiene and domesticity had to some degree established European models of bodily behavior as a form of hegemonic "common sense" for many Africans, causing the content of state propaganda to become more specific and less didactic to some extent.

These subtle differences in tone can be found in the following selection from the 1960 *Bantu Mirror*, from a regular column called "The Home Teacher" written by staff of the Department of Native Affairs. The column detailed the morning routine of an idealized housewife named Mrs. Chamunorwa:

> The two children appeared in the kitchen. A basin of warm water, soap and towel stood ready. Ellen had cut up an old worn towel into four small pieces so that each member of the family had a wash cloth.
>
> Kudzai, her little son, rubbed the soap onto his cloth and began to wash his face and neck with vigour, until it shone with cleanliness. Next he took a small brush and cleaned his nails. His mother reminded him that he must rinse his mouth with clean water after he had eaten and then clean his teeth with a toothbrush. She told him that he must work the brush up and down to get all the food out of the crevices, then rub his gums with the brush; this would stimulate the blood in his gums and keep them healthy. She told him that failure to observe this rule every day would result in his getting toothache as his teeth would decay.
>
> . . . Rudo, Ellen's daughter, observed all the rules of hygiene without having to be told. Every day she stripped off all her clothes and washed her body all over with soap and warm water. Every day she brushed her hair vigourously and once a week she washed her hair with soap and water.[95]

A set of root concerns and obsessions linked the vision of this passage with earlier productions of domesticity. Cleanliness was still presumed by white

authorities and employers to be an ideal form of behavior most Africans ignored. The column still held women responsible for the hygiene and health of
the family. It presumed that women have an innate inclination for this task
(Rudo observes the rules without being told, unlike her brother). Still, the
practices being promoted were more specific and numerous and were tied to
specific commodities. There was less moralization and mystification about the
dangers of dirtiness. There were fewer instances of racist sermonizing about
African bodies and practices in *public* arenas.

"Doing It for Themselves": Women's Clubs and Radio Homecraft

Changes in tone were also evident in new institutions for the promotion of
domesticity that began to appear in the 1940s and grew dramatically a decade
later: the "homecraft" movement and African womens' clubs. At the same time
that messages about hygiene were infiltrating every aspect of everyday life that
colonial authorities could touch, the model of "home demonstration" developed by Jeanes teachers was reproducing itself in new mass institutions. Each
part of the daily routine of the Jeanes women—group meetings with community women, "home visits" and village cleanups, demonstrations of "proper"
domestic practices for female (and occasionally male) audiences—helped form
the basis of activities for a wider postwar network of African women's groups
and societies. The stereotypical images that the Jeanes women and their superiors tried to create around the institution of home demonstration formed the
basis for one of the most enduring archetypes of female personality in most
Zimbabwean communities from the 1950s to the present day: the "respectable,
club-going, Christian wife."

The reproduction of home demonstration and its attendant domestic
ideologies in mass organizations in colonial Zimbabwe mirrored similar efforts
ongoing throughout the rest of colonial Africa. Nancy Rose Hunt, for example,
has described the institution of *foyer social,* "social homes," in colonial Burundi
from 1946 to 1960. The practices and ideologies that Hunt recounts resonate
deeply with developments during the same period in Zimbabwe.[96] The earliest
mass institutions formed to promote domesticity in the 1920s and 1930s were
the *Ruwadzano* clubs for African women in the Methodist Church,[97] and the
Wayfarers, the branch of the Girl Guides reserved for African girls.[98] In all cases,
these institutions were closely attached to forms of hygienic and domestic
training already developing in the missions and their schools. These groups
often acted in concert with township administrators to sponsor classes and
lectures in domestic subjects.

A significant proportion of the supervision of hygienic training and domesticity was gradually assumed during the 1940s by the all-white Federation of Women's Institutes of Southern Rhodesia (FWISR). The FWISR set up an "African Interests" section immediately after World War II and offered a compilation of familiar justifications for this move: "an ordinary housewife superintends and controls African servants and is in close daily contact with them. . . . African women's problems are as much our concern as ever . . . it has become so obvious that the women must replace the men in our domestic lives."[99] A few of the members of the FWISR had been involved through missions in the 1930s with the development of domesticity training; some had actually been teachers or supervisors of the Jeanes women or similarly trained female students. However, the FWISR also created opportunities for a larger group of settler women to observe and intervene in the everyday life of African women. White women who were FWISR members were in some ways subjected to some of the same forces and expectations that affected the lives of African women; the gradual control granted to FWISR members over institutions that shaped domesticity for Africans was largely motivated by an attempt to grant white women a specialized form of colonial power without interfering with patriarchal prerogatives.[100]

FWISR members steadily expanded their role in institutions for training African women in the 1940s and 1950s. In 1942, for example, Catherine Langham created a state-sponsored "Homecraft Village" to train what she saw as clean, disciplined, and domestically able African women. Cleanliness remained a central fascination in Langham's blueprint. She "saw the tremendous need for teaching the African women hygiene," thought "husbands should be free to visit and stay a few days, that they might see the possibilities of a village run on hygienic lines," and formed students into "hygiene squads."[101] The Homecraft Village, with enthusiastic endorsement and support from the FWISR, went from 37 pupils in its first year to 130 by its third year. The students' training was similar to that given earlier to the Jeanes women. Like the Jeanes teachers, the graduates of the Homecraft Village were encouraged to return to their home areas and teach other women to follow the dictates of domesticity, though they were not always paid by the state or by missions to do so. Graduates became the core founders of the "homecraft movement" and the African women's clubs.

African women's clubs appeared in the late 1940s, based on earlier civic organizations like *Ruwadzano* branches; the largest of these clubs was the Mrewa Bantu Women's Club, with a membership of "over 700" in 1950.[102] The size of this club was largely due to the influence of Helen Mangwende, the second wife of Chief Munhuwepayi Mangwende of Murewa. Chief Mangwende was later deposed following a complicated personal and political struggle with white authorities, but he had originally been a favorite of the Department of Native Affairs and was continually cited in the 1940s as the exemplar of a "progressive" chief. Both of the Mangwendes shared a strong commitment to "modernizing" African daily life in Murewa District and elsewhere. Helen Mangwende was especially committed to domestic training. Though she died quite young, in 1955, she had an enormous impact on the organization and tone of domesticity and hygiene in postwar colonial Zimbabwe. She began her efforts by aggressively proselytizing for domesticity and hygiene in Murewa itself, through personal visits, propaganda of her own design, and the organization of a women's club. To pursue these projects, she solicited the assistance of local state authorities and representatives of the FWISR.[103]

Something of Helen Mangwende's personal dedication to these goals came through in the Central African Film Unit's 1949 film *The Wives of Nendi*. Mai Helen[104] appeared in the film, and according to one of the filmmakers, she also functioned as an unofficial director and scriptwriter during its production.[105] The film tells the story of the establishment of the Murewa African Women's Club. It opens with Mai Helen and a female assistant giving a typical lecture to women of a local village. The women share their domestic accomplishments with their teachers. One village in the district is recalcitrant and insists on remaining dirty. The film develops into a sort of hygienic Western at this point. Mai Helen appoints a local woman to "clean up the town" and reform the dirty headman and his wives. This demonstrator strives to change the practices of the chief's wives, showing them again and again proper domestic habits like washing with soap, washing in clean water, and cleaning out dirt from the village; but they refuse to obey and eventually ostracize the demonstrator. Chief Mangwende and his wife are eventually forced to personally intervene, and the recalcitrant headman "converts" to good, modern ways. To signify this epiphany, he starts wearing a suit and tie and holds a cake making competition.[106]

Helen Mangwende's personal commitment to the spread of domesticity through club organizations and the interventions of demonstrators, as outlined in *The Wives of Nendi,* intersected with the FWISR's growing authority over the official and semiofficial network of institutions promoting the domesticity and hygiene of African communities. This produced two major results. The first was the formal creation of the Federation of African Women's Clubs (FAWC) in 1952 and the second was the growth of *Radio Homecraft.* The FAWC knit together the seventeen African women's clubs already in existence in 1952, placed them formally under the control and patronage of the FWISR leadership (women with past involvement with missions and African administration like Catherine Langham, Fay Staunton, and Barbara Tredgold), and sought to create new clubs in all the remaining districts, villages, and townships where there were none. The FAWC seized upon all the characteristic features of colonial domesticity: home visits and traveling demonstrators, classes, competitions, songs, lectures, parties, and social meetings, and gave them a true mass dimension, involving African and white women in an expanding network during the 1950s and 1960s. There were 250 clubs by 1963 and 900 by 1968.

FAWC members wore badges inscribed "Progress" and chose individual names for their club branches designed to reflect the goals of the members— names like *Kufambira Mbera* (go forward) or *Rufaro* (happiness).[107] The white FWISR supervisors kept a careful eye on the clubs, cautioning members that they must remain "nonpolitical." FWISR leaders actively linked the Jeanes era interest in domesticity with FAWC, giving a progressivist account of the "rise of African womanhood" and hammering at the basic preoccupations of domestic and hygienic training.[108] The FAWC membership was further linked with past programmatic forms of domesticity by the presence in the clubs of women trained in schools in the 1930s, including Jeanes teachers: "At any women's meeting composed of leaders amongst the African Women's Clubs, there was always a good proportion of old-girls from Hope Fountain."[109]

The FWISR took over most previous programs for reproducing colonial ideals of hygiene, domesticity, and household management in African communities, accomplishing this objective through the FAWC and through the popular *Radio Homecraft*. *Radio Homecraft*, broadcast in a mix of Shona, Ndebele, and English, was a regular feature in the lives of FAWC members, but it also reached women who did not live near a club, who only attended FAWC events sporadically, or who had not yet joined a club. In particular, competitions for the best craftwork held through *Radio Homecraft* attracted huge responses, initially swamping the FWISR bureaucracy. The shows helped recruit more and more members for the FAWC and helped create a demand for publications like the FWISR's magazine for African women, *Homecraft*, in which the results and rules of competitions were sometimes announced. *Radio Homecraft* popularized the symbols and rituals of "club-going" society, spreading songs and slogans over a wide expanse. A typical song on *Radio Homecraft* went as follows:

> *Murikuenda kupi, Mai?*
> *Murikuenda kupi?*
> *Amai, Amai, Chipindurayi*
> *Murikuenda kupi?*
>
> *Ndiri kuenda kuClub, vana vangu,*
> *Ndekuchimidzaendape.*

Kunowandisa dzidzo yangu,
Ndiri kuendapo.

Where are you going, Mother?
Where are you going?
Mother, Mother, turn back
Where are you going?

I'm going to the Club, my children
I'm just going there quickly.
There my knowledge is increased
I am going there.[110]

Radio Homecraft not only supplemented the growing network of FAWC clubs, but it encouraged most other civic organizations in African communities to form their own special "women's sections": the wives of black policemen formed their own club, Kuyedza, as did the wives of prison officers; most church congregations with an African membership reformed or revived their women's auxiliaries.[111]

Radio Homecraft's programming and the content of FAWC and other clubs' activities continued to reproduce all of the figurations concerning gender, hygiene, and domestic work that had been common in the 1930s, though the new media and structures that these institutions worked through gave them a particular flavor. Changes in official doctrines were also significant. The FAWC and *Radio Homecraft* were frequently cited by the Rhodesian state as success stories in "community development." This was of course specious, a slight variation on South Africa's claims about its "homelands" policy, but institutions like the FAWC developed enough internal dynamism to seem to give some credence to the proclamations of the Rhodesian state.

Hygienic and domestic training and language had strong generative connotations among men and women, especially women, who were affected by the homecraft movement and the FAWC, continuing the process by which colonial discourses about African bodies became disciplines governing African subjectivity. Some of the most powerful evidence in this regard comes from the impact Helen Mangwende had on many of her contemporaries. One of these women, involved in the founding of the Murewa chapter and subsequently the FAWC, now represents Mai Helen as being "as good as Christ himself." For Mbuya Gwata, the time she spent working with Mai Helen to build the clubs is now her richest and most treasured memory. Today, her craftworks from that time are among her most honored and valuable possessions, proudly shown to any visitor who inquires. More importantly, Mbuya Gwata has insisted that the white women of the FWISR played little or no role in the regular business of the clubs and were of no importance to the rank-and-file membership. She claims instead that Helen Mangwende and her assistants founded the clubs largely on their own.[112] Such sentiments surely reflect in part discomforts stemming from the *chimurenga* (the struggle for independence, fought primarily during the 1970s), when some club members were accused of being *vatengesa* (sellouts) because of their associations with whites. Some of the other former club members we spoke with were also uncomfortable with contemporary Zimbabwe African National Union (ZANU) women's organizations that are trying to reinvent the clubs, which probably sharpens their nostalgia for the past. Still, Mbuya Gwata's pride and affection for her role in the Mrewa African Women's Club was broadly reflected in many other conversations we had with women who had attended clubs or competed in homecraft competitions. For many of the members, club attending had been a considerable source of strength, knowledge, and identity, and the ideals of cleanliness and domesticity it promoted were incorporated in many ways into their lives.

At the same time, the storyline of *The Wives of Nendi*, with its recalcitrant "dirty" villages, suggests that the opposition to colonial discourses about the body and household that had bedeviled the Jeanes teachers in the 1930s still had some purchase in the everyday lives of many Africans, particularly in rural communities, in the 1960s. Moreover, the mass character of the club movement and the increasingly commodified and interpenetrated culture of domesticity in the 1960s meant that the emphases of club meetings and activities often generated new desires and pressures that were unprovided for in the basic charter of colonial domesticity and hygiene. Some club-going women, for example, were eager to acquire a knowledge of "high" European manners and fashion. FWISR publications uneasily took note of these interests.[113] In one forum, they began a series of stories on "Manners and Customs" featuring club-going "Mary" and her ill-mannered, embarrassing, "raw" neighbor "Elizabeth Dube." In each story, Elizabeth Dube's aspirant husband and Mary both exhort Elizabeth to learn new stylish interpersonal skills, such as "charm," "sitting nicely," "eating properly," and "decorating her home."[114] At the same time, recent scholarship has shown that rural women attended club meetings, sometimes over the objections of men, to serve a variety of interests and needs that were relatively independent of settler ideologies concerning domesticity and femininity.[115]

During the 1970s, the intensity of the *chimurenga* shut down many club operations, particularly demonstration vans and lectures, which had been among the most popular features of club meetings. The bluntness of the Rhodesian state's propaganda and its preoccupation with warfare during this time similarly led to an ebbing of state interest in domesticity and hygiene. However, membership in the clubs also became much more problematic, as the supervisory role of white women in the movement and the Rhodesian state's obvious interest in claiming FAWC as a successful instance of "community development" made many activists regard African women in the clubs as "sellouts." Many other women deserted the clubs to assist the armed struggle or used the clubs to assist the liberation armies.[116] But in those two decades prior to the intensifying of the war, "club going" had been a common experience for many urban and rural women.

The leaders of the FWISR and the various assemblages of official and philanthropic power that surrounded them continued to regard the FAWC and *Radio Homecraft* from much the same perspective they always had. The movement, in their eyes, remained the Promethean gift of the civilized to the uncivilized. However, the mass character of the membership of the FAWC and the deteriorating situation of colonial rule led to a new note of doubt and concern about the construction of domestic knowledge and institutions. When the Jeanes women were educated, said one FWISR leader, "they were so few amongst the many thousands."[117] As with so many other aspects of colonial pedagogy and socialization, many of the earliest settler promoters of hygiene and domesticity never really imagined that the targets of their campaigns, female and male, would actually change. FWISR publications continually stressed the controlling role of whites and faintly cautioned that FAWC members should consider retaining domestic practices appropriate to their race and class. However, as Sita Ranchod-Nilsson had pointed out, FAWC members seriously challenged the presumptions of FWISR leaders about the nature and interests of African women.[118] Those African women who flocked to the FAWC in the 1950s and 1960s made their own demands on the clubs. Some of them, the wives of the urban elites, demanded "inappropriate" knowledge of "European" manners, needs, and skills, though such demands sometimes also created possibilities for mutual understandings between upper-class white female philanthropists and aspirant African women, through the mutual practice of high-society manners.[119] The FAWC also solidified European-inspired ideals of hygiene and domesticity into a shared, collectively built form of "common sense." All of these institutionalizations of hygiene, manners, and bodily demeanor, from early mission schools to the FAWC, powerfully shaped the cultural and social facts upon which the marketing and use of manufactured toiletries drew heavily from the late 1930s up to the present day.

3 BUCKETS, BOXES, AND "BONSELLA": PRECOLONIAL EXCHANGE, THE "KAFFIR TRUCK" TRADE, AND AFRICAN "NEEDS"

The colonial construction of "African bodies" through hygiene, domesticity, and manners provided many of the "prior meanings" that shaped the consumption of manufactured toiletries in Zimbabwe. The pre-1945 development of merchant capital and its relationships with precolonial exchange and production formed another essential part of this background. This chapter reviews trade, and attitudes toward trade, involving various African communities and small merchants from the nineteenth century to the late 1940s, when a boom in manufacturing reordered consumption and exchange.

The Value of the Foreign: Precolonial Exchange in Zimbabwe

Studies on the Shona states of the fifteenth through eighteenth centuries have stressed the crucial role played by long-distance trade in successive polities.[1] In his analysis of the precolonial Mutapa state, S. I. G. Mudenge agrees with other scholars that various luxury goods (beads, cloth, seashells, porcelain, and other items) traded for gold and ivory were important for expressing the public symbolism of elite power. On the other hand, Mudenge stresses that "some of the external luxury goods, like beads and cloth, had equally valuable internal substitutes."[2] Moreover, he argues, assertions that the Mutapa elite had a strict monopoly on such goods should be regarded with suspicion: "In some areas the peasant way of life may well have quite accidentally, qualitatively improved as a result of external trade."[3] Access to "foreign" luxury items may well have been more common than many have assumed, and such consumption was surely an arena for active struggle between elites and others.

Nevertheless, one legacy of successive plateau states was a persistent cultural association between imported goods and elite power among Shona speakers. Such associations were reinforced during the second half of the nineteenth century. The eclipse of the larger Shona states and the coming of the Ndebele shifted the dynamics of long-distance trade, giving greater though not exclusive emphasis to trade from the south rather than the east. As Ngwabi Bhebe has pointed out, the trade with parties from the south had two dimensions, a "subsistence-oriented" trade in cattle and grain, and the exchange of ivory for manufactured goods. Like Mudenge, Bhebe points out that the southern trade in luxuries and novelties was not solely a royal Ndebele monopoly even to start with, touching a wider and wider circle of people by the 1870s.[4]

Traders in nineteenth-century Zimbabwe (and missionaries, who acted as traders), carried goods similar to the ones listed in the following inventory:[5]

	£ s d		£ s d
Muskets, 20	60-0-0	Single Rifles, 12	84-0-0
Powders, 5 lb. bags	30-0-0	Lead, 300	15-0-0
Tins, 50	5-0-0	Caps, 10000	2-10-10
Blue beads, 100 lbs	5-0-0	Red beads, 100 lbs	7-10-10
Tinder boxes 12/-	0-10-0	Clasp knives 12/-	1-10-0
6 American hatchets	2-0-0	2 larger axes	1-0-0
Copper wire, 25 lbs.	1-5-0	1 doz. Blucher boots	10-0-0
Boer brandy, half arm	12-0-0	1 doz. butcher knives	0-12-0
3 cheeses	2-5-0	2 cases fish	5-0-0
Do jam	5-0-0	Dried fruit	1-0-0
Coffee, 300	15-0-0	Sugar	10-0-0
Tea	5-0-0	Chicory	1-0-0
Chocolate	2-0-3	Blankets	5-0-0
Handkerchiefs	1-0-0	Calico	1-0-0
Canvass	3-0-0	Saddle/bridles	10-0-0
Currie comb	0-10-0	Wagon	130-0-0
Oxen	75-0-0	6 horses, salted	200-0-0

Other goods typically included in such inventories were other types of beads and clothing, including hats, jackets, dresses, and singlets; mirrors; and snuff boxes or tobacco products, though there was also a thriving regional trade in tobacco grown to the north of Bulawayo.[6] Importantly, with regard to the overall subject of this book, soap was not part of such inventories and did not become a regular part of the trader's stock until significantly later, between 1910 and 1920.[7]

Ivan Fry, who traded at Lobengula's court, described in his memoirs a volatile world around trade. In particular, Fry recounted elaborate maneuvers and strategies between traders and their customers over the matter of local needs and tastes: "The beads used by the Matabele were all small beads. The Red

White eye, the Ingazi, was all the rage originally. . . . Then the pink bead came in. I have forgotten the name of it. Then they had the Intutuveyani. These were the white and blue beads. . . . They did not use big beads. A lot of traders found themselves stuck with beads they brought. . . . They used to sell them to the Makalaka, who would not wear the same beads as the Matabele."[8] Mzilikazi, Lobengula, and other ruling elites in both Mashonaland and Matabeleland certainly had preferential access to such imported items; the tastes of the elites were important to setting the pace and trajectory of fashion. Lobengula in particular became known among the expatriate community for his strategic use of manufactured goods to symbolize his simultaneous power and influence over his own society and over the expatriates. Many of the European visitors to the capital noted his carefully chosen mixing of "indigenous" and European fashions and commodities. They were especially interested in his use of manufactured clothes, as well as his reputed caches of other trade items like expensive liquor and perfumes.[9]

However, there was also an increasingly frenzied and growing community desiring "foreign" goods during the 1870s. As the European community's need for food obtained from local cultivators increased, and Lobengula's control over ivory became less exclusive, the desire for and access to the prestige, novelty, and intrigue of manufactured commodities accelerated. Many visitors in the 1870s and 1880s described the experience of being besieged by crowds hoping for "presents," and those who traveled beyond Bulawayo into Mashonaland and the Zambezi Valley met with increasingly larger and more assertive demands from local elites for a variety of goods as the price for their safe passage.

The Establishment of "Kaffir Truck"

As Ngwabi Bhebe accurately points out, Zimbabwean peoples "did not have a sharp and abrupt jump into the market-oriented economy of the colonial twentieth century . . . the change was in many respects one of intensifying some already existing activities."[10] There were, however, some important though subtle changes that came with the formal establishment of colonial rule in 1890. For one, traders were freed to a significant extent from the dictates of local African elites. Moreover, in the early 1900s, traders were gradually drawn into the building of various colonial institutions. However, the inventory and operating strategies of traders changed more slowly, remaining fundamentally similar to pre-1890 activities until the second or third decade of colonial rule.

The role of merchant capital in the articulation between the rural peasantry, settler mining, and agriculture, as well as the growth of urban settlement, has not been examined a great deal in Zimbabwean or indeed southern African colonial history as a whole. However, these traders, often referred to at the time as "kaffir truck" merchants, have sometimes been discussed in studies of the broader colonial political economy.[11] Scholars like Ian Phimister, Charles Van Onselen, and Giovanni Arrighi largely view traders as instruments of primary industry and agriculture who acted to drain off surplus and block rural accumulation, thus accelerating proletarianization. In contrast, Barry Kosmin's 1974 Ph.D. dissertation approaches merchants as ethnic communities of socially disempowered "middlemen" and explicitly takes issue with the arguments of Phimister and others. Kosmin argues that the major legitimate criticism of colonial traders is simply that "there were just too few [traders]."[12]

It is important in this light to emphasize that traders made new goods, with all their provocative cultural and economic possibilities, widely available to peasant cultivators and migrant workers, as well as to the new colonial elites. On the other hand, stringent colonial polices of labor conscription, taxation, forced removal, and segregation implicated manufactured goods and merchants in an increasingly oppressive system. Moreover, the growth of settler agriculture following early crises of viability in mining led in the 1920s and 1930s to the construction of laws and mechanisms designed to impede peasant accumulation while accelerating proletarianization. These efforts added to the already growing legal and social armament of traders during the 1930s and thereafter permanently altered the balance of power between merchants and their African customers.

Retail stores were opened on farms, in rural "reserves," on mines, and in townships in increasing numbers throughout the first third of the twentieth century. The proportional location of licensed general dealers (the license was required for trading in imported goods), shifted during these years: mine stores accounted for 28 percent of licenses held in 1912 but only 6 percent in 1942, while other areas evenly picked up the slack left by the decline of mine stores.[13] The bulk of sales to African customers until the 1930s consisted of clothing and ornaments, followed in descending order by tools and farming implements, patent medicines of various kinds, soap and other toiletries, foodstuffs (especially bread and tinned foods), cigarettes and matches, and bicycles. A significant but by no means dominant segment of "truck" salesmen were either ethnic whites held to be "inferior"—Greeks, Central European Jews—or Asians. Others were white migrants from South Africa, including a significant early group of "public-school boys" and men like Stanley Portal Hyatt.[14] As Hyatt makes clear, in the first decade or so, "kaffir truck" was a common pursuit for many of those who came north, "until the Boom should come. Then, of course, it would be left to the coolies and Germans."[15] Owners of farm stores were also predomi-

"Truck" Store in Murewa, 1906. (Zimbabwe National Archives)

nantly Afrikaans-speaking or British. (Notably, farm stores accounted for over 50 percent of the general dealers' licenses granted from 1900 to 1942.) After the 1920s, "coloreds," Asians, and non-Anglo-Saxon whites made up an increasingly larger proportion of the active traders. Still later, in the 1950s, Africans were a growing proportion of traders dealing primarily with African customers, but by that point the entire nature of the "truck" trade was changing.[16]

The fundamental structures of merchant activity from 1890 to the 1940s were marked by several levels of relative subordination within the world and local economy. Most "kaffir truck" came, until the 1940s, from distant foreign producers. Clothing came from large producers in England and later Japan, Italy, and India who manufactured cheap, low-grade textiles expressly for the African market, in particular West Africa. Likewise, other goods, like ploughs and bicycles, were initially imported from distant sources, usually in large consignments, by colonial wholesaling firms, the largest of which maintained their own purchasing agents in London. Other goods, like soap, came from South Africa, Mozambique, and by the 1930s from Rhodesian producers.

"Truck" retailers therefore depended on distant sources whose products were purchased in bulk consignments chosen solely for their low price. These retail traders were also dependent on local wholesaling houses responsible for the purchase and shipping of these goods. Wholesale firms like Lasovsky Brothers, Landau Brothers, or Latif and Sons exerted tight control over client retailers and often financed new merchants with extensive credit, thus creating "tied" stores. The retailer existed simply as the purchaser of the huge bulk shipments that wholesalers' agents had managed to obtain. Wholesale firms would dump large amounts of goods that they had purchased largely out of price considerations and trust retailers to dispose of those goods however they could. During World War II, for example, the usual sources of cheap clothing were cut off, so wholesalers bought a huge consignment of Argentinian blankets and demanded that retailers purchase a number of these blankets along with any other orders. The blankets were universally loathed and hugely over-valued, but retailers were forced to try to sell them.[17] The son of a Chinhoyi trader active in the 1940s says, "I remember my father saying that to get ten pieces of cotton drill he had to get three or four thousand blankets, secondhand from the war and they were imposed on you, they were forced on you. That trading did exist, a point of conditional sale."[18] The growth of local manufac-turing, new channels of distribution, and a secondhand trade in goods, espe-cially clothing, all eroded the power of the traditional wholesalers beginning in the 1940s, but their dominance over retailers only really began to substantially fade a decade later in the 1950s, when less exclusive "cash and carry" wholesalers began operating in concert with local manufacturers and new retail chains. But from the earliest days of colonial trading until that time, the wholesalers played a role that Hyatt referred to as "rapacious," operating as "moribund" concerns that "stink in the nostrils of all decent folk."[19]

A further restriction on merchant activity came from the general nature of the colonial political economy. Until manufacturing expanded in the 1940s, Rhodesian capital was overwhelmingly oriented toward primary production and overtly hostile to increases in African purchasing power or accumulative capacity, both of which would have opened significant new opportunities for "truck" retailers—though not necessarily opportunities benefiting African con-sumers themselves. The low social status of the profession also limited the political leverage held by retailers and wholesalers alike. White and Indian "truck" traders had a good deal more leverage and power than their African customers, but their collective interests frequently suffered in relation to other factions of local and international capital prior to the 1950s.

Merchants, Rural Communities, and the Shifting Culture of Exchange

The structural fissures in the economic and social positioning of the "truck" trade from 1890 to 1945 combined in interesting ways with the relations between traders and African customers and changing modes of consumption in different African communities. In rural communities, many merchants were initially viewed with a mixture of suspicion and interest. The former sentiment arose out of the general antipathy that many villagers held toward whites and their institutions. The latter stemmed from two sources. First, rural traders did more than just sell goods. Their predominant economic activities actually centered on purchasing or bartering for surplus grain and cattle to supply to mines and urban consumers. For rural African villages near growing urban areas, mines, or trade routes,[20] the presence of traders seemed to offer an opportunity for new forms of accumulation or at the very least provided a ready source of cash for paying annual taxes without resorting to wage labor. Yoshikune's work on early "peri-urban" elites around Harare suggests some of the types of class formation made possible by these early relationships between merchants and local cultivators.[21]

At the same time, the very premise of the rural "truck" trade, even when explicitly racist in the physical separation of goods for white and black shoppers under the same roof, was initially full of attractions. The theatrical nature of shopping, the verbal play of barter, and the discovery of new objects whose cultural potential was open all gave the new stores considerable allure. One detailed settler account of a typical morning in a Chinhoyi "truck" store in the 1920s incidentally captured some of the pleasures involved in shopping:

I watched the trade. It was mainly barter, and the details of the business were ruled by rigid custom. Grain was weighed out, and the price paid in coin to the bringer, who fingered each separate disc and passed them round to the others for inspection. Meanwhile B. attended to other traders. Finally, the bringer of grain laid out the full amount of the coins on the counter, and began to buy. As each article was handed over, B. removed the price from the coins, replacing with change where necessary. Each piece of goods was a separate deal, the customer taking a vast delight in spinning out her shopping as long as possible. . . . The first woman had brought a large basket of grain. . . . She received her payment in cash and gazed around the store, drinking in the pleasure of anticipated purchase . . . the woman gingerly fingered the top of a blanket, slowly unfolding it. She seemed torn between the attraction of the broad red stripe on one and the triple green ditto on the second. . . . The prospective buyer turned the green-banded blanket over and over, feeling its texture.[22]

The attraction that traders and their wares had for local communities in the early colonial period, in both the rural areas and in the growing urban townships, was situational and deeply rooted in precolonial valuation of the "foreign," as well as in more immediate crises in local production of iron and other durable commodities, crises caused in part by the articulation of the colonial labor market with local modes of production. As Nicolas Thomas points out with reference to Pacific societies, "Western commodities cannot be seen to embody some irresistible attraction that is given the status of an inexorable force. Indigenous peoples' interests in goods, strangers and contact were variable and in some cases extremely constrained."[23]

Some of this variability expressed itself through various collective identities and affiliations. For example, "truck" stores were often especially appealing to rural African women. As more and more men were drawn into wage labor and the responsibility for all cultivation fell on women left in the villages, the opportunity afforded by the stores to trade surplus grain for cash and commodities provided women with new leverage over household and family economies. Village patriarchs from the outset of colonial rule sometimes regarded "truck" traders with alarm for precisely this reason. Chief Mtekedza of Sabi Reserve protested to state officials about the threat *to* women and the threat *from* women posed by a new store: "What grain is left by the locusts will be taken to the store. If anything happens to the women-folk by the store servants than I will say I was not consulted as to the store site being granted. There are bound to be complaints to me that there is a shortage of food in the land if the store is erected."[24] One state administrator complained similarly and typically, "a trader will foist an article on a wife while the husband is away and returns later to collect the money from the man, thus causing quite a lot of unpleasantness between husband and wife."[25] These relationships also involved the social lives of the traders themselves. Many of the earlier traders took local wives or mistresses and followed local customs to a significant extent. Indeed, much of the scorn directed at "truck" merchants by the governing elite was provoked by the visible intertwining of the lives of rural shopkeepers and their customers. Two of the most prominent institutions of rural trading, hawking and "bonsella," underscored the uneasy but often fruitful intersection of the lives of early traders and their customers.

There were a number of different sorts of hawkers operating by the 1920s. Those carrying imported goods of any kind had to be licensed. Some of these licensed hawkers operated as independent, mobile traders, but most were the agents of a large wholesale or retail business, carrying goods to townships, rural homesteads, and compounds. If a hawker carried only Rhodesian-made or -grown produce, he (hawkers were almost always male) did not need a license. However, these unlicensed hawkers sometimes surreptitiously acted as the agents of a central organization. They also sometimes bent the letter of the law—by selling bread stuffed with imported jam, for example. And, of course, some hawkers ignored the law altogether—probably more so in the depression of the 1930s than earlier.

Typically, hawkers were sent out from a well-stocked central location to reach a more isolated market, though they also operated in townships where municipal authorities had refused to grant general dealers' licenses. Hawkers who were employees of a centrally located store were usually Africans, though until the 1920s many white and Asian traders also traveled about in wagons or on bicycles. Even earlier, in the 1890s and 1900s, African salesmen from South Africa traveled into Matabeleland and Mashonaland selling knobkerries decorated with brass wire and other similar crafted goods, drawing the wrath of "truck" merchants.[26]

Some white traders and businessmen complained strenuously about more centralized forms of hawking after the 1920s and characterized them as the nefarious work of "undesirables" and as a type of dangerous social disorder. One trader protested during the 1940s: "You get some people sending out Natives on bicycles with stuff for the specific purpose of selling them around the other man's store, in competition. Some send them from the Township to go into the Reserves or to the farms where there are lots of boys. In some cases they get reported to the Police."[27] Aggrieved traders and farmers implied that the use of hawkers somehow broke the unwritten rules of the trading game and suggested that the willingness to employ hawkers was a sign of racial or ethnic inferiority. It was regarded as unfair that hawkers, as agents of these merchants, were able to meet Africans on their own ground, speak to them in their own languages, and offer unusually low prices. Alluding to these unwritten social rules, one trader complained, "We have got any amount of hawkers . . . they are from the Indians . . . these hawkers come to them [African customers] on Sundays, when they are at their beer drinks and aren't thinking of business, and sell them a woman's skirt . . . it is just the idea of it. Talking to Natives at their beer drinks."[28] Merchants who did not employ hawkers reported losses of up to 50 percent of their business when hawkers moved into their operating area: "Some of the natives are employed by Indians and others both of Que Que and Gwelo and do not have licenses—they carry passes to sell Rhodesian products. The natives are doing us untold harm as they are going from village to village, every day of the week including Sundays and irrespective of the distance from stores."[29]

Hawkers and other itinerant salesmen helped bring consumption of foreign goods into the intimate domestic space of the *musha* (homestead). They did not do so altruistically, of course, and this was widely understood by all parties to these transactions. Many of the traders and their hawkers were aware that peasant cultivators had accumulated small reserves of cash which they attached little value to beyond paying annual taxes and state fees. Equally, many villagers viewed these travelers with cautious anticipation and apprehension. Some of this ambivalence was described by Vambe in his account of Mashonganyika village in Chishawasha in the 1920s:

> When the travelling traders sailed into the village, with their gaudy wares and sleek ways, the hoarded coins were reluctantly fished out in exchange for such things as beads, bangles and edibles. But in every case there would be haggling and bargaining and a contest of wit. A very regular visitor of this kind to Mashonganyika was called Pondo, Mr. "One Pound," because whatever he offered for sale, however big or small, started at the standard price of £ 1 until you beat him down. He travelled on a very old, rusty bicycle . . . bulging with an assortment of goods, including makokisi and sweets to give away as tokens of friendship and good will. . . . Almost the first remarks he made were a summary of his general philosophy of life. . . . He said that money was something to spend and not to keep under mud floors. However, in spite of his charm, Pondo showed plenty of signs that he did not trust his customers, for he always kept a sharp eye on his goods.[30]

This mingling of civility and suspicion was even more evident in the practice of "bonsella," a key part of salesmanship employed by rural merchants. Like hawking, bonsella was regarded by some as inappropriate because of its implied social familiarity with Africans. Bonsella was the custom of giving small gifts to African customers—usually sweets or biscuits, sometimes cigarettes, matches, or snuff. Sometimes it also involved sitting down with customers and sharing tea or some other drink with them. Vambe described several instances of bonsella in his memoirs. The traders operating at the Arcturus mines were all given nicknames, and one was named for his variety of bonsella: "one man was called Makokisi, a large, ebullient-looking man . . . so called because he was in the habit of offering his African customers sweet buns which were as hard as toasted bread and were called Makokisi."[31] Another trader, "Pfumandiwe," also used bonsella to encourage sales: "Pfumandiwe used a trick which usually worked very successfully with men from Nyasaland. When they came into his shop, he treated them as guests of honour. He took them to a special room, fed them with plenty of white bread and sugared water (as tea) and inquired about their health, jobs and relations first before any talk of business. Thus physically and psychologically well cared for by this benevolent and friendly Pfumandiwe. . . . they literally opened their wallets in a spending spree. These men bought any rubbish which Pfumandiwe dangled before their faces."[32]

Bonsella was seen by settlers as the characteristic line that divided merchants in terms of their racial status. One colonial official commented: "there are quite a lot of British traders who don't play the game towards the Native who comes to purchase. They kick him around to the back, and he says he will not bother to go to the British, he will go to the alien, where they do get a bit of attention and a little *bonsella*, a few biscuits or a few sweets."[33] An Indian storekeeper agreed: "If a customer comes into the store we have to treat him properly and show him everything."[34] The Bulawayo Chamber of Commerce showed its allegiance by condemning the practice.[35] For many whites, the giving of bonsella threatened boundaries of race and class by promoting social familiarity between whites and blacks: "the average kaffir-store keeper will lower himself to any indignity to get a native to buy his wares. He will set out to make the Native feel that he is honoured by his trading with him and will give the Native a 'bonsella,' in addition to what he has purchased. 'Bonsella' is always demanded and given."[36] A white official raged, "The practices adopted by the trade in order to get money out of a native make one feel somewhat degraded. . . . I have seen a white man sitting on the floor of a Kaffir store eating ginger nuts and drinking ginger beer with native men and women. It is against morality that these people should be allowed to trade."[37]

Bonsella and hawking were both strategies for producing connections between rural communities and the act of consuming manufactured goods. The traders who pursued such tactics were perceived differently by their customers, as Vambe made clear: "We were by now effectively, but erroneously, indoctrinated with the idea that all white men who owned shops were maJuta [Jews]. . . . But these traders turned out to be the most friendly and . . . were closer to the African people. They were more human and humane. . . . They always endeavoured to put us at our ease, to make us feel that we were human beings and the result was entirely satisfactory in both human and commercial terms. They tried hard to learn our language, idioms and customs, which was in itself a good sales-gimmick."[38] The discomfort felt by some whites about these practices was based on a semi-accurate understanding of the actual state of relations that they implied, for the intimate connections between traders and their customers often extended beyond the stores themselves. For example, a Jewish trader named Selesnik was condemned as an "undesirable" for residing within an African village and living under the authority of its headman.[39]

Another prominent "truck" trader of the 1930s and 1940s, G. T. Thornicroft, was said to be the "father of the Coloured community," underscoring the nature of his involvement with rural Africans.[40] In Arthur Cripps's novel *Bay Tree Company*, published in 1913 and based on a forced labor scandal in Makoni in 1911, the two heroic white characters are a missionary and a grain trader. The trader is made sympathetic through his genuine empathy for and understanding of African life. As Terence Ranger points out, Cripps's portrayal, though fictionalized, underscores something important and real about the closely intertwined relations between early grain traders and many rural African communities.[41]

However, the same trader who one day gave out bonsella could tomorrow trick a customer into acquiring cheap junk or alter the accounts of credit payments. As Vambe pointed out, courtesy and kindness were sometimes also perceived as a sinister deception: "Men like these . . . brought into very sharp focus the acquisitive, often ruthless instincts of the European to gain wealth, which we did not have, and we therefore had cause to despise as well as fear him. The shopkeepers used every trick known to them to make Africans spend their money."[42] The motives of shopkeepers were perpetually probed, and their good relations with local communities and with specific factions of those communities, such as patriarchal leaders and the elderly, were always at risk. As one African customer complained, "some storekeepers are not fair men. They read a person first and find out what he knows and whether he knows what he is doing, and then charge some Natives more and some Natives less for the same article."[43] Many traders achieved infamy as notorious cheats from the outset of their activities. Moreover, some traders developed an openly hostile atttitude toward their clients from the beginning, particularly toward those customers whose sense of market exchange was still somewhat experimental. Hyatt, for example, recounted an incident in which he attempted to beat an African whose asking price for a pumpkin was many times higher than the usual.[44]

The 1930s were a watershed in these relationships between merchants and rural consumers due to accelerating changes in the nature of colonial rule and the local stresses produced by global depression. The increased pace of removals to infertile lands and the induction of increasing numbers of impoverished cultivators into the labor force of settler farms, along with the increasing impact of proletarianization and urbanization on rural and peri-urban peasantries, throttled most local accumulation and severely reduced the strategic resources of rural peoples. More importantly, the Rhodesian state in the 1930s completed a comprehensive juridical and bureaucratic program through measures like the Maize Control Acts designed to regulate and suppress African agriculture.[45] These developments effectively gave the traders enormous power to compel near-involuntary bartering of surplus grain for any cheap store goods that the trader wanted to dispose of, especially any consignments of undesirable low-quality goods forced on the retailer by the wholesale houses. Peasants with market crops before the 1930s typically used stores to acquire small reserves of cash for paying state fees and for later discretionary spending, while bartering for the goods that they most wanted with any other grain available. Prior to the juridical protection of white agriculture, African farmers were able to negotiate their relationship with traders and with the world of goods they represented on more equal terms. As the Native Commissioner of Mtoko noted in 1914, "If traders raised the prices of kafir truck, the natives would retaliate by increasing the price of their grain."[46]

But later, under the network of laws built up in the 1930s, "trader-producers" became the appointed agents for the purchase of African produce and were supposedly required to pay cash if so requested. (African farmers could also theoretically take crops to Maize Control Board depots, but so few existed and transport was so marginal that it was not really an option.) In actual fact, merchants rarely paid out cash for grain once they acquired these compulsory powers over their customers. Typically, a cultivator would bring grain into the local store. Its value would then be determined not by weight but filling a paraffin tin used as a standard measure. Once these buckets had gained wide currency as a measure, many traders began to hammer them out to get extra grain and required customers to fill them past overflowing. The overflow was then collected by the merchant without any additional compensation. In payment, the trader might theoretically comply with the law, offering a cash price and a "higher" price in goods from the store. Some were blunter and only offered goods or "good-for" tokens usable at a later date in the store. But even when a cash price was cited, it was often bogus: if the seller then requested this cash, it would suddenly be "unavailable at this time." Returning a week later, covering as much as fifty kilometers on foot, the farmer usually discovered that cash was still mysteriously unavailable. Alternatively, traders played coy games designed to comply with the letter of the law mandating cash payments while violating its spirit: "Some of the storekeepers to cover themselves put the money on one side and say 'Here is your money, what do you want?' When they say they want the money, he goes along and attends to someone else and comes back and says 'Here is your money, what do you want?' and he goes on like that until they take something."[47] Cultivators were thus virtually compelled to exchange produce for store goods. Such goods often had surprisingly elastic values that skyrocketed when they were being offered as compensation for grain, and often the choices in such deals were restricted to the lowest quality goods in the trader's stock. Not only were peasants forced to acquire goods, but they were also compelled to accept grossly disadvantageous terms of barter. Dramatic price increases after 1939 only worsened the problem.

Resentment of virtually every aspect of this system ran high in the mid-1930s and into the 1940s. Many peasants desperately needed cash for taxes or for discretionary spending, and in the wake of a series of painful famines, the global depression, forced removals, and new forms of state extortion, the margin for maneuver became virtually nonexistent and the momentum of proletarianization was greatly accelerated. The accelerating transformation of consumer needs and the attendant near total collapse of the indigenous production of items like agricultural implements and clothing tied individuals in increasing degrees to "truck" merchants. Traders' refusal to pay cash and the toothless enforcement of the law ensured that peasant producers were locked into a system that victimized them at every turn. The native commissioner of Gutu received an impassioned letter from local headmen: "[We] are much clamouring to you Sir, by sake of to be spending all their garden crops to the white men traders stores, without getting any money or pennies for them, but goods only those. . . . We don't want to sell our garden crops to whitemen traders stores, as we were used to some years back because we were blindmen or Hoodwinked . . . now these white traders they are always cajoled us without pity on us. . . . "[48] In Mtoko District, a similar sentiment was voiced: "The older natives say 'our children and women take a basket of grain here and there and exchange it at the stores for something that is worth nothing—we would prefer to sell our grain for cash and by weight and then buy them the clothes they want!' "[49] One African agricultural demonstrator, speaking for Chief Chiweshe, noted: "There was not any way through which Africans could sell their agricultural production but only through local stores. No cash received. . . . No-one's maize was received unless he or she agreed to buy useless and cheap goods from the same store."[50]

It was thus not only the refusal of cash that galled but also that the goods offered in exchange were deliberately overvalued and of the poorest quality: "They take monkey nuts to the trader and he says they can have 10/- in goods, and when they are given limbo it does not look worth 10/-";[51] or similarly, "if a man brings mealies, they say 'your bag is worth 6/6d., but I have got this shirt here,' and the Africans have the feeling at the back of their minds that they are being done down."[52] Even minute aspects of the process were criticized. For example, the hammering out of paraffin buckets was constantly attacked, and for many, it became symbolic of all the larger frustrations suffered by African cultivators. One witness passionately and repeatedly cried out during his testimony to the Godlonton Commission, "I want to know about the bucket."[53]

Traders active in the 1930s and 1940s made a few attempts to dismiss this outpouring of criticism. Often, the supposed instinctive love of Africans for bartering would be cited. The head of Landau Brothers wholesaling firm said, "it is nothing for a native to walk 25 miles to get an article a tickey cheaper,"[54] thus concluding that it would be foolhardy to alienate customers. The Bulawayo Chamber of Commerce even claimed that Africans enjoyed such journeys to the extent that "it is not uncommon in the trade for a man with a concession store actually to open another store in competition with himself to accommodate this peculiarity of the natives who enjoy walking from store to store."[55] Storekeepers also pointed out, with some accuracy, that more distant or peripheral rural communities where no stores were operating were frequently desperate to acquire their own local trader. This was not so much a testament to some universal and uncritical love for consumption or a fondness for the "truck" trade as much as it was a recognition of the growing power and indispensability of colonial capitalism. By the 1940s, the articulations between peasant production, market-driven exchange, wage labor, and the colonial state had grown increasingly oppressive.

Laborers, Credit, and Workplace Traders

Many rural people also had their primary contact with stores through their work on settler farms. The clientele of farm stores was often similar to that of other rural trading operations, in that it was drawn from surrounding communities. However, these stores also channeled consumption directly through a captive workforce. This institution of the "company store," familiar in the history of industrial capitalism, often forced workers to involuntarily expend their wages in a disadvantageous exchange for goods in the store situated on the work site. In this respect, the experiences of some rural African consumers resembled the forms of exchange and consumption encountered by Zimbabwean mine workers from 1900 to the early 1940s. Mine workers, like township residents and rural villagers, encountered the complex mix of relationships and rhetoric that grew up around storekeepers and hawkers—indeed, as migrants, miners themselves often experienced these multiple contexts of exchange and consumption—but miners also had the most intense encounter with another problematic institution of early colonial trading: credit.[56]

Early wage laborers entered the capitalist workplace in Zimbabwe for very specific reasons: obtaining cash for taxes, alleviating shortages caused by drought or relocation, or because of other forms of labor compulsion. As the Rhodesian state and local capital changed over time, such limited contact with wage labor became harder to maintain. Initially at the mines, and later on farms and in urban and rural locations, storekeepers made a variety of forms of credit available and thus helped to create new entanglements with manufactured commodities. Workers, like other African consumers, brought a variety of autonomous cultural and economic interests to the purchase of goods in these stores. More so than most, however, they were captives of labor regimes with considerable power over everyday life and physical survival. Stores could provide commodities necessary for continued life: supplementary food, clothing, equipment. At the same time, mine stores seemed to offer workers, many of them migrants from northern regions where manufactured commodities were still rarities, the opportunity to acquire valuable goods with their labor which could then be brought back to ensure a long-term livelihood in the workers' home communities. Credit seemed to be the key to both perceived needs, while in fact it was the forge on which subtle forms of bondage were hammered out.

One of the earliest and most prominent of these systems of credit was the "box system," which was functioning on many mines by 1900 and elsewhere by 1910. The system worked like this: At the start of a worker's contract, he could buy a box, "the more gaudily coloured the better."[57] Each month, the box owner could then add goods to the box, which was kept in the store. The items could be paid for in full, in part, or not at all when placed in the box. Installment payments were then scheduled with the storekeeper for any items not completely paid for. When the box owner was ready to claim the box (at the end of a work contract, unless the owner was using the box system to pay for supplementary foods), all of the contents had to be paid for or the box and all of its contents, including paid-for items, could be claimed by the storekeeper.

The underfeeding of workers meant that they often had to purchase supplementary foodstuffs on credit, either using the box system or other form of credit, and from there, their involvement with credit usually deepened. Secondly, the ease and allure of placing goods in one's "own" box encouraged many workers to continually add to the box without ever paying off any one item completely. To add to the problem, merchants usually kept no records and workers received no invoices or receipts of payments. Prior to 1912, the traders also had the only access to the box and even afterwards it was possible for the merchant to unilaterally change the contents of the box. Should any fraud occur, it was the worker's word against the merchant's—and authorities usually favored the latter.

As a consequence, many store patrons were unable to recover the items placed in the box and lost considerable sums of money simply keeping up with the installment payments. Sometimes, customers might believe they had paid off the box only to discover that the merchant had a different perception of how much money was owed. If they finally laid claim to the box, they sometimes found that the contents had been changed—higher quality goods initially placed in the box and paid for had mysteriously changed into cheaper equivalents. One trader admitted in 1910 that "only half the boys got their boxes back again . . . some of them got so much into debt that they left their boxes altogether with the dealer."[58]

Mostly, miners and others placed a small but carefully chosen group of goods in their boxes, "discretionary" purchases that they hoped would provide an additional long-term recompense for their work experience. Typically, these goods were various items of clothing, with small items of hardware and toiletries sometimes rounding out the selections. But on the meager salaries available to African workers, even a few such items were frequently impossible to pay off except on extremely lengthy labor contracts. Still, the boxes lured store patrons to commit far beyond their means. Many of the early workers in Rhodesia's mines came from present-day Malawi or Northern Rhodesia and hoped to gain lifetime security from a single experience in the capitalist workplace, a pattern repeated on a larger scale to the south on the Rand, as Patrick Harries has pointed out in recent work.[59] The upper limits of such ambition were demonstrated by the contents of one box left behind by a Yao miner: "20 shirts, 26 clothes, 9 singlets, 3 jackets, 25 short trousers, 1 blanket, 3 sweaters, 1 red fez, 2 scented soaps, 1 knife, 1 scarf, and 1 handkerchief."[60] While this was an extreme case of ambition, many other store patrons pursued equally fruitless and frustrating commitments. The box system was regulated in 1912, but it continued to be used and abused by merchants for some time. One British South Africa Company policeman who patrolled the Charter Road area of Salisbury recalls the box system still being widely used in the early 1930s.[61] However, the box system thereafter rapidly disappeared and could be found only on a few mines by the 1940s.

Another system of credit with origins in the mines was the "ticket" or "coupon" system. Tickets or tokens were used in a variety of contexts in the colonial economy: to record and pay wages or to record "payment" by a trader for grain brought in by peasant cultivators. As store credit, tickets were used in a number of ways, but most typically farm or industrial workers were given coupons on a monthly basis that could be used in the local store concession. These coupons might represent either a portion (or all) of the current month's wages, or they might be a form of promissory note that would automatically commit future wages to the store owner. In the latter case, money would be deducted from wages at the time of their payout by the employer and then given to the storekeeper. One early legislator fretted nervously: "At the end of the month, they [workers] had no money to withdraw and they thought that they somehow had been defrauded. In a country like this, where so much depended on the labour supply, it behooved them to be very careful not to do anything which would make the labourers dissatisfied."[62]

These kinds of automatic wage deductions became common in many stores under a number of guises and though often technically illegal, the authorities usually restricted themselves to expressions of concern and stern warnings.[63] In some rural areas, rather than a pledge of wages, future crops were promised to obtain food in times of scarcity.[64] Unlike the box system, this sort of credit survived well after 1944. Systems involving "coupons" and automatic wage deductions did vary somewhat over time; during both World War I and World War II, the scarcity of goods made traders reluctant to extend credit, and the increasing regulation of the economy forced such credit underground to some extent after 1944.

There is little question that Africans in colonial Zimbabwe objected strenuously to these new forms of usury. Miners at Shamva distributed leaflets in 1920 importuning their fellows not to get credit.[65] The compound manager of the Shabani Mine noted the loathing felt by workers for the box system: "Natives would come and complain to me that they had purchased certain articles, had left them in this box, and those articles had been substituted by an inferior article . . . the blanket was not the same one he had purchased."[66] One Chief Native Commissioner, Hugh Simmonds, acknowledged ruefully that because of credit, Africans "rather get into the hands of some storekeepers."[67] The editor of the *African Weekly* declared, "As the credit system is open to abuse, it should be discontinued at once. The most wicked thing which results from it is the issue of 'paper tokens' which are issued to natives who obtain credit at the stores."[68] As most observers acknowledged, the existence of any complaints at all was remarkable—the local monopolies exercised by merchants over all forms of trade in African communities made complaining to anyone a perilous exercise.[69] Furthermore, traders sometimes further alienated African customers by using violence to collect credit payments, as debt incurred by Africans had a legally uncertain and shifting status. For example, one supposed debtor who was beaten reported resentfully; "I paid £3 some three years ago to Deary. I thought I owed nothing. . . . I only paid the £1.2/- because I was beaten. Had I not been beaten I would have suggested Mr. Deary getting his book and telling me the things I had."[70]

The Touts of Charter Road: Urban Stores and African Communities

Africans living in urban communities had similar relations with traders in the first four decades of the twentieth century. Many of the early elites associated with colonial Harare and Bulawayo relied at least partially on cultivation for their wealth and power, while other urban workers traveled back and forth to rural homesteads. Thus, the kinds of shifts in colonial law that so incensed rural villagers and problematized the experience of shopping in "truck" stores affected most urban residents as well. Rhetorical gestures of courtesy like bonsella were less common in urban stores than they were in the countryside, but they still occurred. The forms of credit experienced in farm and mine stores were also found in many urban "truck" establishments. Thus, urban traders often were viewed with considerable resentment in township society.

However, an additional problem was that many urban traders, whether located in segregated downtown commercial zones like the Charter Road area in colonial Harare or in the townships themselves, were seen as ethnic competitors who were taking away opportunities for livelihood from Africans themselves. Urban Africans often found themselves without access to cultivation and in a situation where the social definition of "necessary" goods was fluid and constantly expanding. Their daily survival and cultural lives were thus very much in the hands of the small group of merchants who catered to African customers. "White" stores in the main downtown areas of Bulawayo, Harare, Gweru, and other colonial towns would occasionally serve Africans, but they would only do so with humiliating strictures and with seriously disadvantageous and randomly shifting terms of exchange. The smaller and more remote the town center was, the more exploitative the "truck" merchants were. Shamuyarira's account of his first childhood visit to a store in Marandellas captures much of the frustration and ambivalence that attended "truck" traders before the 1950s: "My first experience of town-life and its grasping nature came in my second year at Waddilove. I had been given thirty shillings by my father to buy clothes in Marandellas. I went by bus, and called at one Indian and one Portuguese shop. . . . I had been told by friends at Waddilove to try to reduce their prices as much as possible. . . . I argued and pleaded with the Indians for two hours . . . having reduced their figure in the course of bargaining, they still charge a higher price than we would find in African-owned shops."[71]

As Volker Wild has pointed out, African involvement in urban trading was poised to expand along with the rest of the economy in the late 1930s, but was stifled by the surge of segregationist regulation that marked that era. In 1938, there were twenty-two Africans with a general dealer's license in colonial Harare, but that number had been reduced to zero in 1941. By 1953, there were eight. Nevertheless, after the eclipse of "kaffir truck" in its earlier forms, in the 1960s, African storekeepers began to be a more significant part of urban retailing, transforming the relationship between African customers and merchants.[72] In some rural areas, there were similar developments. Leslie Bessant and Elvis Muringai have chronicled the evolution of "business pioneers" in Chiweshe District, African men who "relied first on their business enterprises for their income."[73] These new merchants were often undercapitalized compared to their white and Asian counterparts and depended on a narrow range of established staple goods—mealie meal, soap, clothes, sugar, tea—to sustain their businesses.

The institution that most symbolized the long-term struggles between urban "truck" storekeepers and their African customers before the 1940s was "touting," which was prevalent in colonial Harare and may have been a problem elsewhere. Competition between traders in this area was extremely intense from the early 1900s to the 1940s. Sometime around 1910, "truck" stores began to hire African touts to help secure customers from both the settled African urban population and from migrant workers passing through the railway station. By 1915, this touting had provoked a great deal of resentment from its targets. A conference of the three superintendents of natives commented: "This question has received the attention of both the Police and the Native Department for some considerable time, and was at one time submitted to the Government with a view to legislation being introduced but . . . the representations made at that time (1914) were not sufficiently strong." Still, the authorities noted "continual urging" by "complaints of natives, that some action be taken."[74]

These complaints focused on several aspects of touting. For one, the sheer number of touts employed by some stores created congestion and confusion: one report indicated that five stores had hired thirty-nine touts between them.[75] More objectionable to Africans in the city were the nakedly coercive tactics employed by touts. Workers for the Rhodesian Native Labour Bureau (RNLB) complained in 1915 of being "forced to enter various stores and there compelled to purchase goods they did not require against their will."[76] The British South Africa Company police asserted that these complaints were part of a much larger group of complaints of "assault and fraudulent dealing." The problem continued into the 1920s. In 1922, cases were brought before the magistrate's court against touts operating in the railway station. It was alleged that "touts entered the Native Waiting Room, accosted native travellers using the room, took from them their belongings, demanded that they should follow them to their master's store to retrieve the belongings and on protest by the travelling natives the latter were assaulted with sticks."[77] The problem continued despite occasional bursts of attention from the municipal government and press. In the mid-1930s, some merchants formed the first version of the Mashonaland Kafir Truck Association in order to stop touting, which had continued to create "a tremendous amount of ill-feeling."[78] These efforts largely failed and touting continued to be an important part of the experience of commodification in urban culture.[79]

The high cost of living in urban areas, regular harassment from white authorities trying to prevent permanent African residency in the cities, and the public visibility of white access to goods made the relationships between African customers and "truck" merchants exceedingly frustrating. Many institutions that evolved in township culture in the 1930s and 1940s tried to alleviate these frustrations. Various mutual aid, credit, and burial societies and other small-scale associations were in large measure designed to help members purchase consumer durables beyond their reach as individuals. There also were high hopes that increasing African ownership of township stores would alleviate the severity of relations between customers and traders. Equally, in the system of informal exchange called *tswete*, as described by Nathan Shamuyarira, workers illegally acquired goods from their factories: "A man who works in a bakery brings a loaf of bread back to his room in the evening, to exchange it for bicycle parts which someone who works in a bicycle shop brings along. Other workers [bear] spanners, pumps, shoes, shirts, groceries, meat straight from butcheries—in fact, everything. . . . It was never regarded as stealing. . . . There is frequent laughter in the African townships when people relate to each other the ingenious means by which they succeeded in obtaining these goods without arousing suspicion."[80] The necessity of dealing with traders and their manipulations was thus an inescapable aspect of urban living, but through a variety of innovations, township residents were able to situationally circumvent or reorder the world of capitalist exchange prior to the 1950s.

"It Isn't So Blighted Easy": Merchant Capital, "New Needs," and Colonial Society

Though various practices of traders helped to facilitate the penetration of manufactured goods into the fabric of everyday life and culture in colonial Zimbabwe, how did the "kaffir truck" trade affect the development of new "needs" and "tastes" within African identities and cultures? The shifting balance of power between traders and customers gave an increasing number of merchants in different social settings the capacity to coerce Africans into market exchange and the money economy. The intersection of these two developments with major changes in the political economy of colonial society and the construction of different African communities helped create significant and lasting associations around commodities and exchange that shaped postwar consumption. An equally important influence stemmed from hegemonic portrayals of the relationship between African needs, "civilization," and the role of "truck" merchants.

The issue of the "needs" of Africans and the alteration of those needs was an important part of metropolitan British thought about colonialism. "We must be prepared to take the requisite measures to open new markets for ourselves among the half-civilized or uncivilized nations of the globe," declared Lord Salisbury in 1895. Lord Salisbury's sentiment was echoed around the turn of the century by many key figures in British imperialism: Cecil Rhodes, Joseph Chamberlain, and others. Chamberlain and his supporters particularly invoked the image of "new markets" abroad to woo support for their imperial ambitions among the British working class. Access to new markets, they assured the voting public, would secure the future of British capitalism and thus British jobs. There would always be a place to buy raw materials and a place to sell the products made by the factories.

The imperial imagining of overseas markets rested on the domestic transformation of market culture and modes of consumption in England during the eighteenth and nineteenth centuries. This was a world in which "the commodity literally came alive."[81] Recent historical works have clarified the importance of commodification in the cultural world of the industrializing West, especially in nineteenth-century England.[82] The expansion of "need" was seen by various Enlightenment philosophers, economists, and polemicists as one of the key components of the social engine that would drive the world to "progress" and "satisfaction." As the desire for more and better goods was increasingly understood in the metropolis as a key attribute of a "civilized" society, the societies of Africa, Asia, and Latin America were increasingly understood as societies that lacked desire and thus lacked aspiration and ambition. With Africa, furthermore, desire was understood by nineteenth-century Westerners to be inert, absent, never present.

This perspective was common in England not only among philosophers like Smith and Mill or later policymakers like Chamberlain, but throughout the length and breadth of modern Western thought. For example, Olaudah Equiano, a freed slave writing in 1789, perceived the coming of desire for manufactured goods as the salvation of Africa, resolving the division between colonizer and colonized, metropolis and periphery, slaves and masters.[83] Even when unaccompanied by actual investments, the *idea* of untapped markets had a powerful impact during the last century. By 1902, the British South Africa Company's metropolitan directors had become increasingly aware that their possessions were not a "second Rand." They struggled to defend the Rhodesian colonies as paying concerns, in order to mollify both investors and general critics of British imperial policies. The BSAC director in 1902, Earl Grey, took up the idea of "new markets" and made a speech at the shareholders' meeting in which he promised that the BSAC would stimulate "the desire of the natives to possess themselves of articles of European manufacturers."[84] The BSAC's metropolitan representatives then elicited comments and suggestions on the speech by sending it to various Rhodesian commercial organizations.

Earl Grey's Rhodesian audience of farmers, mine owners, businessmen, and traders perceived his speech in light of their perceptions of the "native problem," which in 1902 was seen first and foremost as a problem with labor. The speech itself invited this connection, as the chairman had elsewhere in its text tried to discover ways of making "the conditions of employment attractive to native labourers." Since 1890, the colonial elite had conducted a raging debate on the subject of whether to use force to obtain workers and what forms such compulsion might legitimately take if pursued. Earl Grey's interest in the "desire of the natives" plummeted into the midst of this controversy. The Rhodesia *Herald* sarcastically framed the debate: "The London Office of the British South Africa Company wants to know how the black can be induced to ape the manners of the white. . . . As the kafir is exceedingly imitative, the correct reply would be: 'Let there be more intercourse between settler and savage.' This may be brought about in two ways. . . . The former would be the result of compulsory service, and the latter alternative a process extending over many years, and relying on a wide spread of colonisation."[85] For those favoring compulsory labor, the "desire of the natives" was usually seen as too slow or even counterproductive. The Chamber of Mines, for example, received a copy of the speech and completely ignored it.

However, it was also taken as generally axiomatic, even by critics of Earl Grey's BSAC speech, that the imitation of European material culture, the adoption of new consuming habits by Africans through the agency of "kaffir truck," would perform the work of "civilizing" Africans. For those who sought less provocative methods of obtaining workers and promoting the broader mission of "civilizing," "the desire of the natives" was potentially an attractive approach. Such a solution rested on the presumption that autonomous desires implanted in the heart of every African would "voluntarily" lead him or her to comply with settler rule and thus avoid the specter of recruitment mechanisms too closely reminiscent of slavery. If Africans "wanted" European things, and "wanted" to work to obtain them, this would preserve the façade of high-mindedness on which British colonialism relied so heavily. For example, the mayor of Salisbury responded to Earl Grey's speech by saying,

> the first thing that should be done . . . was to get them [Africans] out of their rugged hills and put them into . . . well-defined areas where they could be kept under control. They would then be brought into more frequent touch with the white man, and consequently they would try to adapt themselves to a certain extent to civilisation. If they wanted to follow more or less civilised methods of living they would want these items of European manufacture and would want to work in order to be able to purchase them. . . . [86]

A Native Department official similarly noted in 1903:

> as long as the natives have no wants they will not and need not work. They have, unfortunately, unlimited ground for cultivation, unlimited food and unlimited fuel, a few goat skins will provide them with wearing apparel for years, all other clothing required can be obtained from traders at the cost of a few buckets of grain. Their total wants then amount to ten shillings per annum to pay the Government tax. . . . What then can be an induce-ment to make a native work? Create for him as many wants as possible and induce him to adopt more modern methods of cultivation and tear him away from his beer pots.[87]

In the same vein, government official Hugh Marshall Hole wrote, "the only means by which he [the native] could be weaned from this idyllic existence was the creation of a desire for European manufactured goods purchaseable with money."[88]

The Rhodesian state turned both to "truck" merchants and to other colo-nial institutions to try to encourage the development of "new needs" among Africans. For example, a 1934 editorial in the *Native Mirror* typically concluded that the "objective of native education" was "to raise the whole standard of living of the African people. . . . Africans should aim at better homes, better furniture, better utensils, better dress, better wagon carts and ploughs, better stock and better gardens."[89] Traders naturally were key participants in these early colonial characterizations of the role of consumption in "civilizing" Afri-cans, but were divided on how to interpret their own role. Some merchants boasted of their worth as a "civilizing influence," seeing their stores as a form of missionary work that brought Africans into the bosom of the West by altering their material culture. One such trader crowed about the change in African tastes: "The trader changes it for him really . . . we started selling the Natives knives and forks, cups and saucers, china plates, bedsteads. . . . *We* did it for them."[90] Other traders shared these beliefs: "Ploughs, bicycles, donkey carts, sewing machines, mealgrinders, etc. have all been introduced to the Native by outside Native Traders, with admitted benefit to him. The Traders have helped to get him away from un-hygienic skins and the introduction of clothes has done away with the 'muchi'—beads, bangles and other useless ornaments had their day but have been educated out of the native largely by trader influence."[91] Another trader claimed, "When I realised it was impossible to do trade in the Reserve and make a living, I decided to create what you might call necessities for the natives."[92] Representatives of the Mashonaland Kafir Truck Association piously declared, "The trader brought life into the reserve at a high cost to his health."[93] The Salisbury Chamber of Commerce echoed this sentiment: "But for such people the native would still be wearing skins."[94]

However, many traders disagreed with these views, arguing that merchants had little influence over their customers and had to meet needs and tastes generated autonomously within African communities. One sighed, "You may have some dead lines. I had a lot of dead lines before the war, not pounds but hundreds of pounds. . . . If anyone is a storekeeper they will tell you the same. There is always dead stocks."[95] Another merchant agreed, saying, "It is impossible for a storekeeper not to have bad lines . . . you have good designs and bad designs." Bad designs, he said, were those that did not keep up with "change in Native taste."[96] The manager of one of the largest wholesale firms, Kaufman & Sons, commented, "There was such a thing as fashion before the war. They wanted a certain type of goods and when they got tired of it they wanted something else."[97] A "Rhodesian Rhyme" from the 1920s captured this sense of powerlessness: "Dinges and I were trading once/Kaffir truck you must understand/The things that 'take' in this cursed land/Beads of yellow and beads of blue/To tickle the taste of a heathen crew/(Which isn't so blighted easy)."[98]

The fundamental structure of pre-1950s merchant capital did in fact act to channel the development of tastes in specific ways. "Tastes" often had to cohere around what was available from those world producers who supplied the wholesalers, who then forced retailers to accept these consignments. Given that merchants did possess the power by the 1930s to compel Africans to accept certain commodities, this inevitably had an effect on local tastes and aesthetics. The heritage of this kind of coercion and the generally negative reputation that "truck" traders had acquired by the 1940s forged a loose and flexible association for most Africans between consumption and exploitation. The low quality of

The Advance of Commodity Culture: Shona Women in 1896 and 1926.
(Zimbabwe National Archives)

"truck" goods, the segregation of commodities into "African" and "European" types, the mechanisms and daily cheating increasingly deployed by merchants, the discrimination against African businesspeople by both the state and whole-saling houses, all left a bad taste in the collective mouth of African society. Merchants in all areas experienced sudden "dead lines" produced by "myste-rious" shifts in tastes or sudden determinations to resist merchant coercion. Boycotts of stores were apt to appear suddenly, last a few weeks or months, and then fade. During the 1920 boycott of the stores at the Shamva Mine, leaflets distributed by organizers argued, "it is a very good thing to agree with each other. There must be no one who will go to the compound store because we have got no money to buy 1/6 bread and 9d. soap."[99] The Native Commissioner stationed in Wankie District reported in 1943 that "A complete boycott of stores ended after a week's duration. . . . After enquiry it was ascertained that the intention of the Natives was . . . to force the traders to reduce their prices and supply articles of better quality."[100] Sometimes, Africans gave up what seemed to be firmly established, deeply felt "needs" overnight: during both the First and Second World Wars, when "truck" prices skyrocketed, Native Department offi-cials frequently reported that Africans in rural districts had spontaneously reverted to "traditional" clothing and that almost no Western-style clothing could be seen.[101]

However, African consumption of manufactured goods by the 1940s was also anything but a straightforwardly collective phenomenon. Many elite Afri-can spokesmen from both rural and urban communities agreed that European goods had become omnipresent in African daily life throughout the colony. One asserted, "Natives are coming to be very fond of taking the European things now."[102] African representatives of the Umtali Native Welfare Society proclaimed:

The people are growing in wanting European-style furniture. They are buying tables, chairs, cupboards, beds. They buy bedspreads and curtains in the market and table clothes in the store. Hundreds of cups and sau-cers came to Meikles Store. Many native people flocked to buy them until it was said that there were not any to sell to natives. They want looking glasses, combs, baths, scissors, flower vases, picture frames, teapots, cut-lery, etc. . . . The people are eating different kinds of food now. They eat porridge, meat, vegetables, bread, tea, fruit, arrow root, beans, native beans, sweet potatoes, peanuts, pumpkins, eggs, milk, sugar sweets, cake, mineral water, rice, etc. . . .

The people are using better clothing. We can tell this by the great

number of Indian stores both in the towns and outside. They want suits of
all kinds, overcoats, hats, shoes, shirts, dresses, sweaters, marriage out-
fits. . . . The people want cleaner and more sanitary surroundings. . . .
They are making rubbish pits and closets. They beautify their homes with
trees and flowers. Other wants are phonographs, musical instruments,
radios. The people have wanted better things ever since they saw the
missionaries and other Europeans having them.[103]

However, such spokesmen also acknowledged that new needs were felt and
expressed inconsistently in African society: "African people in the Reserves do
not buy furniture for their houses and wear boots, and other things that are
good for civilization, and financially they will not be able to reach that stage."[104]
At the same time, the growing influence of hegemonizing colonial institutions
and spaces like missions, workplaces, schools, and various media, combined
with new types of collective organization and identity formation within African
society, had a major impact on the imagining and enacting of consumer desires
by Africans. Wealth, prestige, and identity, as expressed through goods, had
begun to evolve a fundamentally new grammar by the 1940s. In the meantime,
the spread of the "kaffir truck" trade and the apparent permanence of publicly
expressed needs for manufactured goods in African communities helped to
engender a complex tangle of anxieties and hopes in settler society, anxieties
and hopes that played an important role in shaping the growth of manufacturing and the development of a commodity culture in colonial society.

4 MANUFACTURING, THE "AFRICAN MARKET," AND THE POSTWAR BOOM

Colonialism in Zimbabwe was founded on the anticipation of mineral resources that never materialized and was kept alive on capitalist agriculture, a sector whose fortunes waxed and waned as world prices for tobacco and other produce fluctuated. The prospects of manufacturing were scrutinized with increasing desperation during the 1930s and 1940s, when the colonial economy rapidly experienced successive crises exacerbated by global depression. During this time, secondary industry began a sustained period of slow growth that gave way to a significant expansion after World War II. Ian Phimister has argued that this post-1945 industrial boom in Zimbabwe had roots that "stretched back to the 1890s and the First World War."[1] Some of these roots, notes Phimister, lay in attempts to manufacture a local supply of tools and equipment for mining and agricultural capital. The relative anemia of the mines and the slow and spasmodic growth of settler farms effectively crippled early attempts to constitute a manufacturing sector; competition from South African and overseas producers also posed an insurmountable challenge.

When local factories appeared in increasing numbers in the 1930s, they were mostly involved in manufacturing goods from agricultural materials. These factories still struggled against regional competition, but were able to take advantage of periodic crises that broke the flow of imports into the colony from both South African and metropolitan sources. Furthermore, when these businesses began their slow growth, the post-1923 "responsible government" regime gradually accommodated them with policies protecting them from South African competition while also securing favorable access to the markets of colonial Zambia and Malawi. Thus, these plants were not overly deterred by periods of regional or imperial recession, though the global depression of the 1930s effectively retarded their growth for a short time. Still, many were left in a good position for subsequent expansion in the post–World War II era. As a consequence, the 1950s in British Central Africa were marked by a rapid and widespread boom in manufacturing activity and multinational investment. This expansion ushered in significant transformations of the architecture of local consumption, including transformations in the nature of distribution and the accompanying decline of "kaffir truck" stores. The shift toward mass production and mass consumption in Zimbabwe, accompanied by increased activity in marketing and advertising, continued even after the first blush of postwar economic expansion faded in the 1960s.

Secondary industry in colonial Zimbabwe was also significant in the collective subconscious of settler society, rising anxiously to the fore whenever the subject of the long-term future of colonial rule was raised. Such anxieties grew more acute and were publicly and officially given voice with greater frequency in the wake of the economic crisis of the 1930s and escalating challenges from African organizations in the 1940s. The state, various sectors of capital, and factions in settler society assessed and formulated industrial policy during the 1940s, struggling but failing to discover a formula by which African aspiration could be contained while ensuring industrial growth through the promotion of new secondary industries. The growing divergence of interests between local and multinational capital that followed in the postwar boom added to white anxieties while simultaneously reconstructing the nature and meaning of consumption for all Zimbabweans. Even after Unilateral Declaration of Independence (UDI), the prospects of manufacturing capital remained a contentious and problematic issue for state authorities and capitalists alike.

In this chapter, the structure of secondary industry and the progress of various formal and informal debates about the place of manufacturing and consumption in the colonial economy are examined closely. Much of the focus is on the growth of industries producing soap and other toiletries, partly because the products of these factories were the locus of the intersection of domesticity, hygiene, and manners with the process of commodification. At the same time, the development of these industries in many ways exemplified the general history of secondary industry in Zimbabwe.

Growth and Competition in the Southern African
Soap Industry, 1900–1947

Soap is one of the easiest commodities to industrialize and thus it usually has been one of the first goods produced in an industrializing economy. Its very ease of manufacture, however, also makes it a commodity that inevitably attracts ferocious competition among companies; new players can enter the game at a very low threshold of capitalization. Furthermore, where raw materials are widely available—sources of fat or oil are the most necessary ingredients—soap manufacture can be done in the household with relatively little labor time spent on extraction or production, thus providing further potential competition for manufacturers. The latter problem has never been a serious concern for soap-producing businesses in southern Africa since the beginning of this century,[2] though household production was widespread both in frontier and settled areas of South Africa in the eighteenth and nineteenth centuries. Since that time, organizations involved with domesticity have sometimes tried to promote home manufacture of soap.[3] However, competition between local, marginal soap producers and multinational or larger local capital has been a persistent feature of manufacturing in the region.

The presence of white settlers in South Africa and the rush to industrialize the Witwatersrand attracted active interest among British soap manufacturers in the 1890s. By that time, the first wave of industrial soap producers in England had already been displaced by new companies like Lever Brothers, Gossages, and Pears, whose commercial power derived partly from their new monopolizing strategies and partly from their aggressive commitment to new techniques of advertising and marketing. W. H. Lever, the founding chairman of Lever Brothers, in many ways the most central manufacturing firm to this study, had a stronger interest than most of his contemporaries in sales and production abroad. Part of this interest stemmed from his desire to control sources of supply for important raw materials needed in soap and candle manufacture, like palm oil or whale blubber. Equally important, however, was Lever's unusually forceful faith in the eventual growth of foreign consumer markets for his goods.

Lever believed, sometimes to the dismay of his subordinates, that the whole world was eagerly awaiting delivery of Sunlight Soap. He commented during his 1924 trip to the Congo that the existence of a market for soap among local African communities was "entirely due to the missionary efforts made twelve years ago on the Lusanga . . . when we took some blue mottled [soap] on board with us and tried the natives on it. Blue mottled is always the pioneer in starting the soap habit."[4] Lever was never so foolishly idealistic about this commitment as to build facilities that were completely without any other function. His interest in marginal overseas consumer markets was usually secondary to a wholly pragmatic interest in controlling a local supply of raw materials, as in West Africa and Zaire. For example, Lever's awareness of the practical limitations of African consumer markets showed through in his instructions to the managers of his tiny soap-making subsidiary SAVCO in colonial Zaire in 1924: "the manufacture of Sunlight, Salvator and other specialities for natives in the Congo would serve no useful purpose. I believe that if you had a doublet of yellow household soap, a doublet of blue mottled, and a tablet of Perroquet toilet soap, such as those outlined above, you would be able to supply the demand of the natives and have a sufficient volume of trade to produce economical manufacture at Kinshasa."[5] Additionally, some attractive foreign settings, like the North American market, were protected by other large soap manufacturers, though Lever tried to break into them. The attractions of other settings were equally apparent: the Indian market for *ghee* made from edible oils was clearly huge, and competition for it eventually played a significant role in the merger of Lever Brothers and the Dutch firm Van den Bergh to form Unilever. Even given these provisions, Lever retained an unusual interest in peoples who seemed at the time to be likely to remain marginal consumers in the world-system for some years to come. This interest was in some measure institutionalized into the fabric of his company.

The evident strengths of the South African market plus these inclinations toward developing overseas consumption attracted Lever Brothers into South Africa at an early date. The company was at first content in the 1890s and early 1900s to import its products into South Africa, backed by traveling advertising agents. The growth of local manufacturers like the Transvaal Soap Company and the New Transvaal Chemical Company and even producers in colonial Mozambique were a challenge to Lever Brothers' attempts to dominate the market. These factors, notes D. K. Fieldhouse, in combination with political moves by the South African state to protect local manufacturing, led Lever to purchase several local factories in 1911 and establish a South African subsidiary.[6] Shortly thereafter, in a manner typical of Lever Brothers' global operations and of monopoly capital generally, the firm busily snapped up most of its competition, though as Fieldhouse notes, various local challengers continued to crop up for some time.

In these South African subsidiaries, Lever's personal commitment to his philosophy of salesmanship initially clashed with the strategies of his local representatives. Lever believed that Sunlight Soap and other Lever Brothers products could succeed in any market at higher prices than the products of competitors because of the "goodwill" or brand loyalty that innovative advertising and packaging could help create. When the head executive of the South African operations, Caesar Schlesinger, cut prices in 1921 in the face of vigorous challenges from small companies manufacturing cheap "filled" or "blue mottled" soaps, Lever sent off a blistering reprimand:

The fact is, that there are certain brands of soap which have no goodwill. It has always been my policy to avoid these. Many of the soap makers in England who are now associated with Lever Brothers have built up their trade on soaps without any goodwill. All these companies are more or less suffering at the present time. On the other hand we have not reduced our price for Sunlight nor have any of our associated companies with goodwill soaps, such as Perfection, Magical, Matchless, Puritan and so on, and yet the sales of the higher quality soaps with goodwill has increased phenomenally notwithstanding no reduction in price, and the sales of non-goodwill soaps although selling at competitive prices are not looking at all healthy.[7]

Schlesinger protested, to no avail:

Our trade situation is entirely different from home owing to no goodwill attached to soap except Sunlight. Anybody offering other brands 6 or lower obtains trade for the time. . . . Don't think for one moment that I do not appreciate highly the expression of your views regarding the manner in which to act in case one is faced with severe competition and price cutting on its part. . . . Our position, however, as explained to you in my former letters is unfortunately, different to other countries owing to its sparse population, and the aggressive competition.[8]

Schlesinger's continuing objections, which Lever obdurately resisted to the point of finally dismissing him, obliquely illuminated the nature of the South African market for soap.

Most Southern African manufacturers operating before World War II were sometimes reluctant to visibly or officially take note of the importance of African consumers. Just as African communities were usually understood politically by whites under the confining banner of the "native problem," the considerable significance of Africans in the economy was often painted over or redirected into discourses about the "labor supply." Lever Brothers' South African plants were making soap for a market where the strongest competition was for African consumers. Schlesinger knew that greater productive capacity among those factories making cheap bar soaps and the subsequent low prices of such soaps spelled serious trouble for Lever operations. Moreover, he was aware that available resources for advertising and promotion could as yet do little to instill "goodwill" among African consumers.

Executives in London recognized the same facts after Lever's death. Hugh Greenhalgh, Lever's successor, almost immediately instituted a program for the

South African plants to begin competing in the market for cheap filled soaps as the centerpiece of their development in the 1930s.[9] Nevertheless, Lever had grasped some important truths about the nature of modern capitalism, namely, that the meanings attached to a given brand of a certain commodity, its "goodwill," could prove in and of themselves to be a source of value. Of equal importance were the strategies Lever Brothers used to importune the South African state during the 1920s and 1930s to guarantee and extend the commercial power of the company's flagship line of soaps—Sunlight, Lifebuoy, Lux, and others. During these years, the company gradually convinced legislators to set legal guidelines for the size, shape, and weight of soap that could be sold within the boundaries of the country. Many of the bar soaps manufactured by the small concerns were effectively outlawed by these measures, forcing these producers to seek other markets in the region. Subsequently, Lever Brothers spearheaded state measures to legally regulate the process of soap manufacture by setting fatty acid content, which also worked against cut-rate manufacturers who lacked Lever's global access to raw materials.

These South African initiatives played a critical role in the development of soap production in colonial Zimbabwe during the 1920s and 1930s. The first soap manufacture there began in 1922, when Rhodesian Milling and Manufacturing (RMM), a British South Africa Company subsidiary, began operations. Other local companies like the Progress Soap, Oil and Chemical Company and Express Nut, Oil and Soap Company formed not long afterwards. These companies from the very beginning received encouragement and assistance from the Rhodesian state. For example, as part of the effort to direct African cultivators away from competition with white farmers that ultimately led to the Maize Control Acts, state administrators and agricultural demonstrators busily searched for suitable new cash crops to introduce to Africans. One of the crops heavily promoted by the state during the 1930s was groundnuts, in part with the aim of assisting the growing soap and edible oil industry.[10]

Lever products were also available in the colony and had been from the early 1900s onward. However, the cost of importation meant that Lever soaps and other imported toiletries were regarded as prestige products. Still, the global reputation of Lever Brothers' products, especially among British expatriates, ensured the local manufacturers would be hard-pressed to claim more than a limited share of white consumption. The competitive sector for Rhodesian producers lay among poor white and African consumers, and it was this sector that these firms sought to protect. For example, the Rhodesian state was urged to shield soap and oil manufacturers from South African competition. A representative of RMM informed the controller of customs and excise in 1929 that "factories in Rhodesia are now capable of supplying all the requirements of both Northern and Southern Rhodesia in so far as Household Soaps and Candles are concerned [T]he need for the suspension of duty on these articles no longer exists and we respectfully request that you will be good enough to consider the controlling of them."[11] Such requests were part of the larger campaign waged by industrialists in the 1930s for a renegotiation of customs agreements favoring South African imports, efforts that bore fruit in the 1940s.[12] The passage of Lever-inspired regulations in South Africa in the 1930s also moved the Rhodesian soap makers to action. RMM, Progress, and Express, as well as other local manufacturers producing other goods, all successfully urged the state to pass a weights and measures law in 1926 to protect their products from importers whose goods had been restricted by a similar South African law. No sooner had the law passed than RMM was aggressively challenging the validity of a number of South African brands imported into the colony.[13] The purchases of "kaffir truck" retailers and their African customers were foremost in the minds of Rhodesian soap producers during this legal maneuvering. An amendment to the Weights & Measures Act that permitted thirty-five-bar cases of soap to be shipped was added in 1929, expressly because "35-bar is the size most favoured by storekeepers who cater for the native trade."[14]

Likewise, South African laws regulating the fatty acid content of soap immediately inspired imitation by Rhodesian legislators, again at the instigation of RMM and other local firms. The chairman of RMM, Russell Ridgeway, wrote to state officials, complaining that after the passage of new South African regulations, "we are afraid that manufacturers of cheap soaps there, being deprived of their market, will either send their inferior soaps to Rhodesia or else open up works in Rhodesia itself." Admitting that RMM, like other firms, depended on the manufacture of "cheap soaps," Ridgeway piously and disingenuously explained: "it [cheap soap] is only supplied at the special request of the purchaser who, unfortunately, in many cases is only concerned with the percentage of profit he can make and has no consideration for his customers. The bulk of the cheap soap is sold to natives who are generally unable to discriminate between good quality and inferior soap as the cheaper soaps have quite a good appearance when kept in the original packing case, but soon dry out on being exposed to air."[15] Ridgeway even mustered the support of medical experts for the passage of the Standardisation of Soap Act, on the grounds that cheap soap was deleterious to the health of Africans and other users of blue mottled and household soaps. As in South Africa, the levels of fatty acids, rosin acids, and caustic alkali required by the new act favored manufacturers with superior access to capital and the world-system and fended off a wave of imports from small-scale South African firms knocked out of their own market by similar regulations. Additionally, the Rhodesian law was considerably stricter than its South African model, which once again helped keep South African competitors out of the market. This law also helped alter the nature of the local market for soap. While RMM, Progress, and Express continued to manufacture inexpensive bar soaps, especially "blue mottled," the provisions of the new law upgraded the quality of the locally available product without immediately raising the price.

Protection from competition allowed local producers to increase their volumes of sales and thus keep their prices low with the option to raise them later without fear of being undercut. The Rhodesian firms had a commercial association which acted to regulate price levels and fend off new local firms seeking to enter the market. In 1940, the Rhodesian Soap Makers' Association, made up of six established firms, sought to block the entrance of a seventh soap manufacturer in the Hartley area, with partial success.[16] The consequence of these moves for African consumers were twofold: the nature of soap and related toiletries available in "kaffir truck" stores changed, to some extent, toward products that had acquired the connotation of being more "upscale," and the manufacture and control of soap was significantly centralized.

Prices of soap were also kept depressed by small-scale trade wars between Rhodesian firms and other producers to the north.[17] RMM moved during the 1930s to acquire privileged access to adjacent markets, especially in colonial Zambia. In the middle of the decade, small firms headquartered in Ndola complained that RMM, Progress, and Express were all dumping soap on the Northern Rhodesian market with the explicit intent of putting the Ndola manufacturers out of business. Despite negotiations between customs officials, arbitration between the companies themselves, and ferocious exchanges of dumping in both colonies, RMM and its peers continued their campaign against northern competition unabated, until RMM took over these factories in the late 1930s. This combined trade, still heavily dependent on African purchases, was described by an RMM salesman of the time as "absolutely terrific . . . a tremendous business."[18]

Thus, by the 1940s, Rhodesian soap producers had protected their critical but relatively unacknowledged market—African consumers—from a variety of challenges. They had also institutionalized price fixing and other forms of monopoly control over local supply and distribution. Furthermore, they established dominance over nearby colonial markets and incorporated them. In so doing, these firms had nurtured a legal framework that favored technically advanced production, global networks of supply, and superior access to capital.

They also had begun to take advantage of new forms of mass communication to advertise directly in African communities, as discussed in the next chapter. This left these firms ideally poised to capitalize on the momentum of the postwar boom in manufacturing in British Central Africa. However, much of their positioning had been in response to and in imitation of similar moves by Lever Brothers in South Africa. As a consequence, the business climate in colonial Zimbabwe in the 1940s was ideally tailored to Lever's requirements for the production of toiletries. RMM and its compatriots, having swallowed all the minnows in their pond, were shortly to discover that they themselves were really only little fish in the world-system. In this, their experience typified the experiences of many other manufacturers.

"Africans Do Not Understand the Use of Hats": The Borders of Consumption

Before moving on to a fuller discussion of Lever Brothers and the postwar manufacture of soap in Zimbabwe, I want to discuss in some detail the per-petually contested status of manufacturing and its relationship to African con-sumption in white society between 1900 and the late 1940s. The proposition that Africans would acquire new consumer needs under colonial rule and thus be "civilized," discussed in chapter 3, may have been a vague form of common sense for many settlers. However, it was also a process that many whites vehe-mently objected to, feeling that the boundaries of racial and class privilege were seriously challenged when elite or aspirant Africans began to consume goods that had previously been used only by whites.

One of the most persistent controversies in the colonial press and among state officials prior to the 1950s involved the terms under which manufactured goods were imported through South Africa and the Beira corridor. This con-cern was motivated to a significant degree by a desire to keep the standard of living of whites above that of all other groups. In 1912, the rising cost of imports led the state to convene a commission to study the cost of living among whites. Witnesses consistently testified to the difficulty of affording the minimal stan-dard regarded as proper for settlers and appealed to the state to somehow assist them. As one witness argued, "In a country where there is a black population the standard of living must be high. You cannot have whites living as natives."[19]

"Poor whites" were sometimes said to have "gone native" and, as such, they were regarded with disgust and great offical concern. One missionary commented, "One man who came from a respectable family had gone native and he was a horrible creature; even with all our Christian forbearance we could not sit next to him in church—he stank."[20] For years, the Rhodesian state

kept files on "undesirable Europeans," individuals who lived too closely with
Africans and shared their material culture. For example, a white cattle dealer
and his wife who lived with their customers drew fire from the local native
commissioner: "M. Selesnik . . . is a very bad example to the natives generally. . . . Selesnik is in the habit of having his meals with and at the same table as
Chief Maduna, in many cases the food eaten being provided by Maduna. He
also sleeps in a hut in the Kraal and it is only quite recently that Mrs. Selesnik
was also staying at Maduna's Kraal . . . a European resident in the neighborhood . . . saw Mrs. Selesnik sitting on the ground outside one of the huts
suckling her baby, with her breasts exposed to view."[21] "Going native" in this
manner was often explicitly understood in terms of what goods the offending
whites consumed in their everyday life. As one settler commented, "we called
them white niggers. . . . They relied almost exclusively upon Africans. They ate
African type foods. They didn't live up to European standards."[22]

Maintaining high standards among settlers was only part of what was
regarded as necessary to police racially coded borders of consumption. Africans
also had to be kept away from the goods considered most characteristically the
province of whites. In many ways, the segregated existence and low social status
of the "truck" trade was a consequence of such preconceptions. "Truck" goods
were often originally marked off as "cheap" imitations of "white" goods by
unusually literal devices. It was not uncommon before the 1950s for low-cost
goods manufactured explicitly for the African market to be emblazoned with
the legend "For Natives Only" or "Not for European Consumption" on the
packaging. Africans were thus given their "own" version of manufactured
goods and were told to be satisfied with such items. Africans purchasing
"white" goods posed two dangers: If Africans employed such goods in new or
innovative fashions and styles, settlers fretted about their possible satiric or
critical intent. If Africans used such goods in their "intended" style, they were
depicted as "cheeky" or "arrogant," stealing fashions reserved for settlers alone.
These attempts to segregate the realm of consumption were frequently naturalized by references back to the long-term distinction in Western economic
discourse between "needs" and "wants." Africans, it was acknowledged, had
basic needs for certain goods, but purchases of emblematically "white" commodities could be dismissed as the actions of childish spendthrifts satisfying
unnecessary "wants."[23]

At the same time, attempts to mark off and protect a white, privileged
form of consumption were difficult to pursue wholeheartedly. The rhetoric of
"civilization," of "dual mandate," promised African advancement, though often
in some extrahistorical future. The allegedly commonsensical settler acceptance

The Ambiguous Spectacle of African Consumption: An Ndebele "New Woman" on a Bicycle, 1900. (Zimbabwe National Archives)

of an eventual change in African consumer needs was often confounded when such changes actually concretely materialized—but other visible contradictions were exposed as settlers tried to block any attempt at imitating their own lifestyles too closely. Moreover, as discussed in this chapter, the colonial economy came to depend increasingly on production for African consumers, a development that accelerated in the 1940s. This further complicated any efforts to maintain racially exclusive forms of consumption.

This did not stop some whites from trying. Criticism of African consumption was a major topic of discussion among settlers and was a frequent feature of propaganda and education directed at Africans. Certain forms of consumption were attacked with particular gusto, most notably involving clothing and to a lesser extent toiletries, houses of particular shapes or designs, and liquor—goods whose consumption was highly visible and that were either affordable by many Africans or were produced and distributed by the state. Such battles were related to other demands for public gestures of deference, such as requiring Africans to stay off sidewalks and avoid speaking in English unless absolutely necessary. The increasing consumption of previously "reserved" commodities by Africans was often viewed by whites as one of the most alarming forms of challenge to these codes of deference. As farmer Jeannie Boggie noted with apprehension, "In Rhodesia, fashion is going to be the race of the white man to get away from the black man."[24] The white publisher of the *Bantu Mirror* similarly fretted to a government commission, "Our civilisation is in the shop windows and if we don't guide them carefully, they will come and get it."[25] Urban Africans were frequently chided by both state and mission about their consumption in the 1930s and 1940s, especially after the Industrial and Commercial Workers' Union (ICU) and other organizations began to pressure the government for wage increases. One typical scolding fell back on the division between "need" and "want": "How much did you spend last year on rubbish? On things that were of no use? On concertinas? Gramophones? Cheap boots that were too small? Hats that didn't fit you? Cigarettes that burned you? Tinned foods that were finished in a few minutes? Liquor that made you drunk? Bangles and brooches and beads and things like that? You cry out for more wages, and I agree you need them, but many of the Whites say you do not need more money because so many of you waste the money you now receive."[26] On the whole, many whites felt that "the Africans should, whenever possible, use utensils, etc., made by themselves as, besides being economical, they are often more suitable and beautiful than the bought article."[27]

Clothing was an especially potent focus for these scoldings. A few whites argued that Africans should wear no Western clothing at all, that they were better suited to wear the type of garments they had worn in 1890. Most whites were at least irritated by the adoption of new "Africanized" styles of wearing manufactured clothing or, worse still, the replication of the European sense of "high fashion" by aspirant Africans.[28] The Assistant Native Commissioner of Bulawayo in 1943 sent a memo to his superiors expressing culturally typical objections to the latter phenomenon: "The African is being hybridised and we see the results all around us. We can see how, in their uncertainty, they seize on those traits they conceive to be the distinctive features of a new status and in doing so exaggerate them just as every parvenu does, missing the indefinable emotional attitudes or taste beneath them . . . a blatant, objectional character, acutely aware of himself and his uncertainty, poses down the street in glasses, a loud suit, a cane and a jaunty air."[29] One man involved in transporting workers returning from the Rand in the 1930s commented, "You know, they were very uppish. . . . They probably went away as raw kraal Africans and came back as 'gentlemen.' They dressed most peculiarly. . . . They all wore the same thing— scarves and multi-coloured shirts. All had suitcases of course full of everything that they could buy in Jo'burg."[30] Missionaries often found themselves on the receiving end of settler outrage at alleged transgressions in dress by Africans. As one official noted, "any well-dressed native with a somewhat aggressive manner is almost bound to be called a 'mission' boy although he has never been near a mission."[31] The *Native Mirror* warned educated Africans: "It is because of the clothes and manners of these 'dudes' that so many Europeans dislike native education and native missions. They say 'that is what the missionaries taught them.' "[32]

Clothing was also by far the most extensive focus of campaigns for consumptive propriety aimed directly at Africans themselves. The Morgenster Mission's Standard VI textbook on hygiene for use with African pupils in colonial Zimbabwe captured the flavor of these campaigns perfectly. It instructed students: "Now, because they have seen the Europeans wear clothes, others have come to think that this habit is better than that in which the bare body was the custom. . . . By trying to copy this expensive custom of the Europeans, people have also taken over most of its bad effects on the human body. . . . The body as a whole is weakened by the wearing of clothes . . . the natural (naked) body is more able to resist diseases than the thickly clothed one of the European . . . all kinds of clothing are not necessary." The textbook continues: "Hats! These are worn because they see it done by Europeans. To the African, the wearing of a hat serves *no useful purpose* . . . it is to them an unnecessary *expense* only. As a rule, Africans do not understand the use of hats: they wear them anywhere and everywhere; in the night, outside as well as inside the house; and they wear them especially if they wish to make a show."[33]

The *Native Mirror* was often filled with propaganda designed to discourage the use of fancy clothing by its urban readers, featured right alongside advertisements for "truck" stores that often aggressively promoted such fashions. For example, a photograph of an African attired in a formal dress suit with top hat and cane was featured in regular advertisements for Raizon's Big Store in Bulawayo starting in 1931. The same photograph headed up an article published two years later entitled "Your Clothes! Do Fine Feathers Make Fine Birds?" In it, the author noted:

Natives from all parts of Africa have complained to the writer of the contempt and ridicule they are subjected to by Europeans when they are wearing their best clothes. European men, women and children laugh at them and make audible remarks about the manner in which they are dressed. In the majority of cases the remarks are justified . . . one sees the most absurd clothes . . . the postures, clothes and walking attitude of these "dudes" is enough to make angels laugh, not only the European. Natives who wear such clothes are doing a great harm to their race; they are bringing themselves into contempt and deserved ridicule.[34]

A similar article published in 1937 warned elite Africans not to overdress by caricaturing the alleged excesses of some African dressers:

His glasses are of plain glass and he wears them only because he imagines they improve his looks. . . . His suit costs him more than he can afford, and his gloves and cane are quite unnecessary. . . . We know him well . . . the more we see his type the more we long for the old African gentleman in his skins—or the new African gentleman who has learnt that character, not clothes, makes the man.[35]

Campaigns to discourage Africans from wearing fancy or ostentatious European fashions went far beyond articles in the African press. Many missionaries and township administrators forbade residents and employees under their purview to overdress, though often ineffectually. Home demonstrators and, later, members of African women's clubs were trained to discourage other Africans from wearing fancy clothing.[36] Moreover, any violation of these codes of social deference, whether it was wearing fancy clothing, walking on the sidewalk, or driving an automobile, was often answered with violence by individual settlers or by the police against an African transgressor.[37]

A number of other commodities received similar treatment: negative representations among whites of their consumption by Africans, criticism of those perceived as promoting such consumption, and propaganda and action directed at Africans to discourage trespass onto white terrain. The African consumption of various toiletries was one common target.[38] Boggie's criticisms of her domestic servants included their use of "scented soap."[39] An article in the *Bantu Mirror* similarly ridiculed a young African boy asking for soap "with a strong perfume."[40] Some teachers in domestic science classes in the 1930s were careful to identify cleansing commodities appropriate for African women.[41]

Each of the commodities targeted in this fashion was enmeshed in other complex issues: fears about the reproduction of African communities and African labor; debates on health, productivity, and the role of the state; and the social and political aspirations of African communities. With reference to these debates, whites frequently articulated a connection between their own consumer needs and their social status. African usage of such goods threatened the distinctions between white and black, rich and poor, ruler and subject. Such usage offered the possibility for subversion, satire, and direct confrontation. The deep-rooted desire to protect the boundaries of consumption clashed with the definition by local and imperial institutions of commodification as a basic goal of the colonial project.

"Our Main Market": *Debates over Secondary Industry and African*
Consumption, 1938–1950

Beginning with the convening of the 1938 Economic Development Commission, settlers reopened a perennial debate about the colony's economic future, with an intense new emphasis on the prospects and implications of the growth of secondary industry. Such debates were made all the more intense by the reluctance of many settlers to permit African consumption of "white" goods. These debates continued for the next two decades and were only partially resolved by significant expansion of manufacturing activity during the 1950s and early 1960s. Though the prospects of both mining and agricultural capital had clearly faded in the eyes of the Rhodesian state and the settler community at the end of the 1930s, a turn to manufacturing seemed to open new ground for contradictions and weaknesses in white rule.

It was impossible to credibly imagine that white purchasing power could sustain the expansion of secondary industries, but some settlers had great difficulty accepting the necessary alternative growth of African purchasing power. Many, after all, could not even acknowledge the crucial role already played by African consumption in sustaining colonial society. Successive waves of African nationalists and activists in the 1940s and 1950s helped sharpen this dilemma still further by dramatizing the economic necessity for increased wages and empowerment for at least some African workers. State commissions, policymakers, capitalists, and the settler public all theatrically schemed and dreamed about methods for increasing African purchasing power, nurturing manufacturing, and resolving the contradictions of colonial rule without actually increasing wages, empowering workers, or legitimizing the social aspirations of African communities. Changing economic and political circumstances, the interests of multinational capital, the political and social initiatives of Africans, and shifting white alliances all ultimately impinged on these desperate fantasies during the 1940s and 1950s.

The Rhodesian state and the wider settler community's disinterest in formal economic policy beyond the limits of their momentary self-interest had been marked since the advent of "responsible government" in 1923. Prior to that, under British South Africa Company (BSAC) rule, commercial and state interests had been definitionally equivalent. Much of the momentum toward "responsible government" had in fact originated in the perceived gap between BSAC economic goals and the immediate livelihoods of the colonists. As Phimister points out, the short-term and local economic interests of most settlers before the 1940s made the political economy powerfully centrifugal, defeating any centralizing or conglomerating momentum: "So long as tactical measures sufficed, there was little call for a coherent strategy supporting domestic capital, still less national capital. Indeed the extent to which the white agricultural and mining sectors were disarticulated from the rest of the economy was manifest in the peculiar system of autonomy enjoyed between ministries under [Prime Minister Godfrey] Huggins."[42] The Economic Development Commission of 1938 was the first official body to be charged with formally assessing the state of secondary industry and recommending policies to promote industrial growth. Keeping in mind Adam Ashforth's incisive portrayal of state commissions as a form of public theater,[43] it is interesting to note how this body chose to stage its inquiry into the problems of colonial manufacturing. The structural nature of the internal marketplace of colonial Zimbabwe was largely avoided or dealt with obliquely, and the figure of "the African" was admitted to the discourse of the commission primarily with reference to agricultural production and a critique of the "kaffir truck" trade.[44] Witnesses described about thirty industries that were in operation by 1938, many of them cottage industries operating out of storefronts, as with the small-scale production of jam from local fruit.[45] Most of the industries reviewed and promoted by the commissioners were designed to give opportunities to farmers or to assist mining capital; secondary industry was soothingly imagined as a supplement rather than a replacement for established economic activities.

In assessing the barriers to growth, the commissioners and their witnesses shaped future investigations and debates. For one, officials of the Native Affairs Department reported the general sentiment among their employees and subjects that the practices of "kaffir truck" traders were one of the main impediments to economic growth. In particular, the existence of "forced barter" was regarded as a problem because it short-circuited the flow of cash and locked commodity distribution into what were seen as inflexible and archaic patterns of exchange, though most witnesses studiously ignored the role played by the Maize Control Acts and white agriculture generally in making "forced barter" inevitable.[46] Furthermore, without fully acknowledging the importance of African consumption to industrial growth, the commission and its witnesses took an interest in increasing African purchasing power through indirect measures: encouraging "appropriate" craftwork like basketry, promoting new cash crops avoided by white farmers, or exhorting Africans to be "thrifty" with wages or crop proceeds.

These visions were carried further by the 1938 commission's successor, the 1944 Godlonton Commission on Native Production and Trade. While other bodies, official and otherwise, were also assessing industrial policy during the early 1940s, the Godlonton hearings represented the ultimate development of an obfuscatory quest for a painless way to reconcile white power, industrial growth, and permanent social and economic subjection of most or all Africans. The Godlonton investigators interviewed a wide spectrum of witnesses, including many Africans: urban activists, rural headmen, agricultural demonstrators, police, and others. Picking up where their predecessors had left off, the commissioners investigated the "truck" trade and found many witnesses willing to criticize the traders. Chief Native Commissioner Hugh Simmonds declared, "I think few people will deny that the conduct of the 'Kaffir truck' trade, particularly the retail trade, leaves much to be desired and as I have stated, has not generally been in the best interests of Natives."[47] Administrators from rural districts who supported the traders and denied ever witnessing fraud or hearing complaints were occasionally forced to recant their testimony—in one case, the commissioners visited "truck" stores to investigate stories about fraudulent measurements, leading the administrator of Bikita District to confess, "beaten out dip drums and paraffin tins are used at most of the stores. I regret that this statement does not tally with the evidence given, bona fide, by me before the Commission."[48]

The Godlonton Commission also followed precedent and investigated schemes for increasing purchasing power in African communities that fell short of raising wages or dismantling the strictures protecting white farmers and white workers. The proposition that "most of us are out to increase the spending capacity of the native, to enable him to take things like sugar, tea and coffee as absolute necessities and not as luxuries" was a modest one as far as the commissioners and their witnesses were concerned, but advocating fundamental changes in the political economy of the colony was not permitted.[49] When some Africans and a few white witnesses naturally protested that better wages and working conditions, along with an end to overt economic domination and agricultural price fixing, were the only possible answers to the "problem" of "native production and trade," the commissioners usually protested that they were not permitted to consider any such "solution."[50]

Many of the directions taken by these two bodies were prompted by various official and semiofficial discussions and policies during the same period, and the inability of these commissioners to formulate a successful industrial and commercial policy was reflective of broader fissures and failures in the settler establishment. Alternative propositions for economic growth were advanced from all sides: secondary industry could grow via exports; Africans could be taught to be "thrifty" and thus consume more usefully; white immigrants could swell the local internal market; Africans could obtain more cash through new types of cultivation that did not compete with white interests; primary industry simply should be left alone; the participation of an extremely narrow sliver of professional Africans in the economy could provide a margin of social safety while still excluding most African workers.

The fragility of these schemes was pressed home from one direction by commercial and industrial interests and from another by prominent African activists, professionals, and businesspeople. In the first instance, local capitalists and store owners were careful not to breach the boundaries of acceptable discourse about the place of Africans in colonial society, not the least because they themselves frequently shared an interest in containing or channeling African aspirations and access to power. Nevertheless, various settlers and investors with an interest in commercial distribution or manufacturing, as opposed to farming or mining, began in the late 1930s and 1940s to increasingly advocate for a new industrial policy that recognized the realities surrounding economic growth and the prospects of secondary industry.

First, these individuals tapped into the preexisting strands of colonial "common sense" about the positive connections between an increase in African "wants" and the success of a "civilizing mission" (as discussed in chapter 3). By promoting the growth of secondary industry, these businesspeople and their allies suggested that the removal of social barriers to the expression of new consumer desires for selected segments of African society combined with support for manufacturing could produce what earlier policies and visions had been unable to imagine: a rapid creation of a significant "African market" that would provide a viable living for settlers while creating a compliant buffer of African consumers. The first key to this conception was the identification and targeting of an "African market." William Margolis, a key figure in the Rhodesian soap industry, argued in his 1938 University of South Africa thesis that "the view has often been expressed . . . that the natives' wants are not only limited but static . . . [but] the opportunity to satisfy his wants . . . is often begrudged the native. . . . a higher standard of living is thought to be sought by the natives." Margolis continued by ceding some ground to those believing in the "limited wants" of Africans: "improved conditions on the Reserves, or higher wages, need not affect the labour supply adversely . . . [but] the native . . . [is] to a great extent unprepared . . . this inexpensiveness of their method of living, the limited nature of their wants and the comparative absence of incentive to work . . . led people to believe their needs were stationary and small."[51] Commercial and industrial businesspeople throughout the colony strenuously promoted their capacity to stimulate the "African market" and resolve structural problems in the colonial economy. Nigel Phillip, head of the Chambers of

Industries, argued, "The native market . . . must be our main market."[52] He later added, "I look upon the value of any citizen whatever his colour as his production and consumption. Our natives and our Europeans have been producing here, but consuming foreign goods. The result is the whole economic structure is upset. While we go on like that, we get nowhere."[53]

These industrialists had allies. A growing proportion of the state bureaucracy was becoming convinced of the increasing social and economic importance of the African consumer in any viable colonial society. Officials of the Department of Native Affairs printed a series of articles on the "segregation question" in 1937, which argued forcefully that "no great industries can grow up here unless the Native is to be called in as a consumer."[54] The newly created Department of Commerce and Industry also promoted secondary industry during this era. A foundational analysis prepared for officials of the department in 1944 argued:

> If then, doubt exists as to whether an entirely new basis of earning the colony's bread and butter will be needed, wisdom must prompt us to assume that this will be so. In any case the broadening of our economic foundations must be to our advantage, and so the first postulate becomes true, although we may be administering a preventative rather than a cure.
>
> . . . there are one or two manufacturing industries producing for the European market, and there is probably scope for a few more, but these are and will continue to be on such a small scale that they cannot hope to influence the general economic structure, nor can they be relied on for any serious contribution to the colony's export trade as compensation for any possible shortfall in the export of gold.[55]

It is important to keep in mind that the figuration of African consumers as a homogeneous group whose needs could be met differentially and separately by white-controlled colonial industries fit in well with Prime Minister Godfrey Huggins's segregationist politics during the 1930s and early 1940s. Imagining an "African market" allowed industrialists and state officials alike to support Huggins's notion of the colony as a "small white island surrounded by a black sea." Rather than see the colonial economy for what it was, a welter of articulations, rather than see colonial society as already "assimilated," supporters of exploiting the "African market" continued to imagine that the colony was a "dual" society. They insisted that Africans were mostly outside the logic of the marketplace, that African culture was a unified and *undigested* whole.

However, the identification of an "African market" and its social and economic potential was only the opening gambit in the debate over industrial policy. Promoters of new industrial policies faced another pressing challenge. Other established and powerful factions of local capital, as well as the wider settler community, had to be convinced that a realistic commitment to increased African wages, improved working conditions, and stabilized urban communities were all absolutely necessary for long-term economic growth. These issues remained the core of the opposition to the serious pursuit of manufacturing activity. The head of the Chamber of Industries straightforwardly tackled these concerns by saying "purchasing power rather than cheap food and clothing is what they [Africans] require." Another leading industrialist added: "one thing that industry must provide for . . . is a reasonable and gradual advancement of the native, and the provision of employment at such remuneration that will allow the progressive native to live, feed and house himself suitably."[56] The president of the Salisbury Chamber of Commerce agreed with his compatriots, arguing, "I think the time has arrived when we should make a serious attempt to raise the standard of our native labour and also the standard of the native as a consumer of manufactured and other goods . . . this in turn means better wages."[57] Once again, factions within the state bureaucracy added their voice to this argument. The Department of Commerce and Industry's core analysis concluded: "the European worker in industry has got to face the fact that outside the native markets themselves there is very little outlet for his labour, and unless the native is placed in such an economic position that he can purchase manufactured goods locally produced, his, the European worker's, chances of employment are even less."[58]

Advocates of increased wages and stabilized labor were nonetheless careful to repeat established clichés regarding the "laziness" and incompetence of African workers. A profile of an influential merchant noted, "Like most other businessmen, Benny Goldstein would like to see Natives' purchasing power increase still further—providing the increase is earned."[59] Such advocates drew upon other sociological clichés to support their views, such as the supposed value of a "buffer" middle class for maintaining social order. Not all manufacturers, traders, or state authorities who identified the "African market" as the key to the future and the importance of increased purchasing power for some Africans agreed straightforwardly that wage increases were necessary. This reserve showed through in the guarded comments of the Chief Native Commissioner in the mid-1940s: ". . . . A further question is the extent to which the Native as a potential consumer could support internal (in this case not necessarily Native) secondary industries. The answer probably lies in the extent to which the spending power of the Native can be increased, and this I suggest can best be achieved by reducing the present unduly wide gap between producer and consumer prices by the orderly marketing of produce and the introduction of cooperative buying and selling."[60] Though all such initiatives carefully protected the fundamental core of colonial domination, they nevertheless posed an acknowledged threat to the position of agricultural and mining capital, particularly when wage increases were advocated, and more generally they potentially threatened radically segregationist visions of colonial rule. This sense of vague danger was compounded by pressures from another quarter, from African activists and organizations agitating for better working and living conditions in both the cities and the Reserves during the 1930s and 1940s.

African activists, professionals, and elites with diverse loyalties, cultural affiliations, and goals were all quick to understand the embedded power of rhetorical connections between economic viability, manufacturing, and African consumption. The growing importance of consumption and rising prices compounded the significance of these connections for elites, workers, and peasants alike. Ndabaningi Sithole, while a politically active teacher working at Dadaya Mission, wrote several times to the *Bantu Mirror* during the early 1940s to insist on the importance of increasing Africans' access to consumption for the sake of all members of colonial society. He noted, "The African of to-day needs clothing. He has many needs because of the Western civilization."[61] Later the same year he protested: "The African has been accused of wasting money. . . . Some people think that to raise the wages of the Africans is to cripple the European economy . . . it is just the opposite. The African constitutes a vital force as a consumer of primary and secondary products of this colony, but he can never be an economically efficient consumer unless his purchasing power is increased by earning more money. More money will enable the consumer to extend his demand for commodities of consumption, and this means economic expansion."[62] Another correspondent took up the theme not long afterward, writing, "Money is so-called wealth as it enables us to get things which satisfy our wants. . . . Now then, what chance does the African in this Colony stand of becoming comfortable and progressive under the present hostile conditions?"[63] A representative of the Southern Rhodesian Bantu Congress was asked by hostile government inquisitors, "Won't it happen that many actually will now say 'We need not work so much each year' [if wages go up]?" The representative adroitly parried, "I don't think that would be the case because the requirements of the African people increase. He will spend that money."[64] Another African witness before the 1943 Howman Commission complained with irritation: "If a young man is throwing away his money, he is still enriching the country. We are not Indians who are going to take the money away out of the country. If we waste the money, it will be left in the country."[65]

These and similar appeals were often made with typical deference by early civic activists, but complaints about wages, the cost of living, and the deprivations of urban life were unmistakably and militantly dramatized by the general strike of 1948, which sent massive tremors through the colonial establishment.

Pressured from several sides, reluctant factions within the state began to fall into step with the patrons of manufacturing. From the mid-1940s to 1950, a succession of advisory committees like the Development Coordinating Commission and the Development Advisory Committee existed alongside and in close relationship to other governmental groups like the Godlonton Commission and the Department of Commerce and Industry. These bodies sporadically entertained the same "great white hope" fantasies about immigrant saviors common among settlers: "it is only by an increase in consumption in the country . . . that sales could be greater, output increased and production costs brought down to a low figure . . . a greater output for these industries will only result if we get a greater white population *or* greater prosperity amongst our native population [my emphasis]."[66] However, these boards and committees mostly approached industrial planning with an increasingly steady and sober attitude, extending the state's assistance to manufacturing capital. As Ian Phimister notes, "Those smaller capitalists who opposed measures, aimed at selectively stabilising the workforce and improving housing conditions found the state relatively unsympathetic. This changed attitude on the part of the state reflected not only their declining importance as the restructuring of secondary industry and the wider economy gathered pace after the war. It was also a measure of the greater autonomy enjoyed by the state."[67] Phimister also points out that discussions between South African and Rhodesian officials, particularly in their joint assessment of the new economic power of the United States, gave the need for secondary industrialization an added urgency.[68]

In the aftermath of the 1948 strike and in the face of an accelerating expansion of manufacturing, the Joint National Council (JNC), an assembly of diverse civic institutions including the Associated Chambers of Commerce, the Federation of Industries, the Federation of Women's Institutes, the Farmers' Union, the Chamber of Mines, and others tried from 1949 to 1951 to resolve conflicts over the industrial dispensation that would prevail in postwar Zimbabwe. Government officials working with the JNC warned that the discussion, especially relating to wage increases, "must be treated as most confidential as, if it were to become public property . . . a general strike would follow."[69] Representatives of primary industries—the farms, the mines, and a smattering of other small heavy industrial firms—fought ferociously to maintain the economic order of past decades. One individual questioned "the idea for increased wages. In his business, native employees averaged 90/- a month wages, which included food and housing allowances and, in addition, received one hot meal a day. They all seemed perfectly happy and looked healthy and well fed. He therefore wondered what was behind the suggestion."[70] The representative of the Chamber of Mines muttered darkly at the next meeting, with farcical understatement, "there was some subtle movement in the country to raise the wage standard . . . the Chamber of Mines would like to know who was behind this movement. . . ."[71]

Trying to appease this faction, the minister of Native Affairs and the body of the JNC agreed to oppose "automatic" wage increases, but declared that they would have to investigate "methods which involved increased wages for increased production."[72] This investigation included the first formal attempt to collect cost-of-living statistics through the form of the "typical" monthly consumption of urban Africans. At the key December 6, 1950, meeting, the conservatives directly challenged representatives of Prime Minister Huggins's administration, asking first, "Is the Government prepared to take full responsibility for any increases in the cost of living brought about by any increase granted in natives' wages?" and second, "What are the Government's views on the effects of an increase in wages for African labour on the general economy of the country?" The first question was ignored as "rather political," but the direction of the Huggins administration was clearly indicated in the answer to the second inquiry: "the proposed rise in wages was bound to have a certain effect on the economy of the country . . . what the effect would be on industrialists, they themselves were better able to judge. It was possible that certain of them would find it difficult."[73] Huggins had reluctantly committed his administration to this direction, declaring, "If the Native is improved, his wages must rise and his spending power, apart from his productive power, will increase. It will circulate money and improve trade and make it possible for more people to earn a living in the industry we have. . . . We cannot return to 1939."[74]

Battle lines thus firmly drawn, the December 6 meeting dissolved into a running string of charges and countercharges. Representatives and allies of manufacturing industries declared themselves "quite ready to advance wages for efficient boys" and vociferously opposed the suggestion of the Farmers' Union representative that any additional compensation for African workers be given in the form of food in order to remove "temptations." They commented that "an employer had no right to force food on the natives if they preferred cash."[75] In a more elliptical vein, a representative of the FWISR supported higher wages while suggesting that African women could also ignite industrial growth alongside their worker husbands by "behaving more like European wives and helping in the family economy," thus presaging a predominant theme in the commodification of hygiene and domesticity during the next several decades.[76] Though discussions of wage increases continued at later meetings and events, in the JNC and elsewhere, the desperation of primary industry and the open opposition of other settler factions on December 6 clearly marked a moment of economic and social transition in the history of the colony.

The Postwar Boom, the Federation, and Changing Patterns of Consumption

The head of the Chambers of Industries surveyed the scene in 1945 and described a "revelation":

> In Bulawayo's main industrial area you will see a new blanket factory already turning out blankets for our African population. This concern has, I hear, already decided to double its plant and buildings. We have another large blanket factory in the course of erection and also a knitting and textile mill. A factory will very soon be started for the production of all sorts of yeasts, vitamins, etc. . . . Another factory is being erected by well-known Union industrialists for the production of macaroni and such like. . . . Messers Lever Brothers are establishing large soap factories here. Pottery and glass are being considered with good hopes of realisation. Sweet manufacture is proceeding on the upward grade, and also biscuits. Look at Salisbury's enormous asbestos-cement factory, turning out really vast quantities of roofing. . . .[77]

This expansion of manufacturing was steady during from the late 1930s but accelerated noticeably after 1945. From 1938 to 1955, the net output of secondary industry grew from £2 million to over £30 million. In 1947, there were 430 factories in operation; five years later there were almost 700.[78] The postwar boom affected far more than manufacturing industries. Increased world demand for agricultural produce and raw materials for heavy industry helped boost Rhodesian primary industry as well during the 1950s, and smoothed over the tensions discussed above. Furthermore, general economic expansion, increased prosperity among whites, the growth of an African elite, and a rise in working-class wages all signified the overall acceleration of commodification.

This expansion was not created or primarily shaped by policymakers, of course, but it was acknowledged and nurtured by the support of the state and other organized factions in settler society. As noted previously, the overt and subtle initiatives and resistances of African workers, elites, and peasants were also pushing the political economy and colonial society in particular directions. The postwar boom in manufacturing was also made possible by the slow accumulation of momentum established in industrial circles during the 1930s. The simultaneous interruption of global trade during the war, widespread though divergent forms of dissatisfaction with the "truck" trade in both African and settler communities, local and global trajectories of capital concentration and growth, and the accumulative weight of the social and economic articulations between precolonial and colonial society in the region helped spur a self-sustaining burst of growth at the end of the Second World War.

The role of the global economy is particularly worth noting here. As the end of colonialism elsewhere in Africa and Asia loomed, modern multinational enterprises knew they would have to renegotiate their relationships with the inhabitants of the new nation-states. This spurred, among other things, an intense new interest in metropolitan commercial circles in the economic potential of African markets. Between 1945 and 1965, numerous books, articles, pamphlets, and promotional materials set out to assess and analyze new African national markets for the benefit of interested British, European, and American investors.[79] Colonial Zimbabwe drew a particularly heavy share of multinational interest and investment during these two decades for a number of reasons. First, as noted above, the slow but solid progress of local capital in manufacturing during the 1930s created a reliable infrastructural base for investors. Second, the existence of white consumers, a small but economically active group, was believed to provide a market beachhead for locally produced goods, a guaranteed market niche to help root a growing company. Third, many multinationals had subsidiary operations in, or connections to, the South African market, which provided financial support and trained personnel for opening new facilities in British Central Africa. Finally, and most importantly, the temporary postwar amalgamation of the Central African Federation, consisting of the colonial states of Zimbabwe, Zambia, and Malawi, seemed at first to promise a huge united market full of economic promise and attractive conditions for sustained growth. To many multinationals, investing in the federation seemed an opportunity to get in on the ground floor.

Local capitalists, state policy experts, and some settler civic leaders in colonial Zimbabwe had understood increasingly during in the 1930s and 1940s that industrial growth was heavily dependent on increased consumption by Africans. Businesses riding the crest of the postwar boom transformed this understanding into a reflexively recited commercial doctrine. For example, the reports of successive British trade missions to the colony in 1931 and 1954 went from featuring the "possibilities of the Native trade" as a minor sideline to being focused exclusively on "the African Native market in the Federation of Rhodesia and Nyasaland."[80] Even after UDI, in 1969, manufacturers were still making frequent pronouncements like "the limit of their [secondary industries'] growth is set by the spending capacity of the local population and only as that is expanded through increased wages and additional employment opportunities can expansion of domestic consumer based industries be expected."[81]

Notably, however, the success of manufacturing never fully dispersed the embedded conflicts over the political economy which had occupied white elites in the late 1930s and 1940s. Industrialists continued having to defend the growing power of their enterprises from the attacks of much of the settler community, which grew increasingly committed to the crude racial capitalism of Ian Smith and to withdrawal from the federation, rather than the smoother neocolonial systems of domination preferred by manufacturers and their allies. Businesspeople had to continually reassure powerful interests among the settlers: "this advancement [of Africans] is based on Africans coming up to European standards of living and not Europeans dropping their basic standards."[82] Before the burgeoning politics of white supremacy crushed his administration and his cautious, pragmatic commitment to a controlled process of change, Prime Minister Garfield Todd declared in 1957:

> It is true that not so very long ago Africans were not much interested in progress and, if offered incentives, these would not have made very much difference to them . . . that day is passing—in actual fact, for many thousands of Africans, it has already passed. . . . Native wages are still not tremendously high in some industries. Where Native labour is particularly cheap, there is not a great deal of incentive to improve methods to bring in better discipline and make sure of every possible method of getting better results from the labour force. Other industries have made great progress. . . .
>
> Some firms are way ahead and possibly hope that their competitors will not yet realise what is actually happening. These are recognising the importance of the African wage bill and are doing everything in their power to attract some of that money into their own coffers. Others do not seem to be aware of the size of the African wage bill. Fortunately, very few firms in our history have served Africans through holes in the wall.[83]

The political struggle that cast Todd into the prophetic wilderness and eventually brought Smith into power served as eloquent testimony to the persistent resistance of many settlers to realignments of power in the direction of standard-issue neocolonialism.

The success of commodification was understood by capitalists to be predicated on the acquisition of knowledge about the "tastes" of Africans and the construction of superior methods for penetrating, altering, and controlling these "tastes" once they were known. Thus, the management of the "African market" was rapidly entrusted to specialized sales divisions, market researchers, and advertisers. This understanding of commodification also posed powerful challenges to the growth and autonomy of manufacturing firms. As various Africans from diverse positions and with different access to power negotiated commodification with industrialists, various misrecognitions, resistances, alliances, and appropriations further problematized the growth of secondary industry. The active promotion by businesspeople of increased wages and economic participation by Africans was anything but altruistic, and many Africans were aware that this was so.[84]

Additionally, one of the by-products of these investigations was budgetary surveys, which today offer a useful picture of the transformation and expansion of consumption in urban African communities from the 1950s to the 1970s (see appendix). These budgets had a complex political genesis. In part, they were designed to appease those who remained critical of wage increases by determining once and for all the "true" nature of African expenditures; many whites accepted the mythology that Africans "wasted" their money on "frivolous" goods and thus came up short when it was time to pay for food and shelter. Many urban Africans throughout the federation feared that the budgets would be used by this reactionary faction to permanently fix a low monthly wage, and thus attempts to collect the data in a few cases sparked small-scale riots and unrest and were generally met with suspicion and opposition. Nevertheless, the survey results provide a somewhat useful picture of the proportional consumption of some commodities among different income groups within urban African society.[85]

For the state officials who supervised the surveys and for many industrialists, the surveys simply represented a crucial piece of data toward the acquisition of knowledge about African "tastes." But the surveys also provide eloquent testimony about the extent of the postwar boom. From 1939 to 1970, the list of products used on a quantifiably significant basis by urban Africans expanded dramatically, from about 15 to over 150. This was partially an artifact of different measurements used in the various surveys, as well as the different motivations guiding the surveyors, but it also outlined the real impact of postwar expansion.

Lever Brothers and the Growth of Toiletry Manufacture

Before 1945, the case of soap manufacture exemplifed the path taken by successful secondary industries and the challenges they were called to negotiate. I now return to the production of soap and other toiletries as a typical example of postwar growth in colonial Zimbabwe that also demonstrates some of the important structural limitations and forces inherent in the production of these goods during this era.[86]

In 1943, Lever Brothers (South Africa) leased the soap factories owned by RMM and Progress in the capital, in preparation for establishing a full-fledged subsidiary. In 1947, the company went ahead and purchased all the soap making facilities owned by these companies and then the factory of the Express Nut, Oil and Soap Company in Bulawayo; the latter plant had just recently been modernized.[87] These acquisitions still left a few local soap makers in business; the most important of these was the firm founded by Margolis, Rhodesian Industries, which would later be known as Olivine.[88] Nevertheless, Lever Brothers in one quick stroke became the overwhelming power among soap manufacturers in British Central Africa (anticipating the federation, the company moved to establish facilities in colonial Zambia and Malawi as well). The characteristic approach charted by Lever in England to marketing, advertising, and manufacturing soap products became the essential corporate model for penetrating the "African market" during the postwar expansion. Furthermore, Lever was only the first of a number of toiletry-producing multinationals to establish local subsidiaries in the colony; later investors included Colgate-Palmolive.

After establishing its subsidiary, whose management was primarily culled from its South African operations, the company proceeded to consolidate its interests in the colony, closing some factories and modernizing others. Following its standard operating procedure, the company also began to cultivate and train local personnel, including, by the 1960s, young African sales representatives whose task it was to reach African consumers. The subsidiary, along with its South African counterpart, clearly regarded these consumers as its most crucial though most perplexing target. Continuing problems with the local production of groundnuts gave Lever a further advantage, with its access to African palm oil supplies and the global tallow market. Locally produced tallow had been mostly exported to soap makers in South Africa, as supplies had far outstripped the needs of local factories, but from 1951 to 1953 these supplies were dramatically snapped up by Rhodesian firms. Exports of tallow dropped from 457,000 pounds to only 10,000 pounds. Soap production rose from almost 12 million pounds in 1946 to 18 million pounds by 1952, and though renewed competition from small South African firms with cheaper access to raw materials slowed growth slightly after 1952, the essential trend continued unabated.[89]

The real growth, however, came first through the expansion of the total range of toiletry and detergent products through market segmentation. By the late 1940s, the need for soap was clearly deeply felt among virtually all urban and most rural African consumers; various budget surveys showed nearly a 100 percent rate of purchase. Early surveys also suggested a remarkably uniform and inelastic demand for soap in African households. Such soaps were probably mostly cheap bar soaps like "blue mottled" and in some cases more expensive brand name examples of these, such as Sunlight, used both for washing the body and for washing clothing. Though individually wrapped tablet soaps expressly for toilet usage had been available both from local producers and importers for several decades, such products were not aggressively marketed to Africans until after Lever Brothers began its local operations. Lever and some of its fellow firms believed "the continuing progress of a manufacturing business depends largely on the expansion of its product range when marketing opportunities arise."[90] The range of soap products was expanded significantly during the 1950s to include various detergents and a number of different toilet soap brands. Moreover, new kinds of toiletries like perfumes, cosmetics, Vaseline and face creams, shampoos, toothpastes, shaving creams, and skin lighteners began to appear on the market in increasing amounts by the late 1950s, with the numbers of both imported and locally produced brands growing noticeably each year. There were notable failures and dramatic sales successes scattered throughout these product lines, but the overall momentum toward more products and more overall volume of sales continued unabated. Though the market for soaps continued to be dominated by Lever, Olivine, and, by the 1960s, Colgate-Palmolive, the growing popularity of other toiletry lines created opportunities for large local firms like the Central African Pharmaceutical Suppliers (caps) and other smaller firms like Robins Remedies, Stobard & Wesley, Parfums Rivoli, and a number of others, as well as multinationals like Ponds and Boots. The expansion of demand for toiletries led the federation government to officially identify "toilet preparations" as an "opportunity for industry." The local share of such production, roughly valued at £400,000 in 1963, was felt to leave considerable room for future expansion.[91]

With toiletries, as with many other manufactured goods, imports still provided serious competition for local production, though there was more local activity in this and related fields than in many other lines of manufacture; electronics, for example, were largely imported before 1965. In a few cases, as with soap, local producers were able to supply almost all sectors of the market and thus reduced the need for imports significantly. In any case, goods targeted at Africans—with the necessary accompanying high volume and low sales price—almost definitionally had to be made locally or at least regionally. Many of the imported toiletries available were aimed at the high end of the consumer market, for whites or elite Africans. A few foreign manufacturers sought to straddle the fence by licensing local firms to make their products, thus retaining the market appeal conferred by importation and avoiding the costs of establishing a full-fledged subsidiary. For example, some of the most lucrative toiletries in the late 1960s and 1970s were skin lighteners and similar facial creams, mostly made by small local chemical and pharmaceutical firms, some of them licensed to make authorized brand names like Ambi which were owned by American or British firms.

Overall, toiletry manufacturers were among the most successful firms in the postwar boom. Consumer surveys from the 1970s show that many of the products introduced after the war had gone from having almost no local market or a market only among whites to being widely or significantly used by Africans as well.[92] Toiletry manufacture was also characteristic of all secondary industry in its pattern of development: established local industries from before the war taken over in a wave of international investment, rapid growth during the federation years, active marketing campaigns in both urban and rural areas, and the expansion of product lines through market segmentation by the 1960s. Lever and other toiletry producers also were faced with typical challenges in the period that followed Ian Smith's Unilateral Declaration of Independence in 1965.

UDI, the Closed Economy, and the Transformation of Distribution

Zimbabwean capitalists today nostalgically look back on the pre-1965 economy as an Edenic time, and many regard the Unilateral Declaration of Independence, regardless of their personal feelings about racial segregation, independence, and the *chimurenga* (the war for independence), as a fall from economic grace. Leaving aside the once contentious issue of the full effects of sanctions on the will of the Rhodesian government to stay in power, the perception by contemporary businesspeople that UDI "closed" the economy is at once accurate and misleading. Certainly, even the limited detachment from the world market that followed after 1965 complicated the task facing manufacturers and made expansion a tricky or even impossible enterprise. Some sources of raw materials were cut off or had to be rerouted and concealed. Retailers and manufacturers alike, in a situation of institutionalized scarcity, found themselves able to sell only within severe limits. As one wholesaler reflected, "Prior to UDI, buyers and sellers were more aware of product knowledge, what came on to the market, how it came on to the market. You'd say, I'll try it out, give me a small quantity, and if it sells, we'll repeat it, and if you sold it again, you'd repeat it more. It wasn't a case of not being able to get it. With UDI and sanctions, we came away from that particular situation to a situation where we became thirsty. When I say thirsty, I mean that when a new product came on to the market, and we hadn't seen new products for a long time, we'd say gee, it's imported, isn't it? Let's try it."[93]

This comment also hints at another side to the post-UDI economy. The dramatic shift to local production that became necessary after 1965 provided considerable opportunities for local capitalists, and enabled some of them, in the short term, to retool their efforts away from the increasingly problematic and always disturbing "African market" to white consumers who had been newly denied their customary imported goods. Furthermore, most companies with established businesses, markets, and facilities simply continued to operate as they had before UDI, possibly tabling their more ambitious hopes for regional and local expansion, but often also continuing to actively map out their continuing penetration of the "African market."

In fact, the decade that followed UDI was marked by increasing efforts by a number of producers to reach not just their usual urban markets but also rural African consumers and retailers, aided by the government's cynical "community development" apparatus—until the escalation of the *chimurenga* made these efforts impossible. Also, the essential momentum of commodification as a process was far too established to derail in any significant way, and it has continued to reshape daily life, meaning, and identity for Africans and whites alike from 1965 to the present. This momentum also continued to remake the essential structures underlying manufacturing and consumption. Most noticeably, moves begun in the 1940s to abolish or reform the "truck" trade reached a culmination of sorts with the spread of new types of retail and wholesale distribution.

The targeting of "truck" distribution was a complicated process. A certain amount of official and other criticisms of merchants by settlers originated either from bigotry, social condescension, apprehensions about whites who lived among Africans, or concerns about provoking dangerous public disorder by Africans resentful of predatory merchants. However, by the 1940s, another important source of antagonism stemmed from the growing community of industrialists riding the postwar boom and hoping to cultivate African consumers. The traditional wholesalers had occupied a command position in the "African market" and their contacts and base of power were still primarily invested in contacts with foreign producers of cheap clothing and other goods. Furthermore, they still controlled or influenced retailers throughout the country. For local manufacturers, as well as some metropolitan producers of high volume, nondurable goods, the existence of this system represented a serious impediment to distributing their products; one noted that "most of the large business firms prefer to supply direct to the retailer. The advantages of this distribution channel are quite strong. . . ."[94]

The restriction and increased supervision of "truck" traders in the 1940s represented the beginning of a process that would eventually eclipse "truck" distribution altogether. After 1945, the next significant development consisted of moves by the Farmers Cooperative and a number of independent traders to consolidate their stores into a single corporate trading company, African Stores Ltd., capable of exerting greater control over prices and distribution as well as dealing directly with local and foreign manufacturers, thus cutting wholesale firms out of the process completely. The effort ran into difficulty with officials, who feared that a trading monopoly in the rural areas would only exacerbate the unrest and bad feelings that were already a serious problem for administrators.[95]

The idea of establishing direct relationships with manufacturers, however, had already been gaining popularity among some "truck" retailers before 1945; storekeepers who hired hawkers had long favored local goods, and other retailers in the early 1940s turned directly to local manufacturers when supplies from the world market became difficult to obtain or came attached to the burdensome selling practices of wholesalers. A second development during this same period was a gradual but perceptible change in the process of licensing traders. As noted in chapter 3, licenses were made easier to obtain in townships and were given to more and more African salesmen. The general increase in the number of businesses led to a relaxation of the local monopolies exerted by stores. African traders whose sources of capital lay outside the traditional relations between wholesalers and "tied" stores represented a new retail clientele predisposed to accept direct delivery of goods from importers and local firms alike.[96] Finally, the accumulation of bad feelings, boycotts, and protests that had increasingly affected the "truck" merchants in the 1940s began to have an effect on the way traders conducted their business. Though many continued to price their goods over what was popularly considered fair, the changing nature of demand and consumption in African communities led many traders to seek goodwill by shifting their stock away from the shoddy materials imported from abroad in favor of brand-name products from reputable manufacturers.

By the mid- to late 1960s, more important developments were underway that would complete the eclipse of the old "truck" networks. For one, many wholesaling firms, like Jaggers Wholesalers and African Marketing Services (AMS), began to operate on a "cash and carry" basis, moving decisively away

Township Store, 1959. (Zimbabwe National Archives)

from the kinds of exclusive client-patron relationships that had marked such trading in the past.[97] This change also helped to encourage more people to open stores, especially in the rural areas, by further opening access to the supply of goods. These firms also tried to help support these traders financially, though without the controlling agenda of the older wholesalers. The growth and success of these firms, even under the trying conditions created by sanctions, highlighted the shifting balance of power between manufacturers and merchants; one leading wholesaler today points out that while traditional wholesalers could no longer compel retailers to accept large shipments of unwanted goods, that power simply had been transferred to manufacturers operating under post-UDI conditions of scarcity.[98]

Even more important, while rural stores dropped some, though not all, of the coercive baggage of past merchant activity, they continued to look much like the stores of yesteryear and were usually still operated by a single merchant and a few assistants. In urban areas, supermarkets and chain stores, first established for the benefit of white consumers, began to cater specifically to African customers. While these stores did not displace traditional commercial activity altogether (the Charter Road area is still filled with the same kinds of small clothing stores and bazaars it had forty years ago), the characteristic style of mass distribution typical in large chains took hold rapidly and subsequently commanded the loyalty of most African consumers. Many Africans, especially the professional and business elite, had already been doing as much of their shopping in "European" department stores in the city centers as was allowed when African-oriented supermarkets began to open. Stores like Kumboyedza, Jazz, and Jarzin sprang up in traditionally "African" shopping districts in the late 1960s and early 1970s. Not all of them succeeded; Kumboyedza got caught in a retail downturn and had problems with supply not long after opening, and by 1974 it had closed.[99] Jazz and Jarzin met with more success and still have multiple outlets in operation today. Interestingly, in both of these cases, the chains were established by merchants who had long been active as "truck" retailers but who reinvented themselves and their businesses to meet what they perceived as unmet needs in urban shopping districts frequented by African customers. The change in the style of shopping permitted in a chain store was without doubt a significant development in the relationship between Africans and commodities, though facilities on the whole remained inadequate compared to shops in predominately white areas.[100]

This transformation of distribution completed the restructuring of commodification that had begun with the slow growth of local manufacturing in the 1930s. Aspects of exchange shaped by "truck" trading remained embedded in the relations between African customers and merchants—credit, for example[101]—and the meanings and associations forged by the relationships between goods, stores, and colonial rule continued to weigh heavily on commodity fetishism and consumption after 1945. However, the course followed by the postwar economic boom had major consequences for the structure and trajectory of commodification. Local manufacturers and their allies had argued since the 1930s that the direction of the political economy and the role of the state had to move away from disorganized and small-scale agrarian and mining capital toward modern monopoly capitalism. While such arguments remained a persistent feature in settler culture, they ultimately did not prevent the postwar boom from taking place. Businesspeople believed that the successes they enjoyed during this time were due in significant measure to African consumers. The "African market" at all times remained a major target for any growing company. As such beliefs coalesced, they led these businesspeople into a complicated and interlocked set of projects for investigating, imagining, altering, and producing "African tastes." These projects never quite worked out the way they were expected to, and ultimately led some capitalists to suspect the limits of their own power to remake commodity culture.

5   THE NEW MISSION:
ADVERTISING AND MARKET RESEARCH
IN ZIMBABWE, 1945–1979

In 1959, while addressing a regional conference on advertising, a South African official in the Department of Bantu Administration and Development, Werner Willi Max Eiselen, proclaimed: "I am satisfied that press advertising in the right way will create a better state of mind amongst backward people."[1] Eiselen was preaching to the converted. His audience of advertisers, market researchers, executives, and salespeople was then riding the crest of a wave of self-confidence. Local and multinational firms had for a decade increasingly identified the "African market" as critical to any continued expansion in southern Africa—marketing and advertising were fundamentally conducted on a regional basis—and they were now entrusting their success in this realm to specialized sales divisions and advertising firms. Like the missionaries of a century ago, these professionals discovered in their appointed task a "civilizing mission" which could ensure the viability of colonial and neocolonial rule.[2] Like the missionaries, advertisers sought to investigate what they saw as a mysterious world of heathen practices and orientations, to codify rules and techniques for preaching the gospel of consumption, and to organize and sustain growing communities of those they identified as the converted. Unlike many other colonial institutions, advertisers shared with the missions the desire to actually create new African subjectivities, to manufacture a new African personhood.[3] Partly as a consequence, some Zimbabwean advertisers and manufacturers were to discover to their surprise in the 1970s that the fruits of their labors were not always what they had anticipated.

"Changing Their Way of Life": Advertisers and "Civilization"

Many advertisers in South Africa and colonial Zimbabwe described their work in the late 1950s as "the most important single factor in influencing, particularly our urbanised Bantu, towards the acceptance of at least the outward symbols of our Western civilization."[4] This messianic tone, applied to the Dark Continent of consumption, flowed neatly from the connections established between "civilization" and consumption by earlier generations of settlers. Advertisers argued that they had a methodology for successfully accomplishing this project: they took earlier and more general settler visions about converting Africans to the gospel of consumption and tried to make them real.[5]

Businesspeople were drawn to this new mission by a number of factors beyond their immediate and primary commitments to penetrate the mythologized "African market." For example, the assertion of such a "civilizing" project had become the raison d'être for any group of institutional professionals seeking to exert their influence within the colonial system. Indeed, with appropriate modifications in tone, the assertion of such a mission still lurks at the heart of many similar contemporary neocolonial institutions, from the World Bank to the Peace Corps. Furthermore, many of the institutions that had long claimed the lion's share of such projects under colonialism were in one way or another faltering in the postwar era.

Advertisers' use of the rhetoric of "civilization" had other important sources. Given the omnipresence of issues of race and class in colonial Zimbabwean society, virtually any text or document self-consciously designed to reach Africans was inevitably going to be rife with notions about ethnic and cultural hierarchy. Other factors specific to advertising were also important. Metropolitan agencies, always at least indirectly influential on the practices of firms operating in more peripheral settings (in many cases, they actually owned these firms), had long boasted of their capacity for solving social problems and overcoming the "resistance" of recalcitrant or dangerous populations within the nation-state.[6] Furthermore, the wave of young professionals in South Africa and colonial Zimbabwe who moved into marketing in the 1940s and 1950s were also, in many cases, trained in disciplines and workplaces with their own heritage of "civilizing." Many had worked in African administration departments of the South African or Rhodesian governments, while others were trained in anthropology at South African universities.

One of the more extraordinary statements of the philosophy of the new mission came in an address to a 1960 regional convention by a South African advertising executive named J. E. Maroun, who also held a degree in Bantu studies from the University of the Witwatersrand and worked for the Johannesburg Non-European Affairs Department. Maroun's address on the "African market" was a remarkably open and boldy formulated exploration of some of the fundamental motifs in southern African advertising discourse. Maroun started by challenging his audience, declaring:

Many of us regard the African market as something concrete and tangible, something which has an existence of its own. . . . Don't let's kid ourselves. . . . AN AFRICAN MARKET DOES NOT EXIST PER SE. . . .

The only African markets that do exist are those that have been created, those that have been made through the efforts, conscious or otherwise, of manufacturers and marketers.[7]

Maroun stressed first and foremost that "in the African market we are in a field where . . . we are more interested in changing the African way of life, the African culture, to make it conform to European standards."[8]

In the advertising ideology of Europe and the United States, advertising was often represented during the 1950s and 1960s merely as "information" that helped consumers make rational selections for satisfying their autonomously determined needs. Consumption was economic democracy, a generously provided pluralism of goods and services in which the vox populi was king and "choice" reigned supreme.[9] In southern Africa, such rhetoric was already self-evidently and transparently ridiculous when applied to the "African market." The Central African Federation was a colonial polity and both Rhodesia and South Africa were ruled by oligarchies openly committed to eternal white supremacy. This didn't stop some advertisers in the region from speaking in reverent tones about the freedom of consumer choice, but most others recognized that developing African consumption was at the least implicitly a command exercise. Maroun was more forthright. He argued bluntly, in an amazing and revealing passage:

A fallacy applicable to all marketing, but of particular importance in considering the African market, is the belief that marketing exists solely to satisfy the needs of consumers. What consumer need was there for roll-ons, lipsticks, bleach creams, perfume and so on until creative marketing stepped into the picture? Marketing creates needs, it sells solutions to problems, it makes people desire what you have to sell. . . .

We are more concerned with selling product categories—ways of doing things—than brands. We are not selling toothpaste or soap or laxatives but new ways of cleaning teeth, washing and blood purifying. We are selling new ways of doing old things. It is only when we realise the cultural implications of marketing to Africans that we appreciate how fundamentally different our task of selling to these people is, it is only then that we can design our tools and make our operations more efficient.

From the outset, we must realise that almost all our efforts in the African market should be designed predominately to change culture—the traditional way of doing things—and in some instances even to introduce ideas which are foreign to and contradict tradition and, therefore, will meet with resistance. . . .

We are offering the African new solutions to his problems and in many cases even new problems. What we should be concerned with primarily is his acceptance of the solution rather than the detail.[10]

While some of Maroun's audience were mildly uncomfortable with the implied collectivism of his arguments—the proposition that manufacturers should put aside their own immediate interests and work as a single unit—the fundamentals of his speech simply represented a more articulate and straightforward exposition of themes commonly voiced among advertisers and executives at that time.

Some of the values and powers attributed to commodification by postwar executives were even appended to specific social "problems" involving African subjects. Stabilized labor and "community development" were praised for their commodifying effects by some capitalists. Fears about underemployed Africans gathering in the cities could be addressed by developing new needs, argued the head of the Associated Chambers of Commerce of Southern Rhodesia: "The vagrant Africans . . . are of no use to Commerce and certainly no use to the country. . . . You may ask, what is the advantage of instilling into such people a desire to acquire better food, new clothes, gramophones, and similar things when they could be left to sleep under a tree, or collect a few pence per day by itinerant hawking, but, as I see it . . . they must eventually be absorbed into industry or . . . become an insoluable problem."[11] Similarly, as the liberation war heated up in the 1970s, many advertisers turned their attention to resolving the "problem" of the rural areas. One executive wrote, trying to redress the industry's past exclusive interest in urban consumers: "the growing sophistication of urban areas must be matched by the creation of markets in the rural area. I say 'must be' because we cannot afford not to bring them into the money economy."[12] Another issued a moralistic call to arms: "It is indeed every Rhodesian manufacturer's *responsibility* to motivate our fellow countrymen to create this cash economy. . . . The main problem in the Tribal Trust Lands is the lack of a motivating force [my emphasis]."[13] Many of those advertising specialists who remained in the country during the 1970s designed "hearts and minds" campaigns for the Rhodesian military, campaigns frequently based on techniques developed by commercial firms.[14]

In short, advertisers defined their work as bringing Africans in from the marketless cold into the salvation of commodity relations, creating needs and converting problem populations into functioning parts of the colonial economy—populations whose problems had allegedly been caused by being "outside" the marketplace, though the truth was exactly the opposite. This new mission was still captivated by visions of an unpenetrated, collective "African market" lying beyond the reach of colonial domination but waiting to be "discovered" and brought within the fold. As always, such an ideal brought with it all the hidden and unacknowledged contradictions involved in sustaining such a project. As Megan Vaughan has pointed out with regard to the promotion of soap: "They [Africans] were created as consumers of products for their new, modern bodies at one moment, and at the next they were told to revive their 'traditional' knowledge of soap-producing plants. By relying so heavily on older modes of production for its very success, colonial capitalism also helped create the discourse on the 'traditional,' non-individualized and 'unknowing' collective being—the 'African,' a discourse to which the idea of difference was central."[15] Heedless of such dilemmas, confident of their general goals, executives set out to explore methods for bringing the new mission to its fulfillment.

"What Time Do They Get Up in the Morning?": Investigating Africans

Maroun argued that advertisers were called to "create" needs rather than passively deal with preexisting "real" needs going unmet in African communities. In so doing, he accurately described some of the hidden assumptions of other marketers. Still, many executives also felt that knowledge about the nature of African tastes and practices was a necessary foundation for future campaigns to control and generate consumer desire. Even Maroun agreed that some kind of prior knowledge was essential if the commodification of Africans were to succeed: "We must realise that in marketing to the African we are faced with a completely new situation never experienced anywhere else before. Things like migrant labour, locations, African villages, African concepts of marriage, wealth, success and beauty are peculiar to Africa. If we are to market in this cultural universe, then from the outset we must recognise this fact."[16] This pursuit of knowledge about "true" African tastes, practices, and attitudes was thus a fundamental feature of marketing in southern Africa, and the hidden implications of such pursuits were eventually to pose a serious problem for advertisers.

Merchants and industrialists had debated prior to the 1950s about whether "knowing" Africans provided any commercial advantage. More than a few "truck" merchants would have agreed with the statement, "I don't allow the Native to complain. If he does not like it he can lump it."[17] Still, as noted in chapter 3, others in the trade argued that sensitivity to changing African "fashions" could make or break a merchant. One "truck" wholesaler reported that his firm studied "native tastes seven days a week."[18] British trade missions and commercial periodicals concerned with the southern African colonies also reported to metropolitan investors about the importance of careful study of African culture prior to any venture. As early as 1901, one journal counseled, "Native love of finery is not confined to any particular section of the African continent but those who make a speciality of catering to the wants of the black fellow should certainly in their own interests study the requirements of prospective customers."[19] Later, the 1931 British trade mission also advised, "All native areas have their own peculiar tastes and ideas, and detailed investigation by technical experts is essential; such investigation, would, we feel sure, be well repaid."[20]

In the 1950s, intense focus on the "African market" brought renewed vigor to appeals for the systematic investigation of African culture and African consumer habits. The periodical *Commerce of Rhodesia* moaned, "there is no information available to show what the Native spends his money on, or his preferences . . . nobody knows."[21] One expert concluded, "The astute entrepreneur will not neglect to study the special idiosyncrasies and eccentricities—even mere whims—of the African consumer, especially in the rural habitat and setting. He has, no doubt, his special likes and dislikes. . . . Care must be taken in satisfying these special tastes."[22] An address from a regional workshop conducted in Durban similarly concluded: "the need for deeper understanding of the African at a personal and behavioural level . . . is fundamentally the biggest problem and obviously is brought about by the very nature of our way of life where there is lack of contact, lack of knowledge of Africans' personal motivations, and therefore very little accumulation of that important thing called instinct on which the businessman can work."[23] Such general appeals helped from the start to solidify the drive for "modern" bureaucratic collection of demographic and budgetary data by the Rhodesian state.

Maroun was canny enough to argue in the 1950s that local African culture was complex, plural, and ever-shifting, "something new" that could not be easily pigeonholed into comfortable binary oppositions like "traditional-modern" or "European-African." Most of his peers were initially less willing to see beyond general colonial orthodoxies. For some, investigating Africans minimally consisted of assembling a knowledge of locally peculiar habits and practices through ethnographic manuscripts, various "commonsense" propositions about the habits of Africans, and anecdotal personal observations. Others simply went ahead and designed advertising campaigns, trusting in the alleged simplicity and compliance of most Africans. However, many companies and professionals had invested a great deal of faith in the "science" of marketing and thus had a great need for formal institutional machinery for consumer research. Some of the larger companies, like Lever Brothers, habitually operated sophisticated internal market research departments wherever they invested. Lever Brothers' methodology for such research provided a fundamental model for other firms. It hired researchers, field assistants, and others to conduct surveys, focus groups, and statistical studies. Many of these workers were Africans who eventually became the core of specialized sales divisions in the British Central African and South African subsidiaries. Lever Brothers and many other companies also lent their support to collective efforts to build research institutions. For example, reports on the "African market" by the University of South Africa's prolific Bureau of Market Research were widely used by many companies active in the region.

The more that was spent on media aimed at African audiences, and the more African consumers assumed a vital place in the minds of industrialists, the more vulnerable corporations were to arguments in favor of the standardization and expansion of information gathering capacities. As the number of reports and surveys multiplied, the list of important missing data identified as requiring investigation unsurprisingly also increased: "And then the living habits of the African—what time do they get up in the morning? And go to bed? Is breakfast eaten at all or do they merely have a cup of tea before going off to work? How far has western influence spread concerning table manners, mealtimes, social conversation and so on?"[24] The phrase "little is known" became an oft-repeated mantra, as in this introduction to a Bureau of Market Research report: "A large and ever-increasing share of advertising budgets is spent on advertising to the non-White sub-markets . . . while little, if anything, is known about the buying behaviour in general and the reaction to advertising in particular of these non-White sub-markets."[25] Once urban African consumption and culture were studied, questions arose about the nature of class-based and gender-based consumption in urban areas. And once these were studied, it became imperative to know what rural consumers were doing, and why they did it. By the time that was known, doubtless urban consumers would have "progressed" and so would require more study, and so on.

Do's and Don'ts: Doctrines of the Faith

Still, the Rube Goldberg machine of market research produced many results, most of which helped shape advertising and promotional campaigns. Research-

ers warned that a variety of rules, guidelines, and procedures they had uncovered in the course of their investigations would have to be applied before any
advertisements could truly succeed. Such rules were a bizarre potpourri made
up of the effluvium of ideology about race, gender, and class; miscellaneous
creative inventions; and genuine discoveries about already accumulated historical experiences with commodification in various African communities in the
region. As such, these overlapping and sometimes contradictory tenets became
part of the subconscious folklore of the advertising business in southern Africa
and had a powerful influence on the design and execution of actual promotional campaigns. Most of these doctrines tried to use the data produced from
market research to evaluate the nature of African reception, to study African
audiences. In this, advertisers resembled filmmakers and propagandists working for the colonial state.[26]

A number of these principles addressed the content of advertisements.
One predominant doctrine about content involved the use of color. South
African racial scientists and ethnographers spent a good deal of time measuring
what were variously seen to be physiological or otherwise innate boundaries of
color perception among Africans.[27] Advertising executives absorbed this preoccupation, and as a consequence marketing discourse was full of discussions and
assessments of the "right" colors to use in making appeals to African consumers. Eiselen informed his audience, "There are two factors that influence
advertising quite a lot in our approach to the African market—the one is colour
and the other is pattern . . . you will find that there is a tendency for the Bantu
not to be as colour conscious as we assume that they are, for example blues and
greens and browns and reds, in the vernacular are grouped together."[28] Different colors were picked as being the most important to use or avoid, though
most advertisers agreed that "bright" colors were essential, that Africans enjoyed "display" and "exhibition." Since each supposedly authoritative study
reached varying conclusions about the "right" colors, one executive simply
noted, "Much has been said about colour combinations and the use of colour
in packaging . . . a little market research amongst one's own African staff before marketing a new product is always a useful exercise."[29]

Beyond color, other typical suggestions regarding visual (and written)
content included "refrain from using abstract material"[30] and "illustrate situations familiar to Africans."[31] The latter was considered particularly important
and was a subject of considerable concern. As data poured in, advertisers began
to get a better sense of the anxieties, challenges, and complexities of African
daily life under colonialism and capitalism. Advertisers were counseled to play
to these concerns as much as possible, though their understanding of these

matters was usually inextricably intermingled with their own preconceptions about African lives. Indeed, on many occasions, even though they had obtained information through their research that permitted them an opportunity to understand and penetrate African subjectivity from within, advertisers reflexively fell back on viewing the African as an alien creature who could be commanded by images but not transformed by them. Still, advertisements became one of the few locations in colonial public culture where urban Africans might see some of their own fears, complaints, and aspirations played out. Advertisers were kept alert to the play of meaning in African communities, aware that potentially fatal misunderstandings or sales-crushing mistakes in their representations lurked around every corner.

Another important principle dealing with content involved the packaging of goods and the promotion of distinctive brand images. Advertisers stressed again and again the importance of "brand identification." This stress was common to manufacturing capital worldwide (after all, the genesis of the success of firms like Lever Brothers lay partly in innovative packaging), but in southern Africa, market research revealed that African consumers had developed unusually intense commitments to affordable branded products. This was understood by advertisers to be an artifact of negative experiences with the poor quality of many "truck" products and a by-product of the relative scarcity of a range of different brands of new kinds of products. Marketers thus stressed that the visual form and feel of packaging was essential to success. Even here, feelings about the "innate" nature of Africans played an important role. One businessman thus advised, "Product names should be short, modern and easy to pronounce . . . product names should have the added virtue of sounding 'African' without being 'African,' like the abbreviation 'Soweto.' "[32]

Another rule directed at the content of advertisements involved the type and form of language used. Part of the issue at stake here was whether to use the "vernacular"—Shona or Ndebele in Zimbabwe—instead of English. Many marketers agreed with the sentiments of J. Walter Thompson executive Nimrod Mkele:

African languages are sharing the same fate, for the African realises that they are as yet incapable of subsuming the concepts of a technological culture towards which he is aspiring and in fact hinder him from making the necessary adaptations. . . . He therefore shows a marked preference for English as the Press and Advertising Annual points out and as the Information Service of the Native Affairs Department in Southern Rhodesia has found out.

. . . This naturally leads to the questions of advertising in African languages. Quite apart from the question of the multiplicity of languages and the absence of vernacular media . . . there is the fact that Africans do not know their languages well. . . . Even translations tend to fail, because a lot of words that are common and so on are untranslatable. . . . The results are apt to become artificial and stilted. . . . On the whole, there is a lot to be said for advertising in English, partly because the African attitude to his languages is negative and also because he is not thoroughly conversant with them.[33]

Mkele's arguments were based on a mix of practical assessments of media available in the late 1950s, a political credo that colonial societies could only be kept stable through promoting common imperial languages, and a cultural faith in the superiority of English. The issue was not always resolved in favor of English. Lever Brothers, for one, ran most of their print advertisements in the 1960s in an alternating cycle of Shona-English-Ndebele. Most marketers agreed that when the target population was the urban working class or rural peasants, rather than aspirant or upscale urban Africans, it was necessary to translate to some extent. On the other hand, many multinationals took it as an equally axiomatic proposition that brand names or language on packaging should never be translated under any circumstances. As one Lever executive noted, "The name Sunlight was retained in European translations and rendered phonetically into languages not using the Latin script. Sunlight was to be known as Sunlight everywhere, even though the nearest a Chinese could get to 'sunlight soap' might be 'isunli-iso.' "[34] Another issue connected to language involved the use of idioms or slang in ad copy. "In particular, plays on words should be avoided," was a typical warning in one guidebook. "They may misfire completely."[35]

There was another entire class of tenets reflexively recited by most advertisers from the beginning of the 1950s. Knowing the history of the "truck" trade, researchers warned industrialists to avoid offending the growing middle class by selling poor-quality goods or assuming any African would take whatever was offered. As one executive noted, "He [the African] was turning against Kaffir truck, or any other commodity of lower quality."[36] With theatrical flourish, businesspeople recalled the "bad old days" (days that had not entirely passed), in which "the African customer . . . went to the small window of the big department store and whatever he desired to purchase was chosen for him and brought to the window; or whenever he was allowed to come in, he was confronted with the question 'What do you want?' "[37] Similarly, another expert advised, "The sale of goods manufactured, marketed and advertised especially for the African market . . . stands a particularly good chance of failing . . . the

African consumer from bitter experience has become very quality conscious."[38] Salespeople and marketers constantly stressed that Africans were "discriminating." "Do not try to fool an African customer," warned one journal. "He is a most discerning shopper and you probably won't make the first sale, let alone any repeats."[39]

The most important and universally accepted principle to arise from this understanding of the impact of "truck" sales was that products must *not* appear to have been made *for* Africans. Mkele warned: "Africans tend to reject those products which are advertised as being exclusively for them. For such products carry with them connotations of inferiority which the African meets in his everyday experience whenever he is at the receiving end of segregated services. What happens in this case is that the product becomes psychologically downgraded and the African does not buy it."[40] The advertising manager for Lever Brothers similarly argued in 1953: "It was, in fact, extremely difficult to sell him [the African] any article which was packed differently from that packed for European trade. As an instance . . . any wrapper marked 'Not for European Consumption' was taboo among the Africans, and was even liable to cause some trouble."[41]

However, many advertisers also fervently believed in the flip side of that principle: as one manufacturer put it, "one could probably sell a lot of my product to the African market, but if I were to use my normal method of sales promotion, of advertising to Africans, the prestige of my product would suffer in the eyes of the European buyer."[42] It was considered axiomatic that if a product acquired a "black" image, then white consumers would immediately cease purchasing it. Indeed, if goods previously considered exclusive were marketed to Africans, whites sometimes protested actively to the company involved. Again, the history of commodification—in this case, the protection of the boundaries of privileged consumption—weighed heavily on the judgments made by executives about the viability of potential advertisements.

On some level, these two doctrines were irreconcilable, and advertisers knew it. The usual solution was to fall back on the old colonial trope about the "imitative" tendencies of Africans. "The buying behaviour of the African is mainly dependent on that of Whites," said one expert. "The African deliberately imitates the buying habits because he thinks they are superior and that their goods give him social status among his own people."[43] Mkele similarly argued, "The most important values that influence the African buying behaviour—and this includes advertising reactions—derive from European standards. . . . But the process of identification is not yet complete, which explains why the African still cannot appreciate classical music, modern art or serious drama."[44] Many marketers thus claimed that campaigns should start with white audiences and then move to African ones—but without visibly *appearing* to do so. One odd variant of this advice that recurred with some frequency was the belief that "taste transfer" could take place from whites to Africans through domestic servants. One journal typically counseled, "As a result [of the high cost of living], Mrs. Housewife has to go out to work. . . . A large proportion of the home shopping is therefore done by the house servant. . . . By giving this section of the public courteous attention the results would be astounding."[45] Mostly, marketers made conscious plans to play to a segmented and segregated market. By the 1960s, such segmentation often went beyond the binary opposition of "African-European." As more money was spent on the "African market," as more research was produced, and as the initiatives and reactions of diverse Africans began to have a telling effect, many advertisers and businesspeople also began to target Africans not as a monolithic unit but as a conglomeration of groups and classes.

Of these smaller groups, the most important by far were African women. Just as educators and missionaries had held women responsible for transforming the social content of African life—accepting in their own way the dictum that the personal is political—marketers believed fervently that women were the key to changing the material composition of the African home, that women controlled most purchases and most tastes. Of course, many of the "soft," nondurable consumer goods, especially toiletries, that made up the bulk of business in the "African market" were primarily considered to be *for* women or to lie *within* their appointed domestic sphere. In general, however, women were pegged as somehow holding the key to the successful creation and reproduction of new needs. As a consequence, advertising discourse was full of discussions of the psychology and nature of "the" African woman. "African women," concluded one pamphlet confidently, "are as susceptible to advertising appeals to feminine vanity as their European counterparts."[46] A Rhodesia Ministry of Information publication declared, "In the towns the changing status of women is particularly noticeable . . . today the housewife . . . does the shopping."[47] One of the more interesting statements of this sort was made by Mkele at the 1959 South African convention: "You will have noticed the highly significant role that women are playing . . . they are bringing in new tastes into African homes. After all, it is they who determine what shall or shall not be bought. The role of a hubby is to fork up and smile."[48] A Lever Brothers (Rhodesia) executive similarly concluded, "The woman now features as the major decision maker for the purchase of day-to-day consumer goods."[49]

Alongside this appreciation of the role of women, some Rhodesian executives stressed from the beginning of their burst of increased activity in the 1950s that careful attention to class divisions among Africans—often rendered as "status"—would be beneficial. This was implicit in the attention lavished on urban consumers, where the small elite were concentrated, and it was implicit in the counsel that many goods should be pitched first to whites and then gradually leaked down a hierarchical pyramid to other consumers looking for the prestige of being "Westernized." The ideal of the "imitative" and "status-conscious" African professional was a favorite with advertisers. "The middle class of African was rapidly increasing . . . manufacturers would soon find it worthwhile to spend a good deal more on Native advertising," concluded one Lever Brothers manager.[50] The middle class was valorized not only for being comprehensible—"just like us"—but also for "leading" other Africans to new tastes.

Another significant product of market research was an evolving set of doctrines about the alleged divide between a related set of Manichean cultural opposites: "young" and "old," "modern" and "traditional," "sophisticated" and "tribal." Studies warned that "product resistance" could be created by the negative and "backward" half of these pairings, that "taboos" could mysteriously materialize from them and destroy well-planned campaigns. Marketer-ethnographers offered helpful advice about these hidden tastes and perils, listing forbidden substances and cultural faux pas. Mkele offered a typical assessment, noting that "Today traditional taboos . . . no longer play as important a role as is sometimes assumed," while also acknowledging: "It is true that in Rhodesia Africans would not buy green-coloured radios. . . . I doubt that you could get Africans to eat lobster, crayfish, eels or shrimp under any circumstances. . . . Today the young housewife takes eggs for breakfast while her mother-in-law looks on horrified at the very idea. The well-known taboo about the eating of fish by the Xhosa and Zulu . . . is very much on its way out. . . . Advertising of course can only accelerate the process." He also warned, like many of his peers, of the "resistance of the elderly" and noted that "you will always find young people in the lead."[51]

All of these various doctrines, tenets, rules, snippets of advice, and speculations were the product of a growing research apparatus for surveilling, reproducing, and subdividing the everyday lives of Africans and came to represent an institutional form of folklore. This discourse was then used to structure the design and implementation of promotions, advertisements, and marketing campaigns aimed at African audiences.

Advertising and the Making of African Audiences

"Television, press and radio . . . reach only a fraction of the African population," noted the house magazine of Lever Brothers. "Special techniques are therefore needed to explain, demonstrate and promote Lever Brothers' products."[52] This was a conclusion rapidly reached by many other executives in the 1950s as they mulled over the creation of new needs among Africans. In short order, marketers proposed a panoply of innovations and marketing strategies that included cinema vans, roving demonstrators, fashion shows, contests and give-aways, public festivals, sponsorship of civic groups, spectacular public displays, celebrity endorsements, leaflets, gimmicks, hoardings (billboards), as well as conventional campaigns in media patronized by Africans.

Advertisements, including those directed at Africans, had been a part of the landscape of Zimbabwe well before the postwar economic boom. Early manufacturers and importers looking to make inroads among African consumers took a leaf from the "truck" trader's book and distributed free promotional gifts like mirrors or other novelty items, usually emblazoned with the brand name and product design.[53] Importers sent small signs, marked crates, or hoardings for store owners, including those catering to Africans, for display in their establishments. Handbills and other printed materials were sent through the mail to store owners and some African individuals, and some of these also ended up on display in storefronts.

Such early advertisements often made whites nervous. Rather than purporting to instruct Africans in some approved dogma, they explicitly set out to appeal to African interests and preferences. The Salisbury Chamber of Commerce in 1926 received "a letter from the Rhodesian Women's League . . . asking the Chamber to request members to be careful in the exhibition of suggestive displays in catalogues, posters and in wax figures in windows on account of the native population."[54] Billboards and other public advertisements on display in the reserves attracted particular criticism, and the Secretary for Mines and Works concluded, "there would appear to be no reason whatever why these advertisements should be exhibited in Native Reserves where the advertisers could have no possible right to advertise."[55]

Most of this concern was a subset of larger attempts to control all printed materials reaching African readers. The *Native Mirror*, as noted previously, was founded on such a principle, becoming the first major standard media outlet for advertisements aimed at African audiences. An editorial in the third issue showed that authorities remained concerned over the impact of advertising, as they counseled African readers, "It has been our practice to protect the Natives from buying worthless rubbish and to place before them the names of firms whose articles are good and who deal fairly."[56] The article went on to review the individual virtues of each of the advertisers in the previous issue.

Recent scholarship on advertising has featured what has become known as the "blindspot" debate. Dallas Smythe, Sut Jhally, and others have argued that communications and advertising must be regarded as a capitalist *industry* as well as the carriers of capitalist ideology (or what Raymond Williams has called the "official art" of capitalism). To sum up their argument briefly, Smythe and Jhally contend that as an industry, the media manufactures and sells a commodity—namely, audiences. Moreover, they argue, the audience labors to produce itself; it does the work of watching advertisements and is compensated with programs. The audience's product, time spent watching advertisements, is then owned and exchanged by media capitalists.[57] I find this a useful though not perfectly applicable concept in considering the development of advertising during the 1950s and 1960s in southern Africa. If audiences are a commodity, then the ability of Rhodesian audience-makers to deliver such a product "made" from African raw material was always somewhat in doubt. The heavy demands of colonial domination and racial supremacy meant that the creation of standard media outlets was spurred by the needs of the Rhodesian state rather than by the desires of manufacturing capital: almost all advertisements aimed at Africans appeared within media founded by and controlled by the state. Jhally's and Smythe's work nevertheless effectively suggests that even in this colonial context it is important to examine the complicity of advertising in the making of mass communications, and to consider the ways in which advertisers identified, regulated, and reproduced African audiences in the region.[58]

The number of available periodicals in colonial Zimbabwe multiplied in concert with the growth of secondary industry after 1945 and included locally published newspapers as well as imported South African magazines like *Drum*. As these publications multiplied, so too did the range and number of advertisements. Print advertising was the most reliable for most advertising executives. They knew how it worked. Campaigns could be designed in and sent from a distant home office to be minimally adapted to local conditions, thus ideally connecting distant individuals as part of the same transnational family of consumers. When radio began to spread, it too provided a familiar and comfortable option for southern African marketers.

Newspapers and radio aimed at Africans were both created by the Rhodesian state, and it took some years before the state controllers of either were comfortable with sharing these institutions with capital. By 1943, the white publisher of the *Bantu Mirror*, F. D. Hadfield, who had often clashed with supervisors anxious to control the paper's content, announced gratefully, "We have now succeeded thoroughly in convincing advertising contractors of the value of native markets."[59] Once the newspapers were run by the semiprivate firm African Newspapers Limited, advertising flourished on their pages and became a significant source of revenue. Businesspeople also rapidly learned that circulation figures did not accurately describe the audience for print media, noting that "the conception of one paper per family hardly applies and it has been estimated that there are, on average, from 8–15 readers of each copy."[60] By the 1960s, the African media had increasingly attempted to appeal to new segmented audiences, though the successive mergers and bannings of a number of these publications during the same era put something of a damper on such expansion. In the 1970s, "African" publications available included *World of Sport, Weekly Express, Homecraft, Murimi, Parade and Foto-Action, Prize, Radio Post, Sitima,* and *Umbowo,* as well as South African publications of various sorts like *Drum* or *Bona.* Of these, *Parade* had the largest readership; according to one survey, about 46 percent of the urban population read it. Many township residents also read the "white" newspapers; indeed, by the mid-1970s, the formal circulation figures for these papers showed that only half as many whites read them as Africans.[61] Many of these advertising campaigns in the press received additional circulation when they were reproduced in similar form on store hoardings[62] and on buses.[63]

Viewed in retrospect, press advertisements are the best witness to historical shifts in the practices and discourses of marketing in colonial Zimbabwe. In the 1930s, when the only locally published outlet for advertisements directed at Africans was the *Native/Bantu Mirror,* almost all of the ads published were placed by the owners of large "truck" stores. Some of these ads focused on the virtues of the store itself, while others stressed the availability and low price of a particular item or type of item. By the 1940s, the emphasis shifted to ads placed directly by manufacturers or their agents. The range of ads published varied considerably over time. Shifts in emphasis were partially a product of changes in production and distribution, as chronicled in chapter 4. The eclipse of "truck" traders and their dominant role in commodification was evident even in the period from 1931 to 1939. Advertisers' beliefs about the selection of goods as especially appropriate for Africans were equally apparent in the distribution of ads over this time period. This distribution disproportionately favored nondurable "soft" consumer goods, especially products for the body like medicines, toiletries, and clothing. In fact, manufacturers making durables like furniture who hoped to solicit African consumers often made no effort to appear differentially to them, simply trusting ads in the "white" media to reach interested consumers of all races. Significantly, by 1976, virtually all of the advertisements in *Parade* were beginning to take on more of this relatively "race-blind" feel, presaging the postindependence climate. Surveys done in South Africa by the Bureau of Market Research during the 1960s and the 1970s also suggested that this intense concentration of advertisements around nondurables, especially medicines and toiletries, helped to produce "brand consciousness" of a sort that otherwise was uncommon among African shoppers.[64]

Radio broadcasting in the region began first in Lusaka and was eventually extended to all of British Central Africa through the Central African Broadcasting Station (CABS). In the late 1940s, some colonial bureaucrats in the region had been mulling over the possibilities and challenges involved in reaching African listeners through radio. Encouraging the production of cheap radio sets was one necessary step, while another important mission was to actively encourage Africans to "acquire the listening habit."[65] Broadcasting to African audiences in British Central Africa began in 1952 and grew rapidly. As with newspapers, radio was originally seen primarily as an instrument for propaganda, a tool for the state to accomplish its educational and rhetorical aims while combating what was decried as the insidious power of "rumor." There was some controversy at first as to whether to allow advertisements; the FWISR wrote to CABS and complained: "Is it fair to submit the African to high-powered sales talk?"[66] Advertisers at the 1959 and 1960 regional conventions in South Africa eagerly begged bureaucrats for access to radio. Advertisements were eventually permitted and rapidly multiplied. Rhodesian executives knew that newspapers mostly reached urban audiences, while radio held the possibility of reaching a much larger audience in cities and countryside alike. However, even with the state's attempts to subsidize the production of cheap "saucepan" radios and the placement of radios in workplaces and public places, businesspeople knew it would be some time before there was a mass listenership. By the 1970s, surveys suggested that as much as 93 percent of the urban population and 51 percent of the rural population were listening to radio at least once a week.[67]

Still, advertisers clearly felt from the outset of their interest in the "African market" that other methods for creating audiences would be needed. They once again turned to a communications outlet developed by the state: roving cinema vans and demonstration units. As noted in chapter 2, the Rhodesian government was operating traveling vans by the 1940s that displayed propaganda and films in rural areas and mining compounds. "Bioscopes" in township halls served the same function. Lever Brothers in particular was familiar with this tactic, having operated a small fleet of such vehicles in India in the 1930s.[68] Advertising took advantage of these vans in two ways.

First, a firm named Dillon Enterprises obtained a contract with the government to place advertisements before and after films shown by the state-operated vans. Market research commissioned by newspaper publishers and the Rhodesian Broadcasting Corporation during the 1970s found that 15 percent of urban Africans and only 5 percent of the rural population saw a film in an average month.[69] Dillon Enterprises' own research suggested that attendance was much higher, and also claimed a much higher "effectiveness" through measuring various reactions to film advertisements. This research suggested that only 60 percent of respondents "believed" radio ads, while 78 percent "believed" cinema spots. On the other hand, it also noted that 51 percent of those surveyed agreed that they "understood *all*" of the cinema ads, as against 62 percent for radio, something that the researchers attributed to the difficulties of "inference" in visual ads and the audience's alienation from those ads using European models. The research also claimed that there was a significantly larger portion of the cinema audiences expressing interest in purchasing commodities they had seen advertised.[70]

Second, a number of firms built up their own traveling vehicle divisions and demonstration staffs and set out to promote their goods in both townships and rural districts. These operations rapidly became an important part of marketing aimed at Africans and spawned a host of related techniques for creating and holding audiences. Lever Brothers ran one of the largest and best organized of these operations. The Outdoor Advertising and Promotions Department (OAP) ran a number of cinema vans and demonstration teams. These vans typically covered a regular route, some in townships and some in rural or peri-urban villages. Upon arrival, the team often showed one or several films interspersed with promotional materials. A supervisor maintained the equipment, smoothed relations with local authorities, and controlled his staff of assistants and demonstrators. The films were often followed by a product demonstration, usually conducted by a female employee, who might show "how to improve the taste and texture of sadza by cooking it with Stork margarine" or some other product, emphasizing "audience participation."[71] The van crew also often included an employee who rode ahead to sites checking to make sure that stores there had a complete stock of Lever products. Some other companies using these operatives also tried to circulate their employees among the crowd that gathered, in order to discuss the products on display and, in some cases, start negative rumors about the products of competitors.[72] Other van units offered live performances of music or performed minidramas (usually aimed at communicating a single product-oriented message). In many cases, the demonstrators gave away small packages of the goods they were seeking to promote. In townships, demonstrators also sometimes operated out of stores and supermarkets or set up booths in marketplaces or public gatherings.

The idea of demonstration, of making consumption into a kind of popular and public theater, proved infectious and led to a variety of related promo-

"Mr. Power-Foam."

tions. Companies began sponsoring fashion shows and beauty contests that used products they were trying to sell. A textile salesman described one such effort: "What is the best way to sell a dress fabric? The answer is obvious—in the form of a dress. Thus it was that our first African Fashion Show came into being. We decided to carry out an experiment—to devise and produce a show of the highest calibre, a show which would not disgrace us wherever it was performed."[73] The state lauded such events through the Ministry of Information: "In old African society, there were no such things as Beauty Contests. It would have been regarded as 'vulgar' to say the least. Now it is an accepted form of entertainment. African girls will even parade in bikinis!"[74] Similarly, many manufacturers began to compete for the hearts and minds of African consumers by sponsoring give-aways and contests, mostly in urban areas. After these became increasingly popular tactics, one expert counseled, "competitions are still effective. But again certain competitions have failed because the sponsor failed to appreciate the realities of urban township life. Refrigerators, washing machines and the like are no earthly good as prizes when there is no power available to . . . get them operating."[75]

Another form of public display that had its adherents in the commercial community was the staging of dramatic stunts and unusual events of various kinds. A typical example was the launch of Lever's Omo detergent in 1969, in which marketers used a helicopter to swoop over townships, dropping leaflets. The helicopter would then land and "Mr. Power-Foam," a weightlifter, would disembark and distribute samples of the detergent.[76] Parachutists were used in other campaigns. Another similar nationwide promotional campaign was Colgate-Palmolive's "Operation Destiny" in 1972. Besides these kinds of campaigns aimed at attracting audiences, manufacturers also wooed rural and urban store owners with promotions and special deals, sometimes providing the retailers with marketing experts to advise them on the look and feel of product promotions.[77]

"We Could Be Completely Ignorant": Ambiguities in the New Mission

Advertising to Africans was seen from the beginning as a project involving first the acquisition and deployment of knowledge and second the construction of rhetorical and cultural resources for the manipulation and creation of "needs." From the beginning this implicitly left marketers vulnerable to the reality and diversity of African life. Once the machinery of research was turned on, it sucked in data that necessarily questioned advertising's understanding of what it was learning—a disturbing phenomenon for an industry whose "capital" was and is rooted in its control of information. Advertising's hegemony thus was always partial and interpenetrated. By the 1970s, for a variety of reasons, many advertisers were discovering that all was not as it seemed. The optimism and self-confidence that had characterized their craft in the first flush of the postwar boom faded somewhat at the discovery of the limits of cultural power.

The framing of market research problems and goals had always had a touch of edginess to it because of the assumption that the "African market" was mysterious, opaque, and confusing. The advertisers themselves stood "outside" the lives of their target audience. Africanist academics, missionaries, development experts, and advertisers have all discovered that the quest for intimate social knowledge is in some ways incommensurate with being a stranger. The evidence of this dilemma was already creeping in at the periphery of advertising discourse in the 1950s and 1960s. Maroun's speech acknowledged the issue by alluding to "lucky" manufacturers who passively watched their laxatives transform into "blood purifiers" and sell like hotcakes. As he noted, "There are many instances today which illustrate that some manufacturers are unaware that their products have an African market. . . . "[78] At the same conference, another executive nervously reflected, "We could be completely ignorant about the African market. Post-war I was told this story about Ghana—manufacturers in Britain were amazed at the sales of their carbolic tooth powder in a primitive society where toothbrushes were unknown. They discovered that this most potent-tasting tooth powder was consumed internally to drive out bad spirits.

Likewise the sale of Brylcreem Hair Cream was fantastic. A personal survey revealed that Brylcreem was thought of as a great delicacy when eaten on bread."[79] The cultural meanings of a commodity and the practices connected to it were held by advertisers to be the key to generating new needs among Africans. As Maroun noted: "The African idea of shaving is something completely different to ours. He uses a razor blade in his own way, he does not use shaving cream, etc. . . . the task therefore is to compose and hammer home an image to that African shaver on the correct way of shaving; to illustrate his face covered with lather."[80] The disquieting fears lurking around the edges of advertising in the 1950s were sometimes posed as disturbing questions: Do we know what things mean to Africans? How can we find out what they mean? What if we lose control of the meaning of commodities (and did we ever control them to begin with)?

It is fair to say that a great many southern African advertisers and manufacturers working with the "African market" were relatively untroubled by such questions in the 1950s and 1960s. Some remain untroubled, especially in South Africa. For others, increases in the amount and type of data available have satisfactorily resolved any uncertainty they might have had. One contemporary Zimbabwean marketer reflected, "With further research—with *correct* research—I don't see anything you couldn't sell."[81]

Many market researchers in the region today follow a steady path of subdividing and segmenting African consumers into smaller and more specific groups, while continuing to firmly believe that their inquiries represent an objective and scientific form of knowledge similar to market research anywhere else in the world. For example, a number of contemporary South African firms use detailed "sociometers" to assess African consumption, which include "value groups" such as "traditionals," "resigneds," "modernists," and "I-Am-Me's"; cultural "trends" such as "adding beauty," "aimlessness," "belonging and approval," "black consciousness," "cultural customs," "economy-mindedness," "familism," "national identity," "novelty and change," "rigidity," "secure future," "solidarity," and "violence and aggression"; or distinctive subgroups like "Inkatha," "Pantsulus," "Ivys," "Rastas," "Punks," "Cats," "Hippies," and "Comrades."[82] Another recent South African marketing textbook counseling executives who want to reach African audiences strenuously tries to be politically correct or at least inoffensive in its reinvention of older and cruder doctrines from the 1950s and 1960s. Earlier apprehensions about the mysterious alterity of African culture become in this recent book concerns about the important but mysterious functionings of rumor on the "TNT—Township Network Telegraph." Demonstration roadshows are urged by the authors to set songs about products to popular folk music, referred to as "zululation." The authors inform South African businesspeople that "music is the currency of Africa," that they should "use relevant celebrities," and that they must take pains to "interact" with customers.[83]

These current South African gestures toward the African market are also being paced by another turn of the wheel in the cyclical "discovery" of black consumers in metropolitan markets. In the United States recently, manufacturers have once again emphasized the need to differentially interpret and satisfy "black tastes," though perhaps with a new awareness that "companies that actively pursue the black market run the risk of being criticized for stereotyping black consumers or exploiting them."[84] These recent South African approaches recall Judith Williamson's observation: "Advertisements (ideologies) can incorporate anything, even re-absorb criticism of themselves, because they refer to it, devoid of content. The whole system of advertising is a great recuperator: it will work on any material at all, it will bounce back uninjured from both advertising restriction laws and criticisms of its basic function (such as this one) precisely because of the way it works in hollowing historical meaning from structures."[85]

However, these creeping subdivisions and gestures toward political correctness in current marketing texts from southern Africa are also testimony to a partial breakdown of the easy mastery once rhetorically exercised over a homogeneous "African market." One of the main factors in this development was the growing influence of African corporate professionals on advertising agencies and sales policies. One of the seemingly obvious solutions to the dilemma of missing knowledge during the 1960s was to hire Africans to sell to Africans.[86] For example, the present chairman of Lever Brothers (Zimbabwe), Charles Nyereyegona, came up through the ranks of African sales and advertising and played a prominent role in discursively reframing the "African market" in Zimbabwe during the 1970s. These new African employees were hardly revolutionaries. They collaborated actively in the making and reinforcement of many of the doctrines discussed above. However, they injected a note of powerful skepticism about some of the other usual formulations common in the industry. Marketing failures, for many of these professionals, were the consequence of the cultural isolation of white executives. Reflecting on the practices of the 1960s, one contemporary African executive mused, "There were a significant number of white Zimbabweans who just didn't understand the cultural norms here. . . . They were totally ignorant of what the people were."[87] As another put it, "the aspirations of African consumers and traders are generally similar to those of their White counterparts. The so-called African market is, therefore, simply an extension of the general market."[88] A leader of the African business community also shrugged off the concept of peculiarly African preferences during the 1970s: "I don't think there is any difference between the consumer habits of Africans and Europeans."[89] Nyereyegona was equally dismissive: "The term 'African Market' has its origins in our divided society. . . . In practice this market differentiation becomes a nonsense really!" He argued that "African customers react as any other customers would."[90] Of course, such arguments were to some extent self-interested in the manner they de-emphasized the barriers maintained between black and white elites.

These kinds of comments engendered and accompanied other types of skepticism. In the late 1970s, a Rhodesian businessman named E. G. Tabor characterized thinking about African consumers as the search for an "Excalibur," a magic sword that would cut through the cultural mysteries of everyday African life. Rather than demystifying Africans, as various African elites involved in industry sought to do, Tabor went in the opposite direction to assert that communication with African consumers would always be marked by "a system and tradition of thinking different from that of the European manufacturer." This system, he argued, contrasted with the logical European mind derived from Aristotle and Plato and consisted of "repetitive culture practices in which curiosity was frequently regarded as threatening group solidarity."[91] This kind of surrender to the alleged "irrationality" and "incomprehensibility" of African life was mimicked by other white advertisers, especially as independence became imminent.

However, not all white executives characterized this surrender with Tabor's implicit racism. For some, this was simply an admission of their own limitations and an acceptance of the independent powers of Africans over the production of meaning. As Raymond Williams points out: "It must not be assumed that magicians—in this case, advertising agents—disbelieve their own magic. They may have a limited professional cynicism about it, from knowing how some of the tricks are done. But fundamentally they are involved, with the rest of the society, in the confusion to which the magical gestures are a response. Magic is always an unsuccessful attempt to provide meanings and values, but it is often very difficult to distinguish magic from genuine knowledge and from art."[92] One semiapocryphal example of the inability of advertisers to fully control African interpretations that was cited by many marketers involved a Raleigh bicycle advertisement that ran in newspapers in the 1940s and again in the early 1950s. I have been told by a number of different informants, advertisers and others, that these ads caused a significant drop in Raleigh sales. The advertisement shows a picture of a young boy on a bicycle pedaling frantically

Buy a Raleigh Bicycle and Be Chased by Lions: African Misreadings of Advertisements.

as a lion chases him. The intended message: A Raleigh makes you faster than a lion. The alleged interpretation by Africans: Raleigh bicycles make lions attack you. This story has a folkloric feel to it, but it accurately describes the cultural dynamic behind other known marketing failures.

For example, one executive described to me his discoveries about the word "power" in advertisements aimed at Africans. A great many advertisements from the 1950s onward promised that the product being promoted would bring "power" or "manpower" if consumed. Such a promotion was believed by advertisers to suggest to African men that the commodity would provide sexual or masculine potency if used. This executive's firm had been employed to promote milk consumption and had advertised milk as an item that conferred "power." Nervous about the success of the campaign, the firm organized focus groups and asked if milk enhanced masculinity. Respondents universally derided the idea as ridiculous. Puzzled, the firm continued its focus groups and asked respondents to name goods that actually did confer masculine "power" and sexual potency. One of the most common answers, according to my informant, was "cheese"—a commodity that had never previously had the word "power" attached to it in advertisements.[93] In the wake of this "discovery," ads for cheese suddenly appeared in the late 1970s that were overloaded with appeals to "power," virility, and masculinity.[94] Similarly, a number of businesspeople described their discoveries that their ads featured images that had long been considered irritatingly unrealistic and ignorant by African audiences: young people with too much material wealth, Africans fly-fishing at Inyanga, or local celebrities endorsing products that township residents knew the celebrities did not use.[95]

This process of self-critique was accelerated by the end of the liberation war. The transformation of political culture since independence has had a noticeable impact on the practice of advertising. No one speaks of the "African market" any longer. Instead, they talk of "high-density" and "low-density" consumers. Most white marketers now seem to regard the most exaggerated of their former doctrines about African consumers—regarding color perception, for example—with some embarrassment.[96] However, postindependence advertisers have not escaped controversy: a recent filmed advertisement for Bata Shoes created an uproar with its use of stereotypical images of "tribal" life. Partly as a consequence of consumer activism and these kinds of controversies, the Mugabe administration is continuing to consider some kind of juridical regulation of advertising.[97]

The New Handbag: Advertising and the Commodification of Cleanliness

To help flesh out some of the general narrative I have outlined in this chapter, I want to offer a number of specific examples and interpretations drawn from actual advertisements and promotional campaigns designed to sell toiletries of various kinds. Hygiene and domesticity, as developed under colonialism in Zimbabwe, were increasingly appropriated by marketers by the 1950s. The organizations, sentiments, and values that had been produced in this field and the interlocking of various powers around bodily practice, manners, and identity all became increasingly commodified in the postwar era. In Mkele's 1959 address, his most triumphalist rhetoric was reserved for "the African woman's" discovery of fashion and hygiene products. He happily described the contents of a "typical" African handbag:

> Here is a list of the cosmetics and toiletries which I found on the dressing table of a lady in one of the Johannesburg suburbs. These were Pond's Talc, Pond's Cold Cream, Pond's Dry Skin Cream, Pond's Lipstick, Max Factor Cream Puff, Butone No. 3, Vaseline White Petroleum Jelly, Glostora, Cutex Nail Polish, Go Deodorant, Ingram's Camphor Ice, Glycerine, Olive Oil, Dettol, Vinolia Bath Soap, Lux Toilet Soap, Maybelline, Goya Perfume, Wisdom and Colgate toothbrushes, Colgate toothpaste, nylon hair brushes and Gloria liquid shoe polish.[98]

The Rhodesian Ministry of Information also officially sanctified the commodification of hygiene as the highest stage of various accumulated official and civic projects designed to "civilize" African bodies and behavior:

> The men have adopted the white man's national dress—shirts, trousers, coats and shoes, while the women dress exactly like European women. Cosmetics have caught on very well with almost all African women. Much money is spent on a wide range of beauty creams and lipstick. Polishing of

fingernails is given much attention. Many modern young African women now have their hair stretched so that they look very much like those of European women. . . . It is evident that whatever the white man brought into this country, the African has the highest appreciation for it. What a European is, an African also wants to be.[99]

Thus, by the 1960s, the hegemonic promotion of manners, hygiene, and appearance was increasingly expressed in terms of products and ad slogans.

Demonstration vans, stunt promotions, fashion shows, and other promotions capitalized on forms of mass communication pioneered by the state, but they also grew out of the entire apparatus assembled around domesticity and cleanliness. The idea of "demonstration" had deep roots in domestic education, while the women's club movement had given domesticity and hygiene a mass character before most manufacturers targeted the "African market" with their full energies. Marketing executives rapidly grasped the usefulness of the FAWC and its attendant institutions for reaching African consumers. In short order during the 1960s, the club movement, the content of domestic teaching, and product demonstration became inextricably intermingled. Where official advice had once merely counseled frequent washing, it now often specified washing with specific brands of soap; proper appearance as defined by hegemonizing powers now included the use of particular beautifying toiletries. While the homecraft movement still emphasized women's craftwork, FAWC members also found themselves watching salespeople demonstrate products, as at one 1970 meeting: "Mrs. Rusere of Colgate-Palmolive has given a number of most enjoyable demonstrations of the use of coldwater washing powder, showing how it can even be used for the washing of woolies. . . . Each member then took it in turns to wash a cup and saucer under the eagle eyes of the watching club members who had learned from Mrs. Rusere's demonstration. . . . Each member took home with her a plastic bag containing toothbrush and Colgate toothpaste, a packet of Cold Power washing powder and toilet soap."[100] Textile manufacturers planning a campaign in the countryside similarly recounted, "Wherever we discussed our plans we discovered countless friends—women's organizations of all kinds all over the country came to light with help and advice . . . it soon became apparent that there were many ways and means of finding out what we wanted to know."[101]

"Sunlight Soap Has Changed My Life": Three Ad Campaigns for Soap

These various transformations are best illustrated by examples. I begin with a close reading of print advertisements for three Lever Brothers soaps. The first of these is Lifebuoy soap, which was presented by the company in southern African markets primarily as a "strong" soap suited for washing particularly dirty bodies. As a consequence, the advertised image of Lifebuoy ultimately drifted inexorably toward both masculinity and blackness. Many lasting, subterranean associations were created by these promotions. Lifebuoy has become in southern Africa a liminal substance marking the boundaries of race and gender. Today, southern African whites still speak of powerful associations between Lifebuoy and Africans as well as Lifebuoy and boys' schools. Journalist Rian Malan's recent autobiography, *My Traitor's Heart,* makes the first of these connections a number of times: "The room smelled of all the things I associated with servants—red floor polish, *putu*, and Lifebuoy soap."[102]

Lifebuoy is a brand-name version of a type of cheap soap often sold to African consumers all over the continent from the early part of this century: carbolic or "red" soaps. With the disinfectant carbolic added to them, these soaps all have a distinctive, unforgettable odor. Given the established weight of the settler vision of "dirty" Africans, it is unsurprising that soaps with extra disinfectant, soaps that give users a distinctive odor connoting cleanliness, were thought by white manufacturers to be particularly appropriate for use by Africans. However, Lifebuoy, as a well-known brand name backed by a company with an active advertising division, was actually initially targeted at white audiences in the region. Advertisements from the 1930s in the *Rhodesia Herald* showed whites of both genders using Lifebuoy, usually after engaging in athletics. However, the speed with which a soap with properties of "strength" was firmly attached to African men suggests how much power the white figuration of Africans as "dirty" continued to have throughout the colonial period.

The first soap brands advertised in African media, in the mid-1940s, were actually RMM's Atlas Soaps (which included a carbolic variety). RMM used rhetoric drawn straight from colonial doctrines of cleanliness, telling readers: "Nature's greatest gift is perhaps water—and one of civilization's greatest gifts— SOAP. The day of the witch-doctor's craft and all its evils are over. Today educated Africans know that disease is spread by germs—and that germs live in dirt. Be educated—be healthy—keep your body, clothes and house clean by regular washing with soap and water. If you are wise you will make sure it is ATLAS Soap—the Best. There is an ATLAS Soap for each and every use. Ask for it by name—ATLAS SOAPS."[103] "Each and every use" was the space that Lever Brothers smoothly stepped into with its various well-defined brand images after its acquisition of RMM's facilities in 1947. Men were Lifebuoy's appointed target among urban African consumers, in advertisements running from about 1950 onward. One of the two predominant campaign slogans for the next two decades was "Successful Men Use Lifebuoy," which went through a number of

Atlas Soap Advertisement,
circa 1940.

subtle permutations as time went on; the other major campaign played variously on Lifebuoy's ability to "give health."

As noted earlier, marketers actively sought to give public expression to the anxieties and complaints that characterized African daily life in the cities. Obtaining work, staying employed, fending off crippling sickness, and achieving upward mobility were all major portions of these anxieties, especially during the years when strongly segregationist municipal administrators in Salisbury and Bulawayo took a hard line against stabilizing labor. At the same time, these advertisements also played very strongly, especially in the 1940s and 1950s, to white preconceptions about lazy and indolent African laborers. For example, a mid-1940s' ad for Parton's Purifying Pills features a comic strip in which a worker is castigated by his white boss for being too slow. His friend tells him, "The Boss never grumbles at *me*. If you're feeling lazy, you're probably constipated. Take Parton's Purifying Pills. I did." The worker later tells his wife, having followed this advice, "I'm glad I'm looking better . . . I'm certainly feeling better. And the boss is pleased, too."[104] A similar ad for B. B. Tablets featured an African man sighing, "I feel too weak to work. . . . I wonder if B. B. Tablets will help me?" A doctor fortunately steps in to give the protagonist the medicine, and the worker triumphantly declares, "It's good to feel well and strong again!"[105] These messages were connected with other products as well. The Tea

Marketing Board's advertisements in the 1940s frequently harped on the same themes. In one ad, a son tells his father that he is always exhausted and will lose his job soon. His father wisely suggests, "Do as I do, my son. Drink a cup of refreshing tea when you feel tired." The son's white employer notes the subsequent improvement and tells him, "I am going to give you a better job."[106]

The "Successful Man" campaign for Lifebuoy fit neatly into this strain of promotions. Its first incarnation, in the 1950s, featured the aforementioned slogan accompanied by drawings of an African man dressed smartly in a suit and tie, sometimes holding a pen, beaming from ear to ear. At the end of the decade, these pictures were replaced by an alternating series of photographs featuring African men in various "typical" work clothes. One of the photographs still featured a well-dressed clerk, but others showed a miner, a bricklayer, and a teacher. The slogan remained the same, simultaneously recalling established associations between African masculinity and dirt, between labor and pollution, between professional success within the colonial system and rigorous hygienic purification. This later version of the campaign also featured one image of the "successful man *and* his family" using Lifebuoy, broadening the stakes involved. The other Lifebuoy campaign used in the 1960s and 1970s in African media played in a different way on the same ideas, with the message "Keep Healthy . . . Keep Clean . . . Use Lifebuoy, the *Health* Soap." Again, the ads almost always featured men, this time often already lathered up and in the shower, clutching a bar of Lifebuoy. In the first campaign, the soap was offered as the guarantor of the potential for success under the social rules of the white-controlled workplace; this campaign stressed Lifebuoy's power to guarantee the ability of men to work by securing continuous good health.

Lifebuoy was one of the mainstays of the Lever line, but Sunlight, as noted in chapter 4, was Lever Brothers' oldest and best known soap product, a higher-quality brand-name version of cheap "bar soaps." Sunlight was the product foremost in William Lever's mind when he chastised his South African managers in 1921 for lowering prices. Sunlight, he argued, had "goodwill" attached to it. It had a value located in the meanings consumers associated with the soap, a value that would drive them to pay more to acquire Sunlight. (Baudrillard's idea of "sign value" is important in this context.)[107] In Edwardian England, Lever was able to forge a new position of dominance for his company through aggressive and innovative marketing and classically monopolistic strategies. Sunlight Soap, suggested numerous Lever promotions published at the time, would relieve the burden of domestic work, improve the durability of clothing, make the lives of housewives healthier and easier, and make their relationships with their husbands better.

As I noted in chapter 4, initiatives of the Rhodesian and South African states on behalf of soap producers created a climate in which the makers of soaps like Sunlight were able to prevent low-end competitors from entering the market. However, Sunlight was also one of the most ubiquitously advertised commodities in the region well before the postwar boom, and its familiarity and seeming prestige also had clearly laid the groundwork for "goodwill," even among Africans. Sunlight, like other bar soaps, was used by many Africans as an all-purpose cleanser for the body, clothes, and the home. Unlike poorly made, cheap bar soaps, which were known to shrivel as they dried, losing most of their substance, Sunlight was reliable. Its use could be paced by chipping small pieces off the bar. Early ad campaigns in African media took advantage of Sunlight as a known quantity and showed scenes of young children and adults washing in tubs with Sunlight, accompanied by the slogan "Use it for Face, Hands, Bathing and All Home Cleaning . . . Best for All Washing."[108]

However, by the late 1950s, the company was hoping to teach African consumers to use Sunlight strictly for laundry purposes, and to purchase other tablet soaps for toiletries. Older African consumers were said to resist by continuing to use Sunlight as an all-purpose soap. "The older people reject many products," noted Mkele while discussing Sunlight sales, "which they dismiss as new-fangled notions of the young."[109] The new ad campaigns for Sunlight seized hold of domesticity and the concerns of urban African women in a series of narrative comic strips. These ads aimed in particular at the image of the "good, Christian, club-going wife."

Just as many advertisements for products played on men's concerns about remaining healthy and energetic enough to work, the capacity of women to satisfy male demands and organize their households was represented by advertisements as being under seige in a hostile world. Advertisements for Feluna Pills (iron supplements) ran regularly in African media from the early 1940s to the 1970s and were typical examples of these themes. In one Feluna ad, a mother wishes her daughter good-bye when she goes to her new husband, telling her to "be a good wife and bear strong healthy children." The young woman replies, "I love babies and hope to have some." Her mother then tells her all her own children have been strong because Feluna Pills have "kept my blood strong and pure."[110] Another Feluna ad asked, "Does a woman have to suffer? . . . Because she was a woman, Maureen Mamba thought she *had* to suffer. Her headaches, stomach pains, tiredness and sadness—these she thought every woman had to bear." Feluna Pills give her "new blood, new health, new happiness."[111] Feluna advertisements, and a number of other similar product promotions, usually boiled down to the fundamental message: "A Woman Can't Work If She Is

Don't accept anything else in place of the genuine

Dr. Williams' Pink Pills

Female Health and Beauty:
An Early Patent Medicine
Advertisement.

Sick." Work in this instance was often portrayed in terms of reproduction—
maintaining households, satisfying husbands, and raising children.[112]

The Sunlight ads of the 1960s played on related themes. In one, a wife is
excoriated by a husband who cannot get a good job because his shirts look "so
old and dirty." His wife admits to a friend, "Ben is right . . . I am a bad wife."
Her friend tells her, "You must change to Sunlight Soap." The change is made,
and the wife exults, "Now everything is going right . . . because I wash my
husband's shirts in Sunlight!"[113] Not only is the woman's success as a wife
threatened, but her failure is also represented as threatening her husband's
success. In another storyline, a woman worries, "I'll never find a husband. My
clothes look so old and dirty." Again, a good Samaritan intervenes, sharing the
secret that Sunlight makes clothes look like new again. Redeemed, the young
woman confides, "Sunlight Soap has changed my life . . . and soon I'll be a
happy wife!"[114]

If Sunlight advertisements were striving to make connections with African
women trying to live up to the ideals of club-going domesticity, then Lux was
aimed at women (and men) who were imagined as living with another type of
urban ideals. As with Sunlight and Lifebuoy, Lux ads were part of a larger strain
within advertising aimed at Africans, including products as diverse as pens,
clothing, beer, education, automobile supplies, and especially toiletries of all
kinds. Lux and these other products were from the early 1940s pitched to
African consumers as the definitional essence of glamor and "smart" living—

Sunlight Soap and Domesticity, 1960.

commercial codewords for the African elite. In particular, the word "smart" (or "clever") was believed by advertisers to be as evocative as "power," except in this case it was held to appeal to those seeking economic advancement, "Westernized" lifestyles, and youthful sex appeal. Lifebuoy and Sunlight worked off hygienic imagery linked to Christian morality, labor, and domesticity, while Lux mined another vein of ideology about manners and bodies, the thread that dealt with "modern living" and "civilized fashion."

Lux advertisements in the African media went through a number of changes over time. The earliest of these Lux ads actually appeared in the 1940s. These promotions were reprints of the Lux ads that had been featured in "white" newspapers like the *Rhodesia Herald* and used white models or international female celebrities like Claudette Colbert. The advertisement set forth to explain "why beautiful women use Lux Toilet Soap": "Lux Toilet Soap has been made in all the big countries of the world for many years. This is because thousands of beautiful women have won lovely complexions from using Lux Toilet Soap. . . . Lux Toilet Soap is pure, you can see that because it is white. Lux Toilet Soap has a rich creamy lather that makes your skin soft and smooth, beautiful to look at. Lux Toilet Soap is the simple secret of beauty. Use it every day to keep your skin clear and fresh. And everyone likes its scent because it is like pretty flowers. Remember it is in a pink wrapper."[115] There were many other ads in the 1940s and early 1950s in which white models were used to pitch goods to Africans. While this changed rapidly during the 1950s, such promotions left their mark on the image of beauty—especially since many elite Africans also regularly read the *Herald* and other "white" publications.

Lux promotions later in the decade stressed "smartness" and "color." One advertisement featured the heads of a young couple above the slogan: "Be like smart people, insist on . . . LUX Toilet Soap." The ad went on: "Today, in the homes of the *smartest people,* you'll find Lux Toilet Soap in colour as well as white. You'll see pink, green, yellow and blue tablets . . . so pretty to look at . . . so pleasant to use. And the people who use Lux Toilet Soap *are* smart. Their skin is clean and smooth and sweet smelling."[116] Most other ads in the same campaign featured a "smart" woman rather than a couple, reinforcing the idea that Lux's beautifying effects were more appropriate for females. This campaign was clearly an attempt to appeal to middle-class consumers, or to those who wanted to appear middle class—"smart"—though it was also clearly a manifestation of tenets about the "innate" African sense of color. To be "smart" was to be modern, sophisticated, a member of the social world of comfort, security, and power imagined to some degree in most advertisements. Alec J. Pongweni has noted in his analysis of Zimbabwean advertisements taken from the papers of the late 1970s the continuing prevalence and range of such appeals.[117]

By the 1960s and 1970s, Lux ads in African media returned to stressing the feminine glamor and beauty of the product, even to the point of renaming the product "Lux Beauty Soap." In these ads, an African model held the soap next to her cheek, accompanied by the slogan "a rich, creamy lather to keep your skin soft and smooth . . . expensive beauty oils . . . a lovely new perfume. Three good reasons why Africa's loveliest women use Lux beauty soap."[118] During this time, Lux was also advertised once again as "The Beauty Soap of the Stars," this time using local black South African and Zimbabwean models and singers. For some African professionals like Mkele, this was a welcome development, as it removed alienating and unfamiliar white female stars from African advertisements and disrupted what Mkele called the "cult of the American Negro":

although the Africans do not have all these things they have found the symbols of their aspirations among themselves: they have their own politi-cal leaders, educated and wealthy men, their Miriam Makebas and fash-ionable women. They have become emotionally self-sufficient. . . . Adver-tisers have discovered that African models make as good symbols as the [American] Negro and perhaps even better since the local model repre-sents an attainable ideal. The proliferation of "Misses"—Miss Africa, Miss South Africa, Miss Butone, Miss Palmolive and I see that a Miss Nivea is also on the way—shows that the glamour of the American Negro has been superceded by the glamour of the African beauty queen.[119]

Mkele, as a member of the black urban elite in South Africa, told white advertisers from all over the region that this promotion of local beauty queens was a key to unlocking the aspirations and good favor of his compatriots in South Africa and Zimbabwe. The Lux campaigns took this advice to heart.

Lux advertisements also implied that urban living was perpetually strained by sexual anxiety. The "glamour" of women promoted by Lux ads almost recalled an older English use of the word "glamour," that which referred to the ability of supernatural beings like faeries to dazzle and enthrall human beings. The fate of women in urban areas in Zimbabwe during the colonial era was perpetually tied to their ability to maintain and institutionalize relationships with men. Whether Zimbabwean women controlled these relationships or vice versa has been a subject of hot dispute within African culture since the 1920s. In masculine culture, women were often depicted as sexually powerful, able to marshal commodities like clothing and toiletries to control men through desire. Advertisements for Lux and other toiletries played strongly on these themes. Women were advised that they could use a given product to enhance their sexual control over men, while appeals to male audiences often depicted a forlorn man who has been persistently rejected by women because he is not yet hygienic or "smart" in his use of commodities.

In all of these cases, there were other marketing strategies appended at times to each of these products.[120] Each of these soaps was also at various times distributed in townships, given away at demonstrations, or otherwise promoted outside of the standard advertising media.

"Lighter and Brighter": The Marketing of Whiteness

Just as the rhetoric of various Lever soap promotions was nestled among larger patterns in advertising directed at Africans in Zimbabwe, the promotional language surrounding skin lighteners tapped into a number of established themes. Bu-Tone products of various kinds—talcum powder, creams, lotions—had been around since the 1940s, but Bu-Tone skin lighteners, along with other brands like Ambi and Hi-Lite, only took off dramatically in Zimbabwe in the 1960s. Many of these products contained hydroquinone, a chemical that can cause potentially serious damage to the skin. As a consequence, they were banned in Zimbabwe after independence and have just recently been banned in South Africa as well. The advertising of skin lighteners rapidly became a major feature of popular publications like Parade, and tapped into established marketing themes involving class aspiration, sexual attractiveness, and proper appearance. The most important appeals, however, involved an equally established theme in advertising: the conflation of whiteness, purity, and social power.

Just as Lux advertisements once used white models to define images of beauty for black women to follow, other promotions, especially before the 1950s, summoned up the tastes of whites as the definitional essence of wise consumption. One 1937 ad for Beecham's Pills contained the slogan, "Progressive Africans Relieve Sick Stomachs With Beecham's Pills As White People Do."[121] More generally, advertisements for cleansers and detergents often played on the totemic equivalence of whiteness with total cleanliness. Advertisements for the detergent Persil featured the slogan "Whites become whiter . . . colours come up brighter," a slogan that could easily be mistaken for a metaphoric description of the entire apparatus of segregated marketing in Zimbabwe.[122] Reckitt's Blue was similarly advertised with the slogan "Out of the blue comes the whitest wash." By themselves, these phrases were not exceptional and were found in similar forms all over the globe. But in a climate where the terms *white* and *black* constitute huge sociopolitical realms with power over the most intimate and local details of daily life, such appeals necessarily were supercharged with extra meaning. The images used in these promotions simply added to this effect: Persil ads offered carefully constructed visual contrasts between brilliantly white garments and the black skin of women busy with their laundry. Moreover, European advertisements for cleansing, purifying, and beautifying products had long played on the common symbolic Western conflation of blackness and dirtiness. As Jan Pieterse has noted, the image of "washing the Moor white" was deeply ingrained in the Western visual imagination and was a common theme in nineteenth- and early twentieth-century advertising for soaps, detergents, and similar goods.[123]

Skin lighteners took these accumulated precedents and infused them into a mix of promotional messages about beauty, attraction, and socioeconomic status. One of a number of products that led advertisers to necessarily violate their dictum that goods should not appear to have been deliberately made for Africans alone, lightening creams were by their very nature about the transformation of the "traditional" African self into something the advertisers argued would be more commensurate with "modern" society. Women were disproportionately the targets of these campaigns, though as the products grew more popular, advertisers shifted somewhat toward appeals directed at men, who were seen as a potential growth market. The earliest ads for these products, mostly various Bu-Tone brands, tended not to stress the "lightening" aspects of the creams too much, instead offering a longer list of virtues promised by the

Mr. and Mrs. Ambi at Victoria Falls, Advertisement from the Early 1970s.

commodity. Most of these early ads also focused exclusively on the product's ability to enhance one's appeal to the opposite sex.

By the late 1960s, however, competition among various producers had grown increasingly fierce and lightening was the name of the game. The ads told readers that the closer to whiteness they got, the better they were, in every way imaginable. The texture and feel of lightened skin was praised: "Hi-Lite . . . Chosen by Men and Women Who Want a Lighter, Smoother Skin."[124] More important by far, users were portrayed as having a lock on professional success. The quintessential skin lightening ad was a recurring Ambi promotion in which a handsome and very light-skinned black doctor is reading charts, with a light-skinned black nurse at his side. Off to the corner of the ad is an extremely dark-skinned janitor, with a hunched posture and nervous expression. The accompanying slogan, "Successful People Use Ambi," left little room for interpretation. The ad goes on, "Ambi can give you the smooth, clear, lighter skin you've noticed successful people usually have. . . . Ambi keeps your skin lovelier—gives it an attractive, non-shiny softness."[125] Another Ambi ad from the same period, with the slogan "Wherever they go, Ambi people look great," featured a prosperous light-skinned African couple dressed in brilliant white suits sitting at the luxurious Victoria Falls Hotel, where "Mr. Ambi and Mrs.

Ambi join the big money gambling set for tea. And fellow guests agree: They look great."[126]

The makers of these products, along with some other cosmetics manufacturers, also tried a new tack for promoting their items in the African media. Using the concept of "product placement" now so favored in the U.S. cinema, advertisers managed to insert their products into the narratives of regular installments of "foto-plays" in *Parade*. These were usually generic action-adventures featuring a hero, his girlfriend, miscellaneous *tsotsi* types who eventually came to bad ends. The foto-plays were sponsored by advertisers keen to promote cosmetics and lightening creams. Therefore, invariably, the various girlfriends and femmes fatales who populated the pages of these stories would stop for a panel or two to apply prominently displayed toiletries, thinking, "Thank God for Lip Ice" or with captions like "Josephina had just started applying a cream (to make her skin lighter of course) when she pauses."[127] Marketers for the creams explored other options as well. While, to my knowledge, none of the firms producing these products ever operated demonstration vans, they were very active in new urban supermarkets, setting up "Ambi Bars" and other outlets for demonstrating the products and catching the attention of shoppers. Furthermore, the producers of various lightening creams also took a very active role in promoting beauty shows and hiring teams of models and female singers to represent their products in public arenas. During the 1970s, a hugely disproportionate number of the competitions and give-aways that attracted so much interest from urban consumers were also offered by the makers of lightening lotions.

Bread Spread or Baby Fat? The Case of Stork Margarine

Demonstration vans did play an interesting role in the formal introduction of Lever Brothers' Stork margarine onto the Zimbabwean market not long after the establishment of the company's "outdoor" advertising schemes. At first, it would seem that this product hardly belongs with a discussion of the marketing of toiletries, but in this case, appearances deceive.

Stork's first misadventure belongs to an entire genre of incidents from the 1940s and 1950s common to the history of commodification in southern Africa—and other parts of Africa as well. These incidents always involved the introduction of goods not previously seen within the local marketplace and a subsequent swirl of rumors about and interpretations of the products. One such incident that recurred all over central and southern Africa involved the introduction of tinned food. This event often set off speculation on "pavement radio" that the cans actually contained the meat of human beings, perhaps those taken away to work in mines, factories, or farms, or those who died while under colonial medical care.[128] Similar rumors about poisoned sugar swept through colonial Malawi and the Copperbelt in the 1950s. Even soap was the subject of such rumors: "On the Copperbelt a soap-company gave away free samples to Africans. This had never happened before and Africans threw away the samples, believing the soap would make them lose their will-power."[129]

Each of these episodes was shaped by an immediate political crisis—many of them were clustered around the antifederation campaign in colonial Zambia and Malawi—but all of them were also reflective of deeper misgivings about the form of manufactured commodities. Experience with wage labor, combined with a knowledge of colonial institutions of compulsion and extraction, gave many Africans in the region a reason to regard capitalist production as potentially horrifying and deadly. This knowledge, combined with the appearance of new goods made in unknown places by unknown people, made rumors a potent form of criticizing the concealing powers of capitalism. Other recurring incidents in the history of commodification in the region bear this out.[130] For example, Harry Wolcott's anthropological study of beer consumption in Bulawayo notes that historical cycles of complaints about the "quality" of beer may have roughly coincided with small-scale urban disturbances near beerhalls. These seem in turn to have some correspondence with the introduction of successive changes to the process of beer production, originally done on the premises of the earliest beerhalls by women, then by workers nearby, and finally in modern factories completely removed from the site of consumption. Wolcott also recounts other similar rumors about beer, including one allegedly started by Bulawayo women that a new "European"-style beer was strong enough to cause impotence in men.[131]

Stork margarine was the focus of similar rumors when it was first introduced. According to one of my informants, Stork was first marketed in the same wrapper used in England at that time, a wrapper with a picture of a baby on it, and the margarine itself was uncolored. It is not hard to guess what the consequences were, given that the purpose of this new substance was basically unknown to most Africans, especially in the rural areas. The rumor quickly spread that the margarine was rendered baby fat, proof of the ghoulish practices of the settlers. Lever Brothers moved quickly to substitute a picture of a stork, which helped to quash the rumors to some extent, though some argued that the product had simply been changed to stork fat.[132]

Some time later, another problem materialized with the product. "Smearing," discussed in chapter 2, recreated itself in the 1940s and 1950s through manufactured goods like Vaseline and Pond's Lotion. Margarine has also been among the goods used for this new kind of body smearing,[133] and is still used for this purpose. Reserved animal fat in the past had been used frequently for hygienic purposes, and, especially in the rural areas, Africans examining Stork in the aftermath of the damaging rumors thought it well suited for the same purpose. When word of this trickled back to Lever Brothers, the company was deeply concerned. In short order, cinema vans and township theaters around the country were showing a rapidly produced film that instructed consumers on the "proper way" to consume Stork.[134]

There have been many other instances of Africans putting toiletries and other products to unexpected or novel uses in Zimbabwe. In most of these cases, the company making the item either has remained unaware of any such reconstruction or is aware but unconcerned. Why did Lever Brothers worry about Stork being used as a toiletry? I would argue that this case highlights the importance of meaning in modern capitalism, that Baudrillard's "sign value" is at stake here. It is one thing if consumers add a new value to a commodity without interfering with its utility- and exchange-value as defined by its producers. In this instance, the tampering of African consumers threatened to fundamentally displace the meaning of Stork as it was understood by its manufacturers, thus in some way repossessing the commodity. The appropriation of Stork in this instance threatened to reverse the circulation of cultural power in contemporary capitalism. The flow of "taste," which normally goes from the transnational and metropolitan to the local and peripheral, was perilously poised in this case to move in the opposite direction.[135] Once the meaning of Stork-as-food was secure, the use of margarine as a toiletry no longer mattered. "We don't worry about it now," pointed out one executive. "The people who still do that know that the product is a food."[136]

"Is African Hair Really Different?": The Scientific Design of Shampoo

The extent and nature of the capitalist investment in the collection and control of knowledge, so much at the heart of worries about the use of Stork, has also been demonstrated during the last two decades by attempts by multinational firms with investments in Africa and the Americas to design shampoos especially for black consumers. One African executive who was involved with this project at Unilever[137] explained that these efforts were approached with the utmost caution for several reasons: First, businesspeople operating in late colonial Zimbabwe and other African nations were intensely aware of the negative connotations of goods made expressly for Africans. Second, he explained, "there's a lot of danger with hair"—any new product for hair has to be tested extensively for safety. The project team that the company assembled set out to acquire a collective "doctorate in hair" and to scientifically, systematically answer the question of whether there was "a structural, intrinsic difference" between the hair of Europeans and the hair of Africans. The research team eventually argued that the kinkiness of African hair was caused by breaks in the uniformity of hair layers, breaks that were not found in European hair. The project concluded that designing a special shampoo was actually necessary, that only a special formula could fully condition these breaks and make them "more pliable, less dry."

For me, this project raises a vital question. Why did the company bother? Why was scientific discourse required by companies to validate what most southern African whites already "saw" when they looked at the black body? There are some institutional factors that helped produce the need for such a project. For one, the growing influence of executives like Douglas Kadenhe and Charles Nyereyegona in the local operations of multinationals meant that no decision could be easily validated by an appeal to "commonsense" principles of racial difference accepted in settler society.

The professional style and bureaucratic structure of large multinationals like Lever Brothers or Colgate-Palmolive also generates such projects reflexively. The reason that such a Pavlovian reaction has taken hold, however, is the evolution of Western ideas about "needs" and their reproduction within colonial and postcolonial capitalism all around the world. The commitment to perpetually identify and promote new needs, and its corollary principle, to convince people that they are presently dissatisfied even if they don't yet know it, is ingrained in transnational capitalism. This is the logic of perpetual growth. Like sharks, whose gills cease functioning if they stop moving, multinational firms can never hope to achieve a comfortable stasis. At the same time, the power of particular discourses that criticize "progress" from within historically local ideals of moral economy means that companies must try to justify expanding the realm of need.

Scientific discourse is the most unassailably "objective" form of such justification in modern societies. The fact that Lever Brothers' Zimbabwean subsidiary, in concert with other African and American subsidiaries, turned to such rhetoric is indicative of a discursive shift within Zimbabwean capitalism and Zimbabwean society, a shift from the colonial to the postcolonial logic of power. This shift came late in Zimbabwe's modern history, but its consequence was that racial difference no longer could be easily naturalized by the simple gaze of whites. The outcome of the study of African hair may have been pre-determined, but its discursive logic was not. The relentless march of market segmentation might have produced hair that was "different" in any event. However, the need to generate a scientific explanation of such difference suggests once again that marketers in Zimbabwe have only exercised limited power over meaning within a system of predetermined rhetorical and institutional commitments.

Such constraints are primarily the constraints of history. Here, we must recall Judith Williamson's dictum that it is "prior meanings" in a culture that allow any given advertisement to itself manufacture or recast meaning. Zimbabwean executive Douglas Kadenhe demonstrated his subtle grasp of these dynamics when I spoke with him. He pointed out, with careful emphasis, "There is *now* enough power in advertising [in Zimbabwe] to make somebody want a product which he does not need."[138] Whatever power the rules and texts of advertising have had to affect the interpretation of commodities, after cultural practices, and infiltrate the subjectivity of Africans comes from the weight of Zimbabwean history. This power came from the white observation of African bodies, from the building of colonial hygiene and domesticity, from patterns of colonial exchange, from struggles over the boundaries of consumption, from the particular architecture of secondary industry. Zimbabwe's manufacturers and advertisers became a power to be reckoned with, but they never achieved the influence they hoped for or anticipated. Africans received the gospel of commodification into their homes and their selves, but the converts have to date never behaved as their putative masters and teachers expected. They preached their own gospel about goods and they interpreted the holy writ of advertising in their own multiple and unpredictable ways.

6 BODIES AND THINGS:
TOILETRIES AND COMMODITY CULTURE
IN POSTWAR ZIMBABWE

In Zimbabwean writer Shimmer Chinodya's novel *Harvest of Thorns* there is a lengthy evocation of the 1960s as "a special period . . . an end and a beginning." Chinodya gives a rich, moody description of that time and its special rhythms and preoccupations:

> Those were the days when there were loudspeakers at the centre of the township which were tuned to the radio station. . . . Those where the days of the mobile bioscope, when the nights belonged to Mataka and Zuze and the Three Stooges. . . .
>
> As we grew up, things changed, slowly . . . we began to notice clothes that stole our souls. We saw bell-bottom trousers, "Satan" denim jeans, platform shoes, bold, bright shirts with large, raised collars, checked jackets, massive belts. Dresses crept up women's thighs and women were so much taller, suddenly; we heard of gogo shoes and hot pants. A strange disease afflicted the heads of the women. It made them rip off their hair and don wigs which, when oiled, looked like the fur of drowned cats. Dislodged from the heads, the disease crept over their faces. They scrubbed it out with sponges and fought it off with creams and lotions. The creams wiped off the disease and the lotions left their faces with a bright, healthy shine. Dislodged from their faces, the indefatigable infection went down to their legs and settled there: they fought it off with pulling socks. Then, alas, the disease returned—in a vicious new form, a charry ashiness that settled permanently on their faces. . . .
>
> On the radios we heard a flood of new foreign songs. . . . Manners died with a hiccup. The young went wild. Hordes of youth left the country to seek out trick fortunes in the towns. Men shunned the countryside. . . . Certain words took on fierce new meanings. Old songs shed their innocence. . . .[1]

Chinodya here depicts the 1960s as times full of emotional and uncertain transformations of culture, meaning, and manners, changes in the relationships between people, between people and places—and between people and things. Though the subjective experience of this era by many Africans was supercharged with "newness," with disjuncture and dislocation, the preoccupations of that time were also heavy with history. I began this work with Dambudzo Marechera's mocking description of a magazine's reproduction of dominal ideology about African consumption. Chinodya here speaks to another side of this history, of the ways that Africans themselves saw and experienced commodification.

The previous five chapters of this book have detailed how complicated threads of the precolonial and colonial past wove their way toward the making of Zimbabwe's commodity culture after World War II. Commodification, a process driven by the imperatives of capital and the "civilizing" projects of state and mission, has also been powerfully determined by the African individuals and communities who were imagined as the subjects of the process. It has been determined by their differential access to power and cultural resources, determined by their diversity of interests and priorities. For Zimbabweans, the transformation of needs and changing relationships between things and people were two of the most recognizable consequences of colonial rule. In this chapter, I examine the commodification of cleanliness, domesticity, and the subsequent production of "modern" African bodies through the meanings and practices that arose around toiletries: soap, Vaseline and lotions, skin lighteners, cosmetics, perfumes, toothpaste, deodorants, and shampoos. In each case, I also consider how the specific use of toiletries has been connected to the larger play of commodification in Zimbabwe's African communities and connected to the determinative role of consumption in producing difference and subjectivity.

"Nyamarira That I Loved": Soap, Vaseline, and
the Authority of History

The hygienic needs that have been most fundamental to African communities in Zimbabwe, and yet most invisibly within the realm of "common sense," have been invested in soaps and in toiletries like Vaseline, Ponds Lotion, and Camphor Cream. The complex historical domain of cleanliness and domesticity in colonial Zimbabwe was increasingly embodied in these particular goods after 1945. It may seem peculiar at first to consider soap and Vaseline together, but, in fact, the two types of commodities have existed in close dialectical relationship with each other, constituting separate but closely linked hygienic ideals.

Table 6.1   Reported Toiletry Consumption, Urban African Women

Commodity Type	Income Groups, in Dollars (Monthly)—Percentage Reporting "Regular Use"				Age Group—Percentage Reporting "Regular Use"		
	60+	40–59	20–39	Under 20	16–24	25–34	35+
Shampoo	21.0	16.9	7.2	11.0	18.6	11.6	8.8
Hair spray	21.0	5.6	4.1	6.8	13.7	5.8	6.3
Facial cream	66.1	62.0	50.5	50.7	58.8	59.5	48.8
Eye makeup	11.3	5.6	8.2	8.2	11.8	5.0	8.8
Hair conditioner	4.8	4.2	2.1	-	4.9	2.5	-
Lipstick	12.9	14.1	10.3	9.6	14.7	11.6	7.5
Perfume	71.0	45.1	34.0	46.6	53.9	46.3	40.0
Deodorants	35.5	25.4	9.3	21.9	22.5	24.0	16.3
Face powder	66.1	35.2	28.9	39.7	52.9	39.7	26.3
Talcum powder	21.0	19.7	9.3	20.5	10.8	19.0	21.3
Hair straightener	22.6	9.9	9.3	9.6	20.6	11.6	2.5
Household soap*	67.9	76.1	69.7	54.4	63.1	66.3	69.6
Toothpaste	88.2	80.6	72.4	70.0	79.6	83.9	66.1
Toilet soap	81.8	85.6	79.9	72.1	76.4	79.8	81.4
Skin lightener	67.7	45.1	40.2	50.7	57.8	50.4	37.5
Vaseline	93.5	97.2	92.8	94.5	98.0	90.1	96.3

Source: Market Research Africa (Rhodesia), *Profile of Rhodesia: African* (Salisbury, 1972).
*The survey listed the consumption of household soap, toothpaste, and toilet soap under a separate heading which combined the responses of women and men; I have included the responses for each commodity on this chart instead.

Tables 6.1 and 6.2, along with the charts in the appendix, reveal that soap, from Sunlight to Lifebuoy to "blue mottled," was a nearly universal presence in the everyday life of Africans, rich and poor, men and women, old and young. Soap had by that time been heavily promoted by colonial institutions connected with domesticity and hygiene as the material embodiment of their campaigns to produce "modern" bodies and manners. Though soap had also long been a fundamental part of efforts by missionaries and others to promulgate their vision of hygienic and properly disciplined bodies, its promotion by these groups accelerated dramatically in the 1940s.[2]

Furthermore, local factories produced cheaper, better quality soaps in increasing amounts during and after the 1940s, and with the establishment of a

Table 6.2   Reported Toiletry Consumption, Urban African Men

Commodity Type	Income Groups, in Dollars (Monthly)—Percentage Reporting "Regular Use"				Age Group—Percentage Reporting "Regular Use"		
	60+	40–59	20–39	Under 20	16–24	25–34	35+
Hair preparations	22.4	20.5	12.1	10.7	18.9	14.2	13.5
Shampoo	1.6	.7	.5	-	.9	.4	.4
Aftershave lotion	15.2	4.6	3.4	2.3	1.9	6.2	7.7
Shaving lather	26.4	26.5	20.8	12.1	8.0	23.5	27.8
Vaseline	82.4	86.1	89.4	86.4	87.3	87.2	85.3
Skin lightener	8.0	4.0	3.4	5.1	4.2	5.8	6.6
Razor blades	70.4	74.2	83.1	67.3	38.2	85.0	93.8

Source: Market Research Africa (Rhodesia), *Profile of Rhodesia: African* (Salisbury, 1972).

Lever Brothers subsidiary, soap advertisements and promotions—in African newspapers, on the radio, in the stores, in roving demonstration vans and cinema shows, in the skies and streets of African communities—became an omnipresent feature of both urban and rural life. Soap marketing from the 1940s onward reinforced the image of soaps as the essence of hygiene and began to segment this essence into more specialized "needs"—soaps for male hygiene, hygiene for the sick, hygiene for workers, hygiene in the household, hygiene for "smart" people, hygiene for sexuality and beauty. "Modern" hygiene was further centered in soap by the co-optation by manufacturers of institutions like the FAWC. The public promotion of cleanliness, which had originally been concerned mostly with manners and practices, by the 1950s had empowered soap to wholly embody the "hygienic ideal." By the late 1960s, an advice columnist in *Parade*, trying to promote cleanliness, spoke largely in terms of commodities: "it costs little to wash . . . cleanliness costs little. Soap is cheap. . . . You say you can't afford to look smart? Rubbish. Get cracking and see to it."[3]

As a consequence of these accumulated connections, soap came to be seen as a fundamental need by Africans. In Chinodya's *Harvest of Thorns,* blue mottled soap is depicted as one of the characteristic intimate smells of the human body.[4] During the *chimurenga,* the guerrillas had local villagers procure supplies they regarded as absolute necessities. The delivery of such supplies was one of the most perilous aspects of the conduct of the war for soldiers and peasants alike. Requests had to be kept to a minimum, and it is therefore notable that after food, the item many soldiers regarded as essential was soap.[5] Former school pupils, especially those educated by missionaries, as well as women who were active in the FAWC or remain active in church congregations today, noted during interviews that they had been taught that soap was a basic part of good hygiene. Indeed, for many, soap and cleanliness are the single most powerful defining memory of early schooling. Such training now meets with few objections among those with whom we spoke, though a few complained that they remembered teachers being a bit excessive in their surveillance of fingernails, ears, and other body parts.[6] Also, one man in his late twenties now living in Mbare noted that when he was a child and going to school in a rural district, teachers were sometimes unaware that many pupils had difficulty affording the amount of soap needed to meet school requirements for hygiene.[7]

Notably, however, most of the people we formally interviewed were not part of the professional class, and their social contacts with whites before independence were limited or nonexistent. The outrage felt at the racist perception of African hygiene and bodies by several generations of elites who worked closely alongside or in contact with whites is unfamiliar or unimportant to many of the people interviewed for this study. However, many educated people familiar with white culture retain a strong memory of this particular strain of racism. The white characters who flit through the narratives of black Zimbabwean writers often have been marked by their peculiar obsessions with cleanliness. In *Harvest of Thorns*, when a white bureaucrat briefly enters the narrative, he is glimpsed yelling at a black subordinate: "Dust! Sprinkle more water on the floor, *Madala!*"[8] Peter Fraenkel recalled that his African associates at the Central African Broadcasting Service closely scrutinized the behavior of whites toward blacks and invariably noticed those lapses in interpersonal etiquette that were occasioned by hygienic fears.[9]

Many informants strongly advocated the use of specific soaps to address particular needs. Lifebuoy, for example, was strongly praised by most people, many of whom said they had used it regularly for decades. Informants were able to trace some of the attributes of Lifebuoy to definable sources. For example, most argued that the soap was generally healthy and specifically good for children, especially for "measles" or other skin ailments. This, they said, was taught by clinics, teachers, and parents alike. Former members of FAWC chapters stressed that they had learned of Lifebuoy and other soaps while attending meetings. Also, many people recalled hearing about Lifebuoy and other Lever (or Colgate-Palmolive) soaps in demonstrations. However, on a number of occasions, people of all ages also asserted that it had long been true that "Lifebuoy was good for men" or that "After a hard day's work, Lifebuoy smells nice."

These sentiments were almost direct quotes from forty years of Lever advertising campaigns, and it is notable that none of those who made these statements could identify where or how they had learned these attributes of Lifebuoy.[10] The masculinity of Lifebuoy or its ability to cleanse a laboring body had become part of "common sense," generally known but without specific provenance. Equally, the historical effectiveness of other campaigns for toilet soaps became clear in interviews. Lux and Palmolive were occasionally said to be good beauty soaps, though once again, few were certain where they had learned this. Both soaps were also said to be specifically recommended for washing infants under six months old, a "fact" that people said had been taught by clinics and women's clubs for many years. The separation between household and toilet soaps which soap manufacturers stressed increasingly from the 1950s onward has also been effective; the few who said that they have used household soaps for washing their bodies confessed that they no longer do so, or only rarely continue this practice because it is more economical. Significantly, if any soaps were felt to be "bad smelling," it was generally brown household soaps, especially those made by cooperatives.[11]

Thus soap, while having been seen as a fundamental need by most Africans since at least the 1950s, has been closely tied to the practices of cleanliness and domesticity promoted first by missions and the state and later powerfully and massively reproduced in postwar advertising. Vaseline and other face and body creams, by contrast, were purchased and used in large amounts in the postwar era without any significant initiatives by manufacturers, who realized only slowly and with surprise that they had a ready-made market requiring no commercial sermons, no consumer conversions. Rather than pursuing a project of inventing needs commensurate with supposedly normative desires found in the West, the companies manufacturing Vaseline and body lotions found themselves producing for hygienic needs conceived independently among Africans. Indeed, other companies, like Lever Brothers, even felt it necessary to take steps to protect products like margarine from being misinterpreted as hygienic substances that could meet these demands.

These products were used to reinvent the practice of "smearing" so common in nineteenth-century southern African cultures. It is difficult to say when exactly the mixing of red or yellow soil and some sort of oil as part of a daily hygienic regimen began to disappear in south-central Africa. It has never disappeared completely in the region as a whole, but *none* of the Shona individuals with whom we spoke had ever heard of the practice, including individuals in their fifties and sixties who had lived much of their lives in rural districts. The intense disapproval of missionaries and white authorities and the commodification of both rural and urban culture rapidly encouraged most individuals to purchase products to satisfy their needs rather than continue to make their own mud and fat pastes for smearing. The use of plants like *ruredzo* or *chitupatupa* persisted during the postwar era, and their use remains somewhat common today in the rural areas. These plants are no one's first choice to meet hygienic needs, but many find such plants a useful substitute when they are available. However, the basic concept behind precolonial practices—that a clean, mannerly, and aesthetically pleasing body should be coated after washing with a paste to protect it from dirt, keep it from cracking or chafing, and give it a rich, shiny appearance—has remained strongly rooted in all walks of life in Zimbabwe.

The use of margarine as a toiletry seems to have been part of a larger pattern of consumer experimentation from the 1930s to the 1960s, experimentation that searched for a commercially available commodity that could be used for the purpose of smearing. Cooking oil and the oil that had separated from peanut butter were among the substances turned to this purpose. Soap was also used: one washed with soap, rinsed, and then worked the soap into a lather; this lather was then applied to the skin and allowed to dry. In particular, soaps advertised as having "glycerine" and being good for the skin were turned to this purpose.[12] Today, the use of margarine or food oils on the body still occurs, but virtually everyone, from Mbare squatters to middle-class homeowners, regards these substances as undesirable substitutes for cream or Vaseline and a last refuge for the impoverished. In a few cases, informants also attributed such practices to youthful experimentation that was abandoned when better substitutes emerged. The use of soap lather for smearing is a bit different. Some women, especially women over fifty who had spent significant amounts of time in rural areas, said that they still regularly used soap for this purpose and preferred it to all other products. Most others interviewed for this study had either never heard of the practice or regarded it as similar to the use of cooking oil or butter for hygienic purposes: a known practice, but attributed to necessity rather than choice.

In any case, experimentation with new substances ultimately uncovered wholly satisfactory commodities for use in smearing—both petroleum jelly and body lotions like Pond's and Camphor's. My own interviews as well as surveys and various texts suggest that, by the 1960s, most, if not all, Africans in urban areas used these products as often as they could afford, and that many rural people did so as well. Virtually all our informants agreed that they simply did not feel clean without using some sort of cream or lotion after washing. Everyone had practiced or tried to practice smearing throughout their entire lives.

Many also alluded to the cracking of the skin that follows when smearing cannot be performed and the pleasing feel and smell of skin when it is covered by lotion. In the Tashinga squatter camp near Mbare Musika bus terminal, many have been unable to afford substances for regular smearing for much of their lives and displayed worn or cracked skin to demonstrate what they clearly regarded as an extremely serious form of deprivation.[13] In a recent anthology of testimonies by Zimbabwean women, one woman describes the desperate poverty that overtook her family after her husband abandoned them: "I wrote many letters to my husband but he never responded. He no longer sent money: he did not care about us any more. . . . My feet began to crack because I had no lotion or Vaseline to put on them and I had no shoes."[14] In interviews, preferences for different brands or types of lotions mostly seemed to be idiosyncratic and personal, though many people noted that Vaseline has long been regarded as superior for winter use because it provides a thicker coating. Equally, a number of people noted that in summer, dust sticks to Vaseline. Some also noted that Vaseline has long been preferred for use on children. Few of these suggestions were ever voiced in advertisements prior to independence. Vaseline, for example, was the subject of far fewer advertisements than most other leading toiletries, and the emphases of these campaigns shifted over time from Vaseline as a pomade for hair to Vaseline as a skin lightener to Vaseline as a sexual enhancement to Vaseline as a necessity for healthy skin. Unlike Lifebuoy, Lux, Palmolive, or other toilet soaps, whose suggested meanings were carefully controlled by manufacturers, Vaseline and similar products were sold either without the benefit of advertising or they floated loosely in a sea of constantly shifting messages.

The relationship between the two different hygienic ideals embodied in cultural understandings of these two commodities has been neither a case of irreconcilable contradiction nor blissful coexistence. Rather, the needs satisfied by each kind of product have been in uneasy and subterranean dialogue with each other. One kind of "common sense," a product of decades of colonial teachings and propaganda, has come to rest in a variety of soaps, while another kind, rooted in indigenous aesthetic notions about proper bodies, is still invested in the practice of smearing—for the last five decades a practice largely defined around commodities manufactured by capitalist factories. At various moments in Zimbabwean culture during the postwar era, the subtle articulation of these two practices has flashed into visibility, idiomatically attached to social class, gender, generation, or migrancy. A persistent thread running through such moments, however, has been shifting and contingent notions about "tradition" and "modernity," shadowed by a complex and fractious relationship between several generations of Zimbabweans. In the postwar era, cleanliness (and the commodities that defined cleanliness in Zimbabwe after World War II) has thus often concerned the ambiguous ways in which history—and the authority of history—was constituted.

As noted in chapters 1 and 2, colonial ideals about clean and mannerly bodies had already infiltrated much of daily life by the 1950s, while explicit resistance to this infiltration was becoming more diffuse. For many, especially for people born and grown to adulthood since that time, cleanliness, hygienic needs, and daily bodily practices have been shot full of unvoiced misgivings and uncertainties. Rural life, "traditional" culture, parents and elders, the urban lifestyle of the *matanyera* (the marginal poor, specifically "night-soil" workers), all have been defined at various moments as dirty, repellent, unhygienic, embarrassing, or insufficient. This has been reminiscent of the ways in which the Mayers' "School Xhosa" reproved their "Red" cousins for supposedly being filthy and backward: "Xhosa, themselves, when asked to explain the Red-School opposition, do so in terms of cultural differentia: Red people do things this way while School people do them that way. 'The difference between a Red man and myself,' said a young School countryman, 'is that I wear clothes like white people's, as expensive as I can afford, while he is satisfied with old clothes and lets his wife go about in a Red dress. After washing, I smear vaseline on my face: he uses red ochre to look nice.' "[15] Similarly, postwar Zimbabwean associations between dirtiness and "traditional" or rural life have not focused on smearing with Vaseline or other manufactured goods, but on a more elemental stereotype—one that envisions a complete lack of hygiene.

For example, one of the standard genres of modern Shona literature is the "urbanization" novel, in which "modern" and "traditional" lifestyles come into conflict.[16] Frequently, the ambitious urban character in such works, who may have left his family or his background behind in his search for success, regards his parents or other rural relatives and friends with disgust—a disgust frequently conceived in primarily hygienic terms. J. W. Marangwanda's 1959 work *Kumazivandadzoka* has an incident of a son rejecting his dirty mother at the heart of its narrative, and numerous other Shona novels and plays, such as Aaron C. Moyo's *Ziva Kwawakabva,* feature similar moments.[17] This typical individual often acts like the character of Nelson in one of M. M. Hove's semi-autobiographical stories: "One of Nelson's biggest worries was that his mother was both uneducated and filthy. She was an embarrassment to him. . . . He had always complained that she was not well dressed and, when she was, dirty. He wished she would be as educated, well dressed and as clean as his associates' mothers."[18] The use of these stock characters reveals some of the complexity of contemporary commodified hygiene in Zimbabwe. This archetype is obviously recognizable to many Zimbabweans—Marangwanda claimed his novel was based on an incident witnessed at Harare's train station—and yet, the criticisms of "dirty" people rendered by such a character are seen by all of these authors as unjustified. The "urbanized" aspirant is portrayed in the genre as rootless and deprived, ignorant of the value of traditions (traditions celebrated by several generations of Zimbabwean authors writing under the watchful eye of the colonial and postcolonial state).

In some cases, this criticism has been depicted even more explicitly in terms of class. In Ben Sibenke's play, *My Uncle Grey Bhonzo,* the title character is a rich man who maintains a studied distance from his poorer relatives. Grey is portrayed by Sibenke as a vain and foolish character, isolated from others by the stifling arrogance of his class. Sibenke saddles Grey with a severe hygienic obsession: he walks about his house pursuing minute quantities of dirt and calling servants to completely clean away even the slightest imperfection. When his brother and young nephews come calling, Grey is horrified by their appearance, telling the servants to bring Vaseline for them, telling them, "you must learn to bathe. To play with water. Learn smartness and neatness, you see?"[19] Grey's own ill-mannered children mock their cousins, waving handkerchiefs in front of their noses and complaining that "there is a bad smell in here."[20] For Sibenke, Grey symbolizes the urban elite's odious assumption of the cultural habits and bigotries of the settlers. Sibenke depicts a fixation on "modern," commodified hygiene as one of the most emblematic of these habits. Once Grey receives his inevitable comeuppance, his fellow businessman, Hey Zuwa, reminds him and the gathered cast of characters that the wealthy must not behave arrogantly, and that relatives—even poor relatives—are the keepers of tradition. After all, notes Hey, "Is it not our relatives who will bathe our dead?"[21]

Thus, the contemporary person obsessed with "modern" living has been depicted by writers as negatively judgmental of the hygiene of others, as judgmental of the bodies of people who one way or another fall under the sign of "tradition." However, this picture is further complicated in such discourses by the envisioning of what constitutes "traditional" hygiene. Though the meanings associated with Vaseline and lotions may be invested with a good deal of "traditional" significance, the use of these products has also been viewed as acceptably "modern" by urban and rural people alike. For example, Grey offers his nephews Vaseline to correct their "dirtiness," and despite his condescension, they eagerly smear themselves when he gives it to them. Many of my informants similarly stressed the "modern" acceptability of smearing with commercial products. Similarly, soap can lie within the field of "tradition." Many medicines prepared by a *n'anga* (healer) need to be applied in concert with washing, and many informants agreed that there are precise rules involved about whether and when one should use soap in these procedures. Soap has itself becomes a "traditional" medicine in a few cases; for example, it is sometimes used in witch-finding procedures (*muteyo*), smeared onto the anus of a person undergoing the trial.[22]

The complications of these relations between one kind of "modern" commodity, involved in a "traditional" practice of smearing, and another kind of "modern" commoditized cleanliness, rooted in the use of soaps and "beautifying" toiletries, have emerged in a number of Zimbabwean texts. The cleanliness embodied in soap, especially scented or fancy toilet soap, as well as related toiletries, has suggested complicated transformations of the self, simultaneously desired, criticized, and feared. In Chinjerai Hove's novel *Bones*, one of the peasant narrators, Janifa, muses about "good things" and how "those who have them always want to make rules so that others cannot get to the good things." She continues: " 'Good things are not for everybody,' Manyepo says. Even when he smears good smells on his body, he says not everybody must have those good things because good things are not for everybody. But why does he not allow even the baas boy to smear those good scents on his own body? Or give the baas boy the way to make the rules?"[23] In Tsitsi Dangarembga's *Nervous Conditions*, this kind of desire spills forth even more intensely. The narrator's brother is sullen upon his return home from school, disdaining his surroundings and his family. Among the impositions he objects to is "travelling by bus . . . the women smelt of unhealthy reproductive odours, the children were inclined to relieve their upset bowels on the floor, and the men gave off strong aromas of productive labour."[24] At school, he has clearly discovered a new commoditized body, and he cherishes it, for it confirms his new status and power. When the brother dies, the narrator, Tambudzai, is sent to school in his place. She foresees a similar bodily transformation as emblematic of the larger freedoms she can now grasp:

> When I stepped into Babamukuru's car I was a peasant. You could see that at a glance in my tight, faded frock that immodestly defined my budding breasts, and in my broad-toed feet that had grown thick-skinned through daily contact with the ground in all weathers. You could see it from the way the keratin had reacted by thickening and having thickened, had hardened and cracked so the dirt ground its way in but could not be washed out. It was evident from the corrugated black callouses on my knees, the scales on my skin that were due to lack of oil, the short, dull

tufts of malnourished hair. This is the person I was leaving behind. At Babamukuru's, I expected to find another self, a clean, well-groomed genteel self who could not have been bred, could not have survived on the homestead. . . . This new me would not be frustrated by wood fires.[25]

At her wealthy uncle's house, she finds cleanliness and soap: "The joy of that bath! Steaming hot water filled the tub to the brim. . . . I washed and scrubbed and rubbed, soaping myself three times over."[26]

At the same time, such desires have been tempered by a knowledge of their sources. Another of Hove's narrators, torn by unrequited sexual need, ambiguously implores the woman he desires but cannot have: "Look how I have washed and cleaned myself so that I smell the smells of the white man."[27] In his distinctive manner, author Dambudzo Marechera also addressed the ambiguous racial and cultural history of hygienic products and practices: "I had such a friend once. He is now in a lunatic asylum. I have since asked myself why he did what he did, but I still cannot come to a conclusive answer. He was always washing himself—at least three baths a day. And he had all sorts of lotions and deodorants to appease the thing that had taken hold of him. He did not so much wash as scrub himself until he bled. He tried to purge his tongue, too, by improving his English and getting rid of any accent from the speaking of it."[28] Poet Kristin Rungano has imagined a frustrated "city dweller" who declares, "I am sick and tired of city life / Of racing with the fashions: / hating my hairstyle / hating myself coz the adverts / say I smell."[29] In one of Charles Mungoshi's nuanced stories, a businessman named Mr. Pfende, who is at the end of a day full of complex revelations and subtle tragedies, "rubbed his hands which were once soap-smooth, but now, he realized with uneasiness, were subtly but surely getting chaffed and scarred."[30] Tambudzai, the narrator of *Nervous Conditions,* also discovers that "modern" commoditized cleanliness is more complicated than it first appears:

I was in danger of becoming an angel, or at the very least a saint, and forgetting how ordinary humans existed. . . . The absence of dirt was proof of the other-worldly nature of my new home. I knew, had known all my life, that living was dirty and I had been disappointed by the fact. I had often helped my mother to resurface the kitchen floor with dung. I knew, for instance, that rooms where people slept exuded particularly human smells just as the goat pen smelt goaty and the cattle kraal bovine. . . . Yet at a glance it was difficult to perceive dirt in Maiguru's house. After a while, as the novelty wore off, you began to see that the antiseptic sterility that my aunt and uncle strove for could not be attained beyond an illusory level.[31]

Tambudzai wistfully reflects that the price of living within this hygienic ideal, of bathing in a tub with soap, is losing the joys of washing in the river near her rural home: "Nor would there be trips to Nyamarira, Nyamarira which I loved to bathe in and watch cascade through the narrow outlet of the fall where we drew our water. Leaving this Nyamarira, my flowing, tumbling, musical playground, was difficult."[32] Poet Judith Moyo has celebrated the river Nyamakondo with similar language: "The ever pulsating pool of life / The haunt of young and old, / tired and dirty from work in the fields, / Your balmy waters soothe their bodies; / You give them that refreshing touch / Which, all day, keeps them clean and healthy."[33]

This subtle and contingent awareness of the hidden costs and subtle attractions of each hygienic ideal, "traditional" and "modern," has originated from a diffuse consciousness of the history that has shaped bodily practice and the cultural significance invested in soaps and body creams. The costs of the type of hygiene invested in soap have been constituted by its associations with the bodily disciplines of colonialism, of decades of mission teachings and domestic training—but the desirability of soap also comes from the same sources. The clean and mannerly body of "tradition" has been celebrated by some cultural authorities in African communities—but the "traditional" body has mostly been maintained for the last five decades through the use of manufactured lotions and creams.

The balance sheet of hygiene, with its accounting of bodies old and new, has been a small part of a much larger tally of the losses and gains, oppressions and opportunities under colonial rule. One of the most subtle but powerful personifications of the mingling of bodily practices, of the desires, truths, and restrictions contained within historically divergent regimens of everyday life, has come from the relationship between generations within African families. The politics of family life in both Shona and Ndebele society have been historically one of the densest types of social power affecting individuals. Forms of social control exercised by organized groups of senior kin or by parents are among the strictures some young Zimbabweans have struggled to escape ever since the 1920s, a process that accelerated after the 1950s and again during the most intense period of the *chimurenga* in the 1970s. Colonial authority over African bodies was often consolidated at the expense of elder Africans who had previously exerted domain over that sphere of education and training. In the reckoning of many Zimbabweans, old and young, this loss of authority has had many complicated consequences for the relationship between the generations. In interviews, informants in their twenties often expressed frustration with the attempts of their parents and relatives to manipulate their behavior. Such frustrations included arguments about what constituted proper habits for the care and presentation of the body. Yet, at the same time, many individuals, both old and young, have been equally expressive in their desire to restore respect for elders, respect for "tradition," to reconstruct a world that has been lost. Zimbabwean writing in both Shona and English has been filled with characters and voices expressing such desires.

Perhaps the most richly emblematic and symbolic example of the intertwining of "traditional" and "modern" hygiene with the struggle between the generations in postwar African culture in Zimbabwe lies in the reputed actions of *midzimu* (ancestral spirits) toward their descendents. Each individual *mudzimu* intervenes in the world of the living in different ways, but many people have reported since the 1950s that some of the spirits have periodically forbidden their descendents to use certain goods. A *svikiro* (medium) often permanently avoids certain products or services,[34] but for ordinary people, such restrictions are usually temporary. Some of the items characteristically banned by a *mudzimu* include trips in motor vehicles and the consumption of tinned food or scented soaps. Many of our informants described this behavior by *midzimu*, but were reluctant to discuss the motives behind such restrictions or simply had no theories to offer. A few people suggested that items are regarded as objectionable on the basis of smell. More significantly, several informants, particularly in rural Murewa, argued that each *mudzimu* seeks to symbolically recreate the material culture of the era in which he or she lived. One older man argued that if the ancestor in question was a grandparent, then the prohibition might focus on a recently introduced item or service. But if the ancestor was more distant ("from the time of the Portuguese"), he or she might target something like soap.[35] He also noted that the same process could work positively: a *mudzimu* might insist that some object that had belonged to him or her be used by a descendent.

During the postwar era *midzimu* have thus invoked the power of history. This has been history understood specifically in material terms, through a single emblematic commodity that constitutes a shaky and temporary authority of the past over the present. Churchgoers of many different faiths often proclaimed their disregard for *midzimu*. Many others, especially young adults, have expressed their disbelief of or resentment toward the spirits and have frequently struggled to put the demands of *midzimu* at arm's length. Indeed, even many Zimbabweans who have praised "traditional" culture in one way or another feel ambivalent toward *midzimu* who intervene in the lives of descendents. Many think ancestors should look after their families, but not disturb them, and that the living should respect ancestors in return. In *Dew in the*

Morning, a mother struggles with the desire of a *mudzimu* to possess her daughter. She sighs regretfully to her neighbor, who knows little about such spirits: "You probably have good ancestors. . . . Your ancestors are lying still in their graves and don't believe in causing any inconvenience."[36] However, since a displeased spirit is sometimes said to cause other family members to fall ill, considerable pressure can be brought to bear on anyone who resists.[37] The selection of scented soap in many of these instances—the most "modern" of soaps, the most "fashionable" soap, the soap most associated with white privilege—is indicative of the subtle but pervasive role of hygienic products during the postwar period to incarnate the power of history and the power of those who have witnessed its passage. Soap and lotions, and the interlocking of their cultural meanings, have played a key role in generating the contemporary Zimbabwean body and that body's role in the presentation of self in everyday life. These bodies and the products that have helped to make them have been neither clearly "modern" nor "traditional" but instead have been uncomfortably and indeterminately both and neither.

Fanta Faces: Ambi, Whiteness, and Aspiration

The interwoven meanings of soap and Vaseline were produced within the realms of "common sense," invisibly or sporadically, but skin lighteners, along with hair straighteners, have been the subject of open and explicit contention in the history of Zimbabwe's African communities. Since the 1960s, when local production of brands like Ambi was licensed by overseas producers or initiated by local manufacturers,[38] lighteners have explicitly defined a charged intersection between social aspirations, wealth, race, and political activism among Africans.

The connections between class status and manufactured goods in particular call out for closer examination before turning to the specific case of skin lighteners. In one sense, using the possession of commodities as a marker of class status seems a wholly natural and necessary sociological proposition, readily comprehensible within the general commonsensical worldview of most Westerners as well as the specific practice of most social scientists. Commodification in Zimbabwe has forged powerful links between wealth and manufactured things, but these linkages are neither straightforward nor linear. James Ferguson, in reconsidering these assumptions about the connections between class and consumption, has recently offered a vitally important reconsideration of the "cultural topography of wealth," arguing that the traditional linear models for the measurement of riches used by development experts and applied anthropologists are deeply impoverished. Ferguson argues that wealth in rural Lesotho has been constituted instead by "a complex cultural-economic terrain of channels" following a multiplicity of commodity paths and symbolic constructions of power, rather than a straightforward linear hierarchy mapped against the raw amount of money and goods controlled by an individual.[39]

As commodification took hold in the 1930s in Zimbabwe, a similarly complex topography took shape and continued to structure the experience of wealth for the rest of the colonial era and beyond. It grew up against the backdrop of an equally complex but different architecture of wealth and class that existed within precolonial polities in the nineteenth century. While the possession of cattle and the right to harvest ivory were powerful signifiers of riches and power among both Shona and Ndebele peoples during this era, much of the rhetorical competition between the two peoples turned on the legitimacy of wealth obtained through tribute. Other economic institutions, like the *nhimbe* among Shona speakers (the "beer party" hosted by wealthier cultivators looking for assistance while harvesting), were precariously balanced between satisfying communitarian ideals and fueling accumulation for a privileged class among the peasantry.[40] While the beginning of the "truck" trade and the later acceleration of commodification in the 1930s and 1940s did not eradicate these various older modes of consumption, there was an articulation between the new commodity relations connected with the merchants and older models of wealth, an articulation under which these older formations were increasingly subsumed and atrophied.

The circulation of goods over the *longue durée* within south-central Africa, as noted in chapter 2, had imparted to "foreign" items an association, albeit conditional and shifting, with elite power and privilege. In part, this is what drove many cultivators to initially pursue the acquisition of selected manufactured items from "truck" traders with enthusiasm and interest. The value of such goods, of course, stretched beyond prestige conferred by personal possession. Those with enough cash from the sale of agricultural produce or wage labor could perhaps hope, though often in vain, to acquire enough goods to become traders themselves in more peripheral or informal markets; this was especially true for migrants from colonial Malawi, Mozambique, and Zambia, as noted by Lawrence Vambe: "The main route from Nyasaland and the northern part of Mozambique lay through Chishawasha and these men continued to use it. . . . The ones coming in looked sadly poverty-stricken . . . while those going back home appeared, by comparison, well-fed and prosperous. In most cases, these nouveaux-riches demonstrated their affluence with such possessions as shining new bicycles, with many mirrors, heavily laden with goods they were taking to their countries of origin."[41] As a consequence of the growth of the "truck" trade, cultural definitions of wealth frequently came to involve the possession of manufactured goods and imported novelties.

However, many of the most powerful "traditional" elites—those whose wealth derived in part from their command of precolonial modes of production—continued to perceive wealth through an agricultural prism, as a matter of land, cattle, and wives, even at the outset of the postwar boom. Gelfand wrote in the *Native Affairs Department Annual (NADA)* that the Shona man who had "two million dollars but no children" would not be "rich in the true meaning of the word" among his people, and that peace with *midzimu* was more precious than anything money could buy.[42] In actuality, the division between the two differing modes of consumption in African communities was never so great as it might seem. "Traditional" elites often acquired a few expensive manufactured goods, while many of the new elite tried to maintain rural homesteads and herds of cattle as extra signifiers of their status and power, as a way of hedging their bets against the precarious status achieved by working within settler society.

Still, most new urban elites, both businesspeople and professionals, increasingly from the late 1930s mapped out new kinds of public definitions of their individual wealth or aspirations through the possession and use of a growing array of manufactured goods, and these definitions were markedly different from older "topographies." As M. F. C. Bourdillon described it:

> Wealth in the monetary economy often makes little difference to the lives of older men, apart from the addition of a motor vehicle and a few luxuries at home. Thus one very wealthy man I met lives much like his neighbours: he wears shabby old clothes and his homestead is arranged like others in his line, a small two-roomed brick house for himself facing four rondavels for his four wives. . . . The only external signs of his wealth were an old lorry parked outside his house and a television aerial on the roof of his house. . . .
>
> There were, however, other wealthy men in the area, who were somewhat younger and lived quite differently. These lived in bungalows, wore smart new suits, and drove flashy cars. They quite clearly had access to wealth. . . . Younger men who emulate the life style of the more affluent white society often display their wealth in this way and find themselves engaged in burdensome "competition" with their neighbours.[43]

Beginning in the 1930s, the conspicious and public consumption of things like gramophones, fancy clothing, automobiles, and expensive toiletries challenged white ownership of "high" fashion while simultaneously imitating and satirizing the look of white society. The struggles of urban-centered elites to assume the characteristic "look" of white society were particularly marked, though they were often carefully focused on a small number of highly symbolic, visible, public, and relatively affordable commodities. Vambe describes members of his family who were typical of this archetype in the 1930s and 1940s, when elite aspirations were increasingly defined by an assembly of "modern" objects and ostentatiously expressed tastes: "Uncle John Nyamayaro seemed to typify best of all the adaptable character of most of his generation . . . the gulf between his way of life and that of his parents was vast. He used to have bacon and eggs for breakfast. His favourite drink was Black and White whisky."[44]

None of these incorporations of manufactured goods was indiscriminate, and their pedigree was frequently indirect and complex. For example, the influence of South Africa's township culture, especially of the townships on the Reef, was immensely powerful. As early as 1910, a Rhodesian native commissioner commented, "Johannesburg is a disturbing element in this district. The natives look on it as a kind of Eldorado where fabulous wages can be obtained and all manner of luxuries can be enjoyed."[45] The late Maurice Nyagumbo's memoirs of his travels through South Africa as a young man describe his fascination with the styles of "smart" clothing and fashionable living he encountered while there, a fascination that he and many other migrants carried back to Zimbabwe.[46] The influence of Johannesburg's township cultures on the social definition of "fashionable consumption" in Zimbabwe continued to be felt through such migration and through the availability of publications like *Drum* all through the postwar years and continues in the present.

Budget surveys conducted by the colonial state outline the solidifying and extension of some of the relevant consumer habits of elites as the quest for status heated up during postwar economic expansion, most notably in expenditures on automobiles and the consumption of "European" lager beers; the budget surveys additionally document the importance of major durable commodities in the budgets of those in the top income quintile.[47] Market research reports were even more suggestive: Reported consumption of major durables has a linear correspondence with increasing wealth and age, but interestingly, goods associated with appearance, style, and fashion, like toiletries and clothing, were associated disproportionately with both the wealthiest and most marginal income brackets, fashions pursued both by the relatively comfortable and by those most desperately excluded from comfort.

The formation of the African middle classes in Zimbabwe has received increasing attention in recent years.[48] Many of these works describe the emerging struggle of mostly but not exclusively urban elites for respect and an appropriate measure of economic and social power. The mounting hunger for the status conveyed by "modernity" and "fashion" among elites fueled new needs for commodified symbols of power and respectability. Vambe recalled how elites, especially businesspeople, were drawn into this kind of consumption: "He [the African businessman] was often ostentatious, and talked big. . . . In the local phraseology, he was a Shasha, a tycoon. Such Harare men were judged by objects that the eye could see and the hand could touch: cars purchased for cash, women whom one could lure out of anywhere and last but by no means least, the fiery liquor of the white man that one could obtain from the Arcadia suburb. All these items were expensive and many an individual businessman went broke or into heavy debt because he could not curb his appetite for them."[49] Mostly, the respect they sought was denied to educated or ambitious Africans by settlers who fiercely protected the prerogatives of whiteness. In this increasingly embittered struggle, commodities were singular and intense symbolic battlegrounds.

In their wake, African elites defined a template for the presentation of self and community which some other Africans, though they might not *structurally* be members of the middle class, nevertheless found a seductive model for the definition of their own identity. Judith Williamson has criticized "the false assumption that workers 'with two cars and a colour TV' are not part of the working class,"[50] and equally, not all of those Zimbabweans who pursued the consumer habits of the African middle class were actually part of that class. Disjunctions between relations of consumption and relations of production have been common in the history of local commodity cultures transformed rapidly by transnational capitalism and provide good reason to be cautious in attempting to interpret class directly through consumption.[51] Just as Boggie commented that Rhodesian fashion was the race of whites to get away from blacks, fashionable consumption among the African middle class was an equally desperate race to get away from the *matanyera* and other "lower" classes. For example, Vambe's desire to distinguish himself and other educated professionals from the "rough, illiterate and disgruntled," the "least healthy and the least stable" of the "lower levels" of urban African society, formed a major part of his observations of township culture. He drew many distinctions between the two groups, among them the difference between selective, careful consumption of fashions derived from whites and the "wholesale, child-like imitations of white culture" he saw as characteristic of "waiters and cooks" and other working-class people.[52]

It is important to note that such fashionable consumption had a limited reach. Some working-class migrants, particularly those existing wholly outside the formal economy, situationally used certain goods connected with "modern" living to publicly signify their aspirations. These individuals often based their consumption habits on their observations of the African elite. However, many other urban residents, as well as most peasants, constructed the use of their manufactured goods with less of an eye to embodying a total lifestyle. For these consumers, individual goods may well have possessed powerful implications, but not as a linked ensemble. The value placed on many goods among such individuals was often centered straightforwardly on affordability. In our interviews, many of the goods mentioned by informants were praised for combining low cost with a long lifespan. Moreover, nationalist political initiatives in the postwar era played a powerful role in reinventing and strengthening critiques of commodity culture with roots in various older Shona and Ndebele conceptions of "moral economy." In many cases, this produced gestures of consumption leading away from "modern" goods. The National Democratic Party (NDP), which existed from 1960 to 1962, "encouraged its supporters to value these things which were African—customs, names, music, dress, religion and food and much else. . . . Sadza . . . has even become a symbol of nationalism. . . . The last meeting of the NDP . . . was proof of the emotion that had been evoked. . . . Youth Leaguers ordered attendants to remove their shoes, ties and jackets, as one of the first signs in rejecting European civilisation. Water served in traditional water-pots replaced Coca-Cola kiosks."[53]

However, various uncertain and shifting desires for the lifestyle of the settlers were nevertheless understood by Africans of all classes and communities after the 1940s in terms of commodities and commodified practices. Attempts to understand and master the material culture of settler society were found in other refracted and contradictory forms far beyond the boundaries of elite culture. Among Shona speakers, for example, a new type of *shave* appeared sometime around the middle of this century. *Shave* spirits are archetypical personalities or incarnated aptitudes who possess their subjects and cause them to act out a particular stereotype or talent. The *shave* of interest in this case was a *murungu shave,* a spirit that caused its subjects to act out "being white." During ceremonies, these *shave* required a hat, trousers, a white shirt, a tie, tennis shoes, a teacup and saucer, a walking stick, and a pipe. The possessed medium, noted Gelfand, "speaks in 'kitchen kaffir.' . . . She uses such phrases as 'my boy, my boy.' She asks for eggs, bread, cakes or any kind of European dish or article bought at a European store. She does not sit on the floor as her *shave* wants a chair."[54] This ceremony contained politically charged elements of mockery and role reversal, but it was also an attempt to break European *people* down into component *things* for the sake of understanding and thus mastering colonial subjectivity. What the *murungu shave* suggests is that interest in acquiring the tastes and objects viewed as characteristic of white life was never indiscriminate among Africans of any social class. Though such interests were particularly potent and concentrated among the growing population of elites who worked within or depended on the colonial system, they were more generally a part of all the lives touched by the spread of commodity relations.

Elites and members of other social classes seeking to acquire elite status, whether mapping out their aspirations against "traditional" agrarian norms or through the acquisition of manufactured goods, found themselves dealing with constraints and resistance from other Africans critical of postwar commodity culture. Gelfand was not the only Rhodesian ethnographer or observer studying Shona society (or Ndebele society) to note the prevalence of discourses chiding unseemly displays of wealth and privilege and restraining the behavior of the rich and powerful. Alfred Gell has noted a similar phenomenon among the Muria of India, in which "rich Muria accumulate wealth they dare not spend and would have no real idea how to spend had they the inclination. . . . Acts of conspicuous consumption not falling within the framework of traditionally sanctioned public feasting and display are seen as socially threatening, hubristic, and disruptive. Consequently, the rich are obliged to consume as if they were poor, and as a result become still richer."[55] In Zimbabwe, one aspect of similar restraints has involved the surveillance and regulation of the public consumption of elites or status seekers. Another has involved demands on the generosity and responsibility of the elites, especially independent businesspeople, by impoverished relatives, neighbors, and even strangers. Clive Kileff has described the effects of these latter demands on businesspeople in the 1970s:

> Mr. Mugudzu . . . says all the businessmen are in this same position; people see their shops and buses, assume that they are rich, and come to ask for help. This morning's visitors include a policeman who asked for $100 for school fees for his children, a man from Wedza . . . who asked him to build a bottle store and let him manage it; a woman who asked if he could find a place in school for her ten-year-old-child; and a man who asked for six quarts of beer. Mugudzu refused all these requests. During the afternoon two of Mr. Mugudzu's rural relatives visit him. They have come to ask for money; one of them must pay a fine to the family of a girl he has impregnated. Mugudzu speaks with them for a time but gives them nothing.[56]

Though new elites in Zimbabwe were aggressively redefining the public symbolism of status, they found themselves uncomfortably hedged in by the adept reinvention of "traditional" cultural restraints and other demands on the wealthy.

A typical example of these resources involved *zvidhoma*, malevolent, diminutive spirits associated with *varoyi* (witches). Witchcraft accusations of all sorts generally have involved claims and counterclaims about jealousy, greed, and the provenance of material wealth, and *zvidhoma* have been only one part of this larger pattern. *Zvidhoma* were said by some of those I spoke with to be both peculiarly modern and urban, in some cases even to have been imported by migrants returning from South Africa (where similar entities are known as *tokoloshe*). Some of the same informants also argued, however, that *zvidhoma*, or something like them, have been around for a long time.[57] *Zvidhoma* are said to have attacked those who have ostentatiously displayed or hoarded their wealth—especially when that wealth has been in the form of luxury goods such as imported, expensive clothing, fancy cosmetics or skin creams, or new automobiles. Rich persons who have failed to show generosity toward those in need are also said to have been historically at risk for such attacks. A sudden stabbing pain or ache has been said to be the sign that a person is being pummeled by the usually invisible *zvidhoma*, who are usually dispatched by a *muroyi* (witch), often at the behest of a jealous or critical person who knows the victim. In one recent case, *zvidhoma* were even said to be busily engaged in sabotaging and slowing development projects initiated by the Zimbabwean state.[58] However, *zvidhoma*, or something like them, are also said to have been "stuck" to objects. In one case, I was told a story about two people finding a bag of money by the roadside. They later discovered that the bag had been abandoned because *zvidhoma* were infesting it and would bring down a curse upon any owner unwise enough to keep the bag.[59] A warning to beware ill-gotten or unearned gains is clear in this particular anecdote. It is this kind of action that some informants claimed *zvidhoma* were known for in the precolonial past, *kare kare*—being "stuck" to the livestock of obnoxiously wealthy individuals, for example. Moreover, *zvidhoma* were only one of a number of sanctions that could be invoked around consumption: in Richard Werbner's work *Tears of the Dead*, one man recounts how the tent in which he kept all his accumulated consumer goods was struck by lightning sent at the behest of jealous relatives.[60] Bessant and Muringai report that a successful rural storekeeper in 1950s Chiweshe recalled being accused of "using magic to increase his wealth."[61] In the 1970s, while *zvidhoma* and other "traditional" sanctions and types of cultural obligations clustered around the display of wealth remained highly potent, the use of goods to signify wealth was also increasingly threatened by political activists armed with the tools of socialist discourse and practice.

Thus, though many elites or their imitators hoped to symbolize or confirm their aspirations through constituting forms of privileged access to the public ownership and display of certain prestigious manufactured goods from the 1920s onward, the implicit criticism and enmity of others in the community has acted as a considerable impediment, especially from the 1950s to the present. In many cases, status seekers have spent a good deal of time trying to discover or invent commodities that could connote modernity, fashion, success, prestige, and power but that would not be desired by other Africans. If no one but self-identified "fashionable" or elite consumers desired a commodity that nevertheless somehow successfully embodied the essence of contemporary living, then the jealousy and resistance that gave the threat of *zvidhoma* and other sanctions their force might not come into play. W. L. Chigidi's play *Kwaingovawo Kuedza Mhanza*, though set after independence, features a fifty-eight-year-old businessman, Matonisa, whose uneasy and incomplete desire for "modernity" and power seems wholly archetypical of the experience of others like him in the past. At one point, Matonisa sings the praises of his whiskey, which, he stresses, is imported. Whiskey is superior to beer, he declares, for beer you must share with all the "parasites" who come calling using familial and neighborly connections.[62] No Africans want whiskey except the powerful, which is what makes it Matonisa's favorite drink.

The desire for Ambi and other skin lighteners has been particularly tied to these intersections between commodification and a racialized consciousness of class, status, and power. The need for skin lighteners has been a need to purchase fashionable respectability in order to confirm power already possessed—or equally, to escape the colonial margins of powerlessness by appropriating fashions associated with the wealthy. The history of lighteners demonstrates the fragility of attempts by the professional and business elite in African communities to preserve their own exclusive claims to certain goods and, more importantly, suggests how the general cultural critique of the behavior of some of the African middle class worked itself out through contention over a specific product. Skin lighteners demonstrated the inextricable intertwining of consciousness about class and race, but gender also played a significant role. Lighteners, like cosmetics, were used disproportionately by women, though they were not quite as "gendered" as cosmetics. Instead, the use of lighteners attracted vociferous criticism both from political activists and from self-identified "traditionalists," who viewed the products as the most potent and immediate embodiment of elite collaboration with the colonial project, as a blatant submission to the corruptions of "modernity," as a distillate of the oppressive colonization of the self.

Many of the women between the ages of thirty-five and fifty-five reported in interviews that they had used Ambi, Bu-Tone, Hi-Lite, or another lightener in the past. Virtually everyone I encountered had relatives or acquaintances who had used these products. With the products having been banned by the Zimbabwean state in the 1980s, discussions of lighteners were ostensibly about a "dead" issue. As a consequence, most users and nonusers alike found the subject wryly amusing. Unfinished business, unsettled scores, and unrequited desire still frequently lurked behind the laughter, however. In Tables 6.1 and 6.2, it is notable that respondents reporting that they used lighteners circa 1970 disproportionately hailed from the highest and lowest income brackets. This pattern was also described more idiomatically in most of the interviews conducted for this study. Past Ambi users were described as having been successful or desperate for success, wealthy or marginal, people comfortably inside the sphere of colonial power or people utterly outside it. Lightener users were also disproportionately among the young at the time that they used the products.

Former users are today remarkably matter-of-fact about their use of the products. The deep-seated nature of their need for lighteners was amply demonstrated in discussions by many former users, who spoke urgently of their continuing desire for effective lightening creams. Hydroquinone in skin lighteners causes potentially serious damage to the skin, damage many still refer to as "burning." Some former users we spoke with showed us blemishes or marks left by the creams, and yet most of these users proclaimed their readiness to continue consuming such "strong" lighteners. Users demurred when asked if they were angry at the manufacturers. Instead, many expressed anger with the government for banning the products, and several confided that the products were still available from smugglers. For some informants, the damage caused by lighteners is precisely why they still feel a need for the products, as "burning" often causes the skin to develop unusually dark patches. Some people thus spoke about lighteners as a kind of endless addictive treadmill: once burned, one is forever in need of lighteners to prevent becoming even darker than when one began. Some also argued that damage was caused only by improper or excessive use. In one case, one regular consumer said that she was burned only when she continued to use Ambi after returning to her rural homestead to work in the fields for a time, which she said was an improper location in which to use such things—an explanation charged with larger metaphorical significance.[63]

Though lightener users freely discussed their own consumption of the lotions and described the community of Ambi users in the language of class and aspiration, most were notably reticent when the subject of their own motivations for seeking lightening arose. The nature of these motivations lay at the heart of the controversy surrounding Ambi and similar products in the 1960s and 1970s. Critics argued, though from different vantage points, that to lighten one's skin was a betrayal of the essence of African culture and a submission to white domination. In the 1970s, lightener consumers were derisively referred to by many others as "Fanta [orange drink] faces with Coca-Cola legs."[64] They were said to be people who tried to become white in their faces but who could never escape their essential Africanity, their blackness. The rest of their bodies would betray them twice over: to the whites they wanted to join and to the blacks they strove to leave behind. A few of our informants claimed that the epithet was merely whimsical or descriptive, but most others, including former lightener consumers, agreed that it had been intended as a derisive criticism. For example, for Zimbabwean poet Musaemura Zimunya, a woman who uses lighteners serves as a symbol of the simultaneous degradation and desirability of urban life: "She was his death sentence./They called her fanta-face and coca-cola legs./'Ah! but beneath the orange and the coke/I am all fire-furnace!' she retorted."[65]

The blatant equation in lightener advertisements of whiteness with aspiration and success particularly angered nationalist activists, who came to their activism in the aftermath of the work of critics of psychological and cultural colonialism like Fanon, Cabral, Memmi, and Ngugi. As the war escalated, Zimbabwe African National Union (ZANU) officials urged women to "reject the practices of stretching their hair, wearing of wigs, painting of lips red, use of skin bleaches (in order to look less African) and showing a general love for white culture. . . . Some would say 'revolution has nothing to do with wigs, lipsticks and straight hair.' Of course these things have a lot to do with our revolution. Those who try in any way to make themselves resemble white people do so because they adore white people. They hate to be African."[66] Activists seized on the language of the ads for lighteners and argued that purchasing the products was a clear instance of "selling out" to the white regime. As one Zimbabwean scholar has recalled, "The layman does not know what 'AMBI' stands for, but the preponderance of adjectives in the comparative inflection (softer, clearer) has in the past been responsible for the 'mischievous' interpretation by young political activists who claimed that the letters A-M-B-I stand for 'Africans Must Be Improved.' "[67] In her recent autobiography, Zimbabwean nurse Sekai Nzenza has voiced similarly critical sentiments regarding lighteners and other similar "beautifying" products: "it would be so nice to be free in body and spirit. The multinational companies have started dressing us up in weird knickers and bras; they have started putting us in bubbly baths advertising Lux or some other soap. Besides putting our black bodies on the market, they have started telling us how we can change our black skin to nearer white. . . . They will go on to make anything to make us feel beautiful and make money out of us."[68] Resentfully, Nzenza remembered, "I knew I was too black to be considered beautiful. I had tried to use some lightening cream on my face to make it brown, but the cream was no good because I got brown in patches and looked really awful. So I had accepted that I was black and could not do anything about it."[69]

Those who identified themselves as "traditionalists" in late colonial texts and in contemporary interviews also decisively rejected the use of lightening creams. Adherents to some evangelical sects, including *vapostori* (literally "Apostles"; members of an independent religious sect with loose ties to Christianity), have also spoken out against the use of any beautifying toiletry. In *Harvest of Thorns*, church members are depicted hissing "Eve" whenever a "perfumed woman with a made-up face and plaited hair" passed them.[70] Some have insisted that the users of such goods—those who were particularly flamboyant—had been the targets of *zvidhoma* and other forms of retribution, often including restrictions from their *mudzimu*. Some of the rhetoric typical of this position once again crops up in Nzenza's account. For example, she recalls an incident from her childhood in which her mother criticized a town woman who used cosmetics and lighteners as being too "modern": "Then I saw a woman with a very short dress, wearing almost nothing. She had her short hair all straight and oily. Her face was almost white . . . her lips red with lipstick and her legs were her natural black colour. . . . 'She stinks, mother,' I said suddenly. Mother laughed out loud and said to me, 'They call it chimanjemanje—modern way of living. We are losing fast—losing everything.' "[71] A "town woman" who returns to live in the countryside in the novel *Dew in the Morning* eventually gives up her skin lighteners and is depicted as being somehow more at peace with herself and her children, "tamed" by the countryside.[72] In my own discussions, the impact of these two overlapping critiques was particularly reflected in the overwhelming rejection of lightener consumption by many young Zimbabweans and by those over sixty years old. (However, many contemporary urban youth use hair straighteners, which have been criticized in much the same way as lightening creams.) In fact, old and young alike, activists and "traditionalists," often used much the same terminology to condemn those who had bought Ambi and similar products, using the word "betrayal" in particular—though what exactly each speaker felt had been betrayed often varied.

Former users themselves usually offered ambiguous or limited explanations of why they had found the advertisements and promotions appealing and why they had felt (and in some cases still feel) a need for lightening creams. Most users said that lighteners had "beautified" people (-nangaidza) and left it at that. In one suggestive case, a middle-aged woman whose friends had used lighteners considered the matter for a short time and then recalled at some length how they often spoke of the advertisements they remembered seeing in their youth, advertisements that always featured white models and celebrities. "When they thought of beauty," she concluded, "they thought of white women."[73] In another case, a young man recalled that his older relatives, male and female, who had used the creams had later attributed their use to the relentless pressures of social fashion. "People here in Zimbabwe would say that we are very competitive," he observed.[74] Mostly, however, lightener consumers dismissed their critics, arguing that the people who had not used the products had refrained because they were too poor to buy them, not because they were too principled to use them. Complaints were attributed to malevolent jealousy. One successful African businessman with a lengthy involvement in the toiletries industry strongly cautioned me that "all of that criticism was quite wrong, lightness and darkness have independent connotations in traditional culture— it's not about becoming white, it's simply that light-colored black skin has always been seen as beautiful around here."[75] As early as 1959, South African marketer Nimrod Mkele offered a similar defense: "I know that in some of the surveys carried out on face creams, Africans have said they prefer certain creams because they want to be 'light' which might be interpreted as meaning that they want to be white. . . . I hold they meant the improvement of the 'glow' of their skin."[76]

Unlike soaps, cosmetics, or other toiletries, lightening creams were advertised with claims that went unmistakably and unambiguously to the heart of the symbolism of power in a colonial state ruled by white settlers, in which even the wealthiest and most accomplished of the African elite were always kept at arm's length from even the lowest rung of the white hierarchy. The material nature of lightening creams offered little room for cultural invention. The promotion of lighteners promised to erase the differences between white rulers and "modern" African subjects, to enable social aspiration, to differentiate elites from other Africans by allowing them to cross the racial divide and join with the settler elite. As with other goods turned to these ends, elites found that lighteners could not carry them past the boundaries of their own racial identity. Furthermore, they found that they could only imperfectly constitute their own class hegemony over other Africans by becoming "lighter and brighter." Some of the poorest and most marginal residents of the townships muddied the symbolic waters by seeking status themselves through the use of lighteners. At the same time, many others fought back against what they regarded as a supreme symbol of racial and cultural betrayal, calling upon a wide range of tools used in the wider struggle over the construction of "relations of consumption" and the symbolic expression of wealth in Zimbabwean culture.

"God Meant Women to Be Beautiful": Cosmetics, Perfumes, and Femininity

The expression of a need for cosmetics—lipstick, eyeliner, and similar items, as well as related goods like perfumes—clearly rose out of the same headwaters as the desire for skin lighteners and was thus intimately tied up in the symbolic construction of class relations. However, to an even greater degree, these beauty-enhancing toiletries have been powerfully linked to the image of African femininity and the regulation of women's behavior, women's labor, and women's mobility.

As a process, commodification affected African women differently from the very beginning. As men were drawn into wage labor and migrancy, female cultivators became the primary actors in rural exchange, the people who grew, transported, and sold cash crops to "truck" merchants.[77] Thus, rural women were among the first Zimbabweans to discover the cultural potential of new commodities and the first to participate in the intriguing public theater involved in examining and purchasing store goods. Moreover, during the years that these goods were moved through informal trade to the peripheries of the colonial economy, women were able to get access through commodification to new potentials for accumulation. As settler agriculture grew in power and "truck" merchants acquired new juridical mechanisms for controlling commodification, such opportunities were strangled and the potential of commodification turned to deprivation and oppression, once again experienced differentially and disproportionately by rural women.

Moreover, as the histories of domesticity and advertising in Zimbabwe amply demonstrate, colonial institutions most concerned with generating new subjectivities among Africans or altering aspects of African culture often considered women to be the crucial population for the success of their project. As one of the Rhodesian Ministry of Information's pamphlets trumpeted: "It would not be an exaggeration to say that, in her eagerness to learn, for her husband's sake, the ins and outs of Western social behavior, the woman often leaves the man at the post. The African woman has an asset which should make her the envy of more worldly women; she is entirely free from self-consciousness, thus she learns and absorbs simply, unaffectedly, as a child does. She has, too, the confidence inbred in a woman reared in complete security."[78]

Various hegemonic projections of "civilized" manners, bodies, and consumption were located within the female sphere, just as the strictures of "tradition" and "traditional" consumption were frequently invoked by African men with regard to women, in the interests of maintaining and extending their patriarchal control over the feminine. Marketer Nimrod Mkele, for example, was striving in his address to the 1959 advertisers' convention to speak for the African middle classes of southern Africa. In this light, the extraordinary reversals, contradictions, and anxieties of the following passage about African women reveal some of the larger ambiguities felt by urban African men toward women's role in commodification:

> Women . . . are bringing in new tastes into African homes. After all, it is they who determine what shall or shall not be bought. The role of a hubby is to fork up and smile. He has a say in purchases that involve a heavy outlay of money but the final decision is the little woman's and the man must only sign on the dotted line. This, of course, is a European value stereotype.
>
> It is the women who egg on the men to keep up with Joneses—or should I say the Kumalos?—for the African man, like his European counterpart, tends to be conservative. This conservatism is of course encouraged by women.
>
> Even in smoking, it is the women who are in part responsible for smoking becoming the last preserve of men . . . they are invading even this sphere, although the cigarette-smoking woman is still regarded as a woman of doubtful morals in African society. This of course is derived from the stiff Victorian morality of the whites . . . via the missionaries, for women do smoke in traditional African society.
>
> Women are therefore the vanguard of progressive ideas in today's emergent African Society in so far as the home is concerned. This factor is so important from a marketing and advertising point of view that it cannot be overlooked. To paraphrase a well-known idiom: Never under-estimate the size of a woman's thumb—there may be a man under it![79]

All toiletries, from soap to skin lighteners, disproportionately contained "feminized" elements, a legacy of the differential focus of colonial domesticity and hygiene on women. Women bore the burden of ensuring their own purity and the purity of their households through the use of soap and cleansers. Equally, the "beauty" of their bodies and the "modernity" of their manners were a major subject of domestic and hygienic training. Cosmetics and perfumes have thus been the most but not the only "gendered" toiletries. However, as the *most* gendered toiletries, these goods in many ways have embodied, for both white and black men, various perceived dangers and challenges from African women, while also expressing a diverse spectrum of feelings involving disgrace, subversion, freedom, sexual power, and style for many women. These dangers and values were especially centered on the presence and activities of female migrants to the cities. These migrants, whose numbers grew as the rural areas became increasingly less viable during the 1920s and especially the 1930s, in some cases were able to achieve a limited degree of autonomy and economic power through entrepreneurial activity, including beer brewing and various forms of prostitution and concubinage.[80] While urban women first used cosmetics in a style that deliberately recalled and exaggerated the look of white femininity in southern Africa (which was in turn regarded by many metropolitan visitors as already exaggerated),[81] many of these women began to use cosmetics such as powder, eyeliner, and lipstick in a distinctive new manner: in large amounts of outstanding colors. Cosmetics were thus one of that selection of commodities chosen by Africans to simultaneously imitate and satirize white fashions.

Therefore, white attempts to protect privileged consumption, previously discussed in chapter 4, often focused on the increasing use of cosmetics by urban women in the 1930s and 1940s. One township administrator railed, "The females dress in most cases within reason . . . [but] rouge and face powder is now being used much to their detriment."[82] Another administrator drew a similar picture: "A visit to any Municipal beer hall during sessions or dances and concerts at which the aping of white men and women reaches its peak, will show a large number of over-dressed, heavily rouged and lipsticked cigarette-smoking females wearing the latest ball-room apparel . . . this type has been the bane of my life in the Location."[83] This critique was particularly caught up in the wavering regard of the Rhodesian state toward African prostitution, which was encouraged, forbidden, and ignored—often in one breath. After the postwar boom, local manufacturers and multinational agents were slow to market cosmetics for Africans precisely because of this intense sense that the use of cosmetics was inappropriate for Africans. A 1955 article in *Commerce of Rhodesia* raved, "African women are already aware of cosmetics, but few could be called cosmetics conscious. The horrible sight of a dark-skinned woman whose face is plastered with cosmetics which obviously were manufactured exclusively for use by a light-skinned person is a common sight in Rhodesia. . . . African woman frequently lacks taste . . . she entirely ignores the final effect in her anxiety to do what the white woman does . . . any natural good looks the African may have are lost. . . . "[84] Manufacturers often argued, with some justification, that many available cosmetics were designed to enhance white skin, though African women were busily making their own rules for cosmetics use which did not correspond to aesthetic codes governing the application of cosmetics among settler women. Businesses were also slow to consider manufacturing cosmetics that were specially designed to enhance black skin—with the exception of skin lighteners—because of their fears about the negative impact of goods made exclusively for Africans. The heavily commodified forms of domestic training and hygienic propaganda prevalent after the 1950s eventually began to promote cosmetics use as a fashionably "modern" habit, but remained heavily constrained by settler notions of what constituted "good taste." An advice column in *Parade* typically warned African women: "Next time you go to a dance or a party take a long look at the girl who is attracting the most attention . . . she will be simply but cleverly made-up . . . her make-up will be slight and well-concealed, not blatant and clown-like. There will be just that suspicion of powder. Her face won't look as though it has been stuck into a flour bag. Lipstick will be specially chosen to suit her and will be finely applied, not in thick daubs as though a house painter has been at work . . . she will wear just a trace of good perfume, and on no account will be loaded with cheap and nasty scent. . . . "[85] Even with manufacturers and domestic trainers accepting cosmetics use among Africans in the 1960s and 1970s, albeit according to their own sense of "taste," many Rhodesians remained adamant about the unacceptable "cheekiness" of African women who used such products. In one of our conversations, a former domestic servant said that "hard" whites who visited the home harshly criticized her employer for allowing the servants to wear beauty products.[86]

Similarly, many African men have regarded cosmetics as the material expression of their fears about women's mobility and women's power. The stereotype of the socially destructive woman, the woman who defies patriarchy and its rules, has long had great currency among Zimbabwean men. Such a woman has usually been seen as urban, often as a prostitute, and frequently has been portrayed as a social climber obsessed with "modernity." As Rudo Gaidzanwa notes, many male Zimbabwean writers have used these stereotypes, disproportionately envisioning women as "adulterous, prostitutes, domineering and evil."[87] In this respect, judging from the content of archival files of complaints by African men throughout the colonial period and testimony by African men before state commissions investigating the social conditions of urban communities, as well as other similar historical documents, these have been typical views within the broader sphere of masculine culture in twentieth-century Zimbabwe.

Since many "shebeen queens," prostitutes, and independent urban women adopted cosmetics as a badge of their identity in the years immediately preceding World War II, descriptions of urban women by African men frequently targeted cosmetics use as the central symbolic attribute of "dangerous" femininity. Vambe, for example, described what he saw as the corruption of the women of Chishawasha in the 1930s: "Many of the local women too had emerged from their traditional role of excessive modesty and subservience. Some of them had a highly developed sense of dress, the result of European influence, and those who had taken to using sopa and other beauty aids were hard to resist."[88] In another passage, he recalled that "Exactly as the village sages had anticipated and feared, these emancipated ladies greatly excited local men of the younger set whose simple work-worn wives and sweethearts were not as alluring nor as well-washed, sweetly scented and finely dressed as their arrogant rivals from town."[89] The image endured into the postwar era. Zimunya's poem, "Be Warned," declares, "Not every woman plies a trade/of lipstick and shaven eye-brow pencilled/or eye-lashes wetted to the point of spikes/not every woman wants your drink/for the love of warmth of a hotel suite."[90] Similar comments were made in a number of interviews. Women who used cosmetics were described by men either as prostitutes or at the very least as dangerous seductresses. "Women will stop at nothing [to seduce men]," said one middle-aged man.[91] Another echoed the comments of a number of other informants by saying that he had long decried the use of cosmetics by decent women and good wives pursuing fashion, because this made it difficult to know who the prostitutes really were.[92] Another proclaimed that married women have been the worst offenders, chronically overusing cosmetics.[93] Some men also criticized African women for using cosmetics from other perspectives; a few of these were reminiscent of the scoldings of settlers, as they excoriated African women for inappropriate or vulgar use of beautifying toiletries. "If you put red lipstick on a black background, it looks terrible, it looks like the smallpox," said one man.[94] Many of these comments were also tinged with an overtone of the type of political or social criticism that lighteners attracted.

At the same time, the masculine comparison of independent femininity with voracious sexuality[95] has endowed cosmetics with a powerfully attractive sensuality as well. If black and white males alike[96] have wanted to be able to distinguish African prostitutes from African wives, this has often not been with the aim of avoiding the former. The woman with lipstick, eyeliner, and powder on her face has thus often been as much an object of desire as the subject of criticism. One Murewa man scoffed at other men who criticize female users of cosmetics: "God meant women to be beautiful, God means them to wear makeup. How could a person have objected to this?"[97] Even male critics acknowledged that cosmetics users were sexually alluring. Still, whether it was desired or attacked, the patriarchal view of cosmetics use in African culture has most commonly been that either the wearing or avoidance of cosmetics is something that women should do to satisfy masculine sensibilities. In these visions, the need (or lack of need) for cosmetics is not rooted in women's being, but instead comes from men. Cosmetics have been viewed in this sense almost as secondary commodities, as the packaging of women-as-goods.

Not surprisingly, many Zimbabwean women have seen the matter differently. Discussions of cosmetics in interview groups inevitably provoked passionate debate among women, debate that invoked a wide-ranging cultural universe of female roles, expectations, and aspirations. Many women in our interviews generally approved of cosmetics use, or at least conceded it to be a matter of individual taste requiring no social opprobrium, while a number of other women were vociferously critical. In any case, from either perspective, cosmetics posed a challenge to each woman's *own* identity and sense of her femininity, each woman's personal struggle to construct what she considered to be a superior lifestyle for coping with the challenges of "modern" life and the demands of men. In some interviews, when men and women were in a group together, many women, regardless of their views on cosmetics, seemed to regard men's contributions as unwarranted interventions into a strictly female issue.

Much of the female criticism of cosmetics has come from those women who have tried to follow the ideal of the "Christian, club-going wife" in their own lives. These women frequently echo the argument of some men that the products have elided the difference between wives and prostitutes. The lifestyle of simplicity and modesty promoted in FAWC branches and the example provided by well-known African women like Helen Mangwende constructed a new kind of "moral economy" for some women. The ostentatious use of cosmetics by those they view as wastrels or, worse yet, prostitutes or home-wrecking seductresses has been a serious provocation to their sensibilities. Other female skeptics have included educated women who agreed with the radical or political critique of cosmetics use. As one young woman said, "You want to look natural. I don't see why you should be more beautiful than you really are. Lipstick makes you look like somebody else, somebody who you really are not."[98]

Defenders of cosmetics were often equally impassioned. One woman, Mai Mbiriri, proudly displayed cosmetics given to her by her former employer, a white woman who emigrated to South Africa. She described how her employer taught her to use cosmetics "properly" during the 1970s and concluded, "Because I stayed with whites for a long time, I learned to like that way of life . . . those who say cosmetics are bad are ignorant."[99] Like other users of cosmetics as well as skin lighteners, Mai Mbiriri accused critics of being motivated largely by jealousy. Additionally, many women have praised cosmetics for their fashionable modernity and considerable allure. When advice columns and advertisements in *Parade* and other African media finally began to embrace the "selective" use of cosmetics, these items were imbued with a powerful aura of elegance. A significant number of female informants also pointed out that while they themselves did not use cosmetics, they could not see why anyone else should feel inhibited about doing so. This may have been partially attributable to a subtle characteristic of social relations in Zimbabwean culture. Many are reluctant to openly criticize the actions of others; such criticism is sometimes redirected in diffuse ways, into witchcraft accusations or into the verbal play connected to threshing at harvest time. At the same time, many women have also clearly situationally accepted a philosophy of individual freedom which includes the freedom to define one's own lifestyle.

Underneath this debate, other divisions between women were being played out. As Table 6.1 demonstrates, cosmetics of various kinds have been disproportionately the province of wealthy women, desired just as lighteners and other luxury products were, for their ability to signify wealth and privilege, though more women from all income groups than Market Research Africa's study indicated probably used various cosmetics. The low reported usage of cosmetics overall in Table 6.1 may have been partially a consequence of social stigma and partially a consequence of the language used by the survey. Few women have used cosmetics "regularly"; many more have probably used them on just a few occasions. A. K. H. Weinrich's complex study of Mucheke Township in Masvingo (Fort Victoria) in the late 1960s and early 1970s testifies to the convoluted intersection of social status with cosmetics use among women in Zimbabwe. In the township hostel known at the time to Mucheke residents as the home of the township's prostitutes, many women relied heavily on cosmetics. However, even within the hostel, the so-called professional prostitutes and those Weinrich terms "semi-attached" women were divided, with the latter group making sparing use of lipstick and eyeliner.

Among the diverse lower-status families in old three- or four-room houses in the township, Weinrich notes that "tradition" was generally valued. Particularly among the women who settled permanently with their husbands in town, cosmetics were the object of intense criticism. However, such criticism was not directed at prostitutes, but instead at the women in Mucheke's wealthier households: "Some of these women are very critical of the rich families in Mucheke and their superior dress and westernized manners. They often comment that, since their dress and behavior are so sophisticated, they must have been prostitutes in the past."[100] A related point was made by one of our informants: "It used to be thought that only loose women used cosmetics, because those who were poor and could not afford these things thought that the women who used them must be getting their money from some other source."[101]

Other lower-income women in Mucheke aspired "to move into the upper social strata." Like other socially ambitious Zimbabweans of the time, they signified their goals by adopting selected "Western behaviour patterns," including landscaping their small yards or practicing fastidious household and personal cleanliness. Additionally, many of these women wore some cosmetics. Significantly, however, Weinrich notes that "they do not use lipstick, for their neighbours think that this is affected and befitting only prostitutes."[102] In 1991, in the Tashinga squatter camp near Mbare Musika, women were still debating the wisdom of using cosmetics in similar terms. "We are always accused of being prostitutes by others," said one woman; "those who use cosmetics are prostitutes, they are the main users of these things. If we all wore them, men would no longer know the difference between prostitutes and other women."[103] Others in the camp argued that the beliefs of other township residents were irrelevant, that critics were only jealous, that even poor women should be free to look however they want—even like prostitutes, if that suited them. To reduce these tensions over social status, one of our informants explained, her church simply prevents members from using cosmetics: "We have so many people from different economic backgrounds, it keeps anyone from feeling inadequate."[104]

In other sections of Mucheke in the 1970s, Weinrich records that women with higher social status—nurses, wives of professionals, the wives of railway workers—all used skin lighteners, hair straighteners, and cosmetics, including lipstick.[105] The highly organized railway workers' wives were particularly interesting. Women in the lower-status neighborhoods, pursuing the role of the "good club-going wife," criticized cosmetics use among the wealthy, suggesting that this marked them as former prostitutes. In contrast, the Women's Guild for the Wives of Railway Workers "declared war on concubines," seeking to expel them from their well-off and "respectable" neighborhood.[106] The implicit desire here was to make the fashions and lifestyles on display in the railway section "pure," free of the taint of prostitutes and their imitation of high-status women, although these high-status women were in turn strategically and selectively imitating white women. Women in similar positions elsewhere in Zimbabwe— the wives of professionals and other men in the urban middle-class elite—

increasingly patronized "charm schools" and "beauty seminars" at the behest of their husbands and peers in the mid-1970s. Some of the teachings in such schools involved the "proper" use of beautifying toiletries so that a fashionable woman could distinguish herself from prostitutes and other supposedly "vulgar" cosmetics users.

Alongside this capacity to define differences in social status and ambition among women, cosmetics also have acquired the power to symbolize for women (and men as well) the cultural divide between "urban" and "rural" lifestyles. James Ferguson has recently criticized scholarship on urbanization in Zambia, arguing that mobility between rural and urban communities has been far more widespread and continuous than scholars have previously understood.[107] As Ferguson has pointed out elsewhere, Zambians, Zimbabweans, and other southern Africans themselves strongly view urban and rural life as separate domains, with distinctly different cultures and lifestyles.[108] While they may move back and forth from place to place and identity to identity, most Zimbabweans have described the city and the country as two different lifeworlds that call for different kinds of behavior. Cosmetics have acquired liminal status with regard to these urban and rural lifestyles. Female migrancy—the process of becoming a "town woman," learning to live with urban codes of femininity, going, in the parlance of settler discourse, from "raw" to "cooked"—has been powerfully symbolized by cosmetics and other beautifying toiletries with qualities of "modernity." In the "Shoe shine city," writes Zimunya, there is "the heaven of clean clean daughters/I won't marry a country girl/who vaselines dusty legs/and burns her forehead with hot stone/pressing wetcat smelly hair."[109] In another poem, he cries, "O Harare! your lipstick and beauty soap and perm salons/have courted the little girl now she blushes and touches/kisses and turns endlessly before the mirror."[110] A young woman arriving from town at a rural homestead stands out for her "blue-painted eye-lids, false eye-lashes, red lips/bangles gritting in her hands/with a European hair-wig above an Ambi-proof face."[111] In Chinodya's Dew in the Morning, city people move in next door to the young narrator, who is familiar with city ways. He instantly identifies one of the neighbors as an urban person by her use of toiletries: "Even from far we could see the easy comfortable features of a city woman—the vibrant skin of her body, lightened on the face by Ponds Vanishing Cream, the stockinged legs, the shiny dark wig. We even caught strong whiffs of her perfume."[112] In interviews, the same fetishistic associations with urban and rural life emerged. In one case, two younger women living in Murewa said that they would never wear cosmetics at home because they would be regarded as prostitutes, but when visiting their friends in Harare they had always worn lipstick and eyeliner to look fashionable and attractive. They said that using cosmetics in Harare was for them an important sign of freedom.[113]

The uncertain and situational meanings attached to cosmetics in Zimbabwe have ranged from a prostitute's form of personal advertisement to an elegant modern feminity to asserting stylistic freedom from male dictates. Cosmetics have been invested with the capacity to separate urban and rural lifeworlds and to define the status and aspirations of women. Makeup has thus become an important recurring feature in struggles over the role of women in African communities. Lipstick, eyeliner, and related products, surrounded by a complex history of advertising, domestic training, white aggression, and patriarchal power, as well as the changing social and historical conditions governing sexuality, labor, migrancy, and production, have both symbolized and produced female identities.

Colgate for Cuts: Commodities, Imagination, and Cultural Ownership

The growing use of toothpaste in Zimbabwe has engendered no controversies, overt or subtle. Toothpaste has been used more by elites and others pursuing "modern" hygiene, to be sure, but whether used regularly or not, toothpaste never has particularly been used to define identity or social difference or to mark the boundaries between various lifestyles. However, the history of toothpaste is part of another side to the historical experience of commodification for Zimbabweans. New goods, new needs, and the processes involved in the acquisition of same have presented, in the eyes of many Zimbabweans, a bewildering, oppressive, and disconcertingly fast-paced challenge. At the same time, each new item sold by merchants has offered a new surface for cultural inscription, a new opportunity for imagination and creativity. Not all manufactured items have been seen as equally interesting by Africans. Both the hidden boundaries of cultural ownership maintained around objects and the fundamental material composition of various commodities have strongly influenced the degree to which new goods have been perceived as open to cultural reclamation, to new interpretations and uses.

In this respect, commodification has been subject to the larger sweep of Africans' creative responses and initiatives in the face of confinements and opportunities presented by colonial rule in southern Africa. As scholars such as David Coplan,[114] Jean Comaroff,[115] and Terence Ranger[116] have demonstrated in diverse ways, within the powerful strictures and oppressions of capitalist production and state domination, diverse southern African peoples have reordered and reimagined their own personal and collective universe of practices, values, and meanings. The interpretations of goods proffered by advertisers and manufacturers have certainly been variously ignored, altered, or purposely misrecognized by Africans, but the specificity and intent of such reactions is difficult for an outsider (and perhaps even an insider) to illuminate. Commodification as a whole process has frequently been regarded in southern African societies—and in the African diaspora as a whole[117]—as a negative and oppressive experience, a tragic form of submission to white authority. In *Harvest of Thorns,* Chinodya retells Zimbabwe's history in the style favored by ZANLA soldiers holding *pungwes* (secret nighttime events for mobilizing and educating rural villagers). Among the lessons preached by the guerrilla leader is that the tragedy of colonial rule is partly a consequence of Africans allowing themselves to be "bought off with bits of cloth and mirrors."[118] In the 1950s, an African radio announcer similarly warned his colleagues:

> "Never satisfied. . . . Ah, these Europeans. . . . And watch out . . . they're making us that way too. When I was a youngster the men from my village went to work for the Europeans to earn a blanket. Then they came back to rest and were happy. To have a blanket was a wonderful thing in those days. But we? As soon as we have one thing, something new starts to worry us. We want a bicycle. And after that clothes, and then dresses for the wife, next a wireless set. There are some already wanting motor-cars these days . . . are we any more satisfied? Are we any happier?"[119]

The prevalence of such nagging doubts in Zimbabwe's history was at the least vaguely oppositional to the wishes of most colonial rulers. In these sentiments, there is also an awareness, which at least some Africans shared from the onset of intensive commodification after 1945, that consumer culture is a trap, a perpetual-motion machine that must infinitely generate new forms of dissatisfaction in order to prod consumers into moving toward the next frontier of desire. More specifically, many objects were effectively rewritten and advertising messages actively reinscribed in the face of strong dictates from advertisers, state officials, and missionaries about their "real" meaning and utility. Yet, many acts of cultural invention with regard to manufactured commodities or marketing campaigns have not been known to corporations, or if known, have not met with any opposition. Furthermore, the degree of reinterpretation imposed on commodities has varied considerably. Some goods have been accepted into practices of everyday life largely as their manufacturers intended they should be.

During the heyday of the "truck" trade, especially in rural districts, most manufactured goods were somehow reshaped to meet relatively indigenous tastes and needs. The merchants themselves mostly acknowledged that they had a limited degree of influence over their customers, while the actual manufacturers of "truck" goods were generally spatially and socially distant from their markets at the periphery of global capitalism. During this same era, many settlers, state officials, and missionaries advanced an interest in regulating consumption and producing "civilized" manners, but these projects were only partially, gradually, and disjointedly effective. While the growing African middle class attempted to selectively acquire "modernity" and "high fashion" through certain items in the 1930s, it also, alongside the rural peasantry and urban working classes, increasingly shaped manufactured clothing, toiletries, ornaments, beer, and other items toward independently determined tastes and aesthetics. Kosmin, for example, points out that "red limbo" sold by "truck" merchants was incorporated into ceremonies involving *shave*.[120] Photographs of *n'angas* from the 1940s and 1950s now held in the Zimbabwe National Archives show a wide variety of imaginative uses of items of manufactured clothing with the aim of crafting a striking appearance. Vambe described one of his Chishawasha relatives calling on the same imaginative resources with regard to clothing:

> For Jakobo, for instance, one article of clothing, except perhaps a skirt, was as good as another. He saw no inconsistency in wearing women's blouses, men's tailcoats without matching pairs of trousers, or riding breeches without boots or leggings. Hats had a special fascination for Jakobo, and he had a whole selection of them, men's as well as women's. . . .
>
> Clad in an incongruous collection of finery, he would be swept into an imaginary world of pomp and circumstance. He would don a hat, for instance, or pick up a walking-stick and jauntily stride about, telling people that he felt like the Governor on his way to a game of golf and tennis. He was a compulsive showman, and while possessed of considerable knowledge of the white world, he admired many of its features in a childlike manner. Sometimes he would contrive some sort of a uniform and cast himself as a military general—one of the calibre of a Napoleon at least. Then he would dream of marching on Europe to give the white races a taste of their own medicine. . . .[121]

Another example of the creative appropriation of clothing, the use of bedsheets for headgear or for the padding of buttocks by "fashionable" women, has been recounted by M. M. Hove.[122] All of this recalls Terence Ranger's analysis of colonial dances in which participants, such as the Beni ngoma in East Africa, appeared to be imitating Europeans in their uniforms and dance steps. Ranger concluded that such dances could not simply be dismissed as "derivative and parasitical" but were instead "deeply rooted, creative and versatile."[123] Of course, some of this innovation was also expressed in negative terms through the rejection of or disregard for new items of clothing or other new goods. In 1919, messengers employed by the Native Department complained that they were being mocked by villagers because their uniforms made them look feminine.[124]

These creative interpretations of new goods and new capitalist practices of distribution and marketing continued after the 1940s, despite the increased density of Zimbabwe's commodity culture in the wake of postwar economic expansion. Some of this invention was mothered by necessity. Particularly with expensive durables, it was essential for African consumers to derive maximum and multiple utility from a product. This engendered numerous skills, mostly self-taught, for the recycling and repair of goods like refrigerators, automobiles, bicycles, radios, furniture, record players, stoves, and sewing machines. In *Harvest of Thorns,* a mother recalls her son's childhood skills for these kinds of repairs: "On that same morning of the day he came back he fixed a plate on the stove she said she had paid an electrician to repair and he said 'You shouldn't throw money to the sharks' and went out to help Peter water the cabbages in the garden while she watched them from the window trying to reclaim from the shaggy-haired man in brown boots the Benjamin she once knew who had repaired a burnt-out pressing iron element with a piece of tin foil and operated the old gramophone on a car battery and disused transformer. . . . "[125] People were also quick to turn these skills and durable commodities themselves into sources of cash. Repair shops, especially for bicycles, were characteristically among the first businesses started by African entrepreneurs as early as the 1920s. Similarly, some appliances were desired not just because of their prestige or their time-saving utility, but because their owners could make money from them. This was particularly true of sewing machines, which are still an important source of cash for their owners, especially in rural villages.[126] In a similar vein, Zimbabweans have learned to reuse the discards of manufacturers and elite consumers to new ends. Some of these uses are familiar in communities living at the margins of capitalism across the world—in the building of shanties, for example. Other uses of garbage or surplus are more specific to Africa as a whole or to southern Africa in particular, and have been less marked by desperate necessity. Toys built from old tin cans and wire are a particularly well-known and prevalent example of this kind of cultural invention in the townships of southern Africa.

Zimbabweans (and other Africans) have continued to more generally in-
scribe new meanings and uses onto manufactured goods since the 1950s. Durable goods have frequently been painted or decorated in new styles—as in the
case of some houses in southern Africa, or even more famously in West Africa,
with the elaborate redesign of buses. However, inexpensive nondurable goods
present a more malleable and affordable canvas for such reinventions. Toiletries
are a major part of this class of commodities and have been the subject of numerous adaptations as a result. I have already mentioned several examples: margarine, cooking oil, and soap lather for smearing; soap in *muteyo* ceremonies;
and cosmetics applied in distinctively township-created fashions. In parts of
southern Africa, instant coffee has been used for color in mixes for smearing.[127]
Liquid fabric softeners have been used as shampoos or perfumes.[128] Blue mottled soap has often been used as fish bait.[129] There are many other examples.

The case of toothpaste is particularly interesting. Like soap, toothpaste had
been imported from South African and metropolitan sources for some time
prior to the 1950s. Colgate was particularly well-known, and even today, the
word "Colgate" is used by many older Africans to refer to all toothpastes. Other
firms, including Lever Brothers, have found it difficult in the postwar era to
make inroads on Colgate-Palmolive's share of the toothpaste market, especially
once Colgate-Palmolive established a local subsidiary. However, even with
more toothpaste advertising and increased promotion of dental hygiene in the
FAWC and similar institutions, toothpaste never achieved the universal indispensability of soap. Table 6.1 shows that 88 percent of upper-income urban
consumers report regular toothpaste use, but only 70 percent of those in the
lowest bracket use it. In rural areas, the figure is certainly much lower. However,
at some point since the 1950s toothpaste consumers discovered an important
new use for the substance. Urban and rural residents alike, both men and
women and of all social classes, reported during interviews that they have used
toothpaste as a skin medicine for many years. When it is rubbed over cuts or
ringworm, toothpaste has been said to soothe a wound while protecting it from
insects. Many also claim that it speeds healing, partly by drying out the cut
or sore. The practice has been widespread for at least the last two decades and
even apparently has included adherents of religious sects who avoid medicines
(*vapostori*). Unlike the case of margarine, in which a new use found for a
product worried its manufacturer, this adaptation of toothpaste has not seemed
to concern Colgate-Palmolive, though they are quite aware of its existence.[130] At
the same time, the company has not incorporated this new use for toothpaste
into its advertising—this goes too far toward recognizing the power of local
culture over the meanings and uses of commodities.

What determines whether (and to what degree) goods are reconfigured in this manner? There are practical considerations of paramount importance. The material nature of goods restricts and directs their imaginable uses. There is an indeterminable but real limit to the malleability of utility. The use of toothpaste as a salve on cuts is predicated on the thick and spreadable quality of the product, as well as its medicinal nature. The use of blue mottled soap as fish bait has come about only because it works. Other important pragmatic factors in determining the extent of experimentation with new goods are cost and accessibility of a given commodity. More nebulous and yet equally significant is what I have referred to above as "ownership." This is determined by the complex historical biography of a given commodity and by the cumulative weight of interests and associations pressing upon it. The sheer amount and thematic intensity of advertisements and the structure of the commodity's production are two crucial factors, as is the manner in which Zimbabweans have seized upon the commodity to produce and define social difference. By these criteria, skin lighteners and cosmetics have had very weighty claims to "ownership" placed upon them, soap only slightly less so, while Vaseline, body lotions, and especially toothpaste have been at the other end of the spectrum.

Does Coke Add Life? Audience Skepticism and Marketing Failures

The limits to the overall hegemony of capital and colonialism over African consumption in Zimbabwe can be further demonstrated by an examination of the history of African reactions to advertising and marketing, particularly with regard to failed or stillborn campaigns. The African audiences organized by postwar media have often turned a skeptical or at least cautious regard on the process of commodification, using critical skills developed during decades of colonial rule to break down or examine the motives and interests of manufacturers and their allies. In the United States and most European countries, such skepticism, when it exists, is often located around the phenomenon of advertising itself, though it also frequently targets specific products or whole categories of commodities. The annoying banality and deceptiveness of marketing is a cliché of modern life in the West, though the presumed universality and inevitability of advertisements, deeply interwoven into everyday discourse and popular culture, blunt any effort at a sustained mass critique.[131] In contrast, most of the Africans in Zimbabwe who have been targets for advertisements and promotions have had little interest in the phenomenon of advertising, with the exception of demonstrations, point-of-sale promotions, and similar public spectacles. Instead, when concern and skepticism have flashed into public visibility, they have mostly originated from intense popular surveillance of the form and packaging of products.

In both Harare's townships and rural Murewa, few people during interviews expressed any direct opinions regarding advertising in general. Ads have long been a ubiquitous feature of everyday life, everyone agreed, but most were noncommittal about their significance. Referring particularly to advertisements taken out by supermarkets and department stores about current sales and discounts, many described print and radio ads as informative. The only advertisements consistently singled out for condemnation were those selling underwear or clothing and featuring semidressed models. A few people were able to recall their reactions to particular campaigns. One older man in Murewa reported that he remembered when hoardings went up in the village center proclaiming "Coke adds life." This was a false claim, he complained, because he felt no extra vigor or energy after drinking a Coke to test the slogan's veracity.[132] Another informant pointed out that you always have to test advertisements to see if they are telling the truth.[133]

As noted in chapter 5, advertisers themselves increasingly found in the 1970s that some of their print and radio campaigns had been widely scorned or ignored for their utter failure to correspond to the realities of African life, so clearly, the individual representations contained in ads have been the subject of much popular interpretation and critique. The ephemeral nature of individual advertisements accounts for some of the contemporary disinterest in the subject. (Personally, coming from a cultural environment saturated with ads, I find it difficult to recall without some assistance advertisements that I viewed twenty, ten, or even two years ago.) Recent AIDS awareness campaigns by the government were mentioned a great deal in interviews, showing that current advertisments often do engender popular discussion. More importantly, however, most Zimbabweans may not have an opinion on advertising as a whole because they do not regard advertising as a coherent and unified phenomenon.

This matter is intimately tied up in the larger issue, discussed recently by Megan Vaughan, of how southern African peoples perceived the visual products of colonialism, such as films, educational theater, or printed propaganda, from the 1950s to the 1970s. Vaughan notes how the employees of the Central African Film Unit and other regional producers of educational films and plays tried, with little success, to assess the reception of their visual materials among African audiences. Some of what these officials and professionals regarded as the most telling anecdotes about African reactions circulated among advertisers as well. For example, Vaughan records one professional's story about an antimalaria demonstration in which a large papier-mâché model of a mosquito was used. The villagers were said to have dismissed the message because their own local mosquitoes were so much smaller. During my interviews with advertisers in 1991, I was told the same story on several occasions.[134] Like Vaughan, I also found that this mosquito story was part of a larger genre of narratives about the supposed inability of Africans to interpret "modern" visual messages. For example, I was told in a number of instances that rural populations viewing their first movies reacted to the images as if they were "real," fleeing in fright from the screen. Rather than documenting some kind of interpretive failure, however, I would suggest that whatever incidents actually lie at the root of this genre of stories represent a persistent though shadowy type of critical distance between the visual imagination of colonial image-makers and African audiences.

The main exception to the lack of explicitly formulated interest in advertising as a whole has come from a small but vocal proportion of the educated elite—especially teachers and activists—who were also the disproportionate targets of print and radio advertising. With considerable passion, one middle-class Harare resident spoke with disdain and loathing for ads, decrying them all as "completely unnecessary."[135] In his own idiosyncratic voice, author Dambudzo Marechera also criticized the presence of advertising in the social milieu of the early 1970s:

There I drank heavily but something was wrong and I couldn't get drunk. It was the place: all garish colours and lights and a band of half-naked girls dressed up in leopard skins and gyrating out some coarse smanje-manje. The big man at the microphone was not so much singing as farting out in an unnatural bass voice. The walls were all plastered with advertisements for skin-lightening creams, afro wigs, Vaseline, Benson and Hedges. There was one in particular of a skin-lightened afro-girl who was nuzzling up to her coal-black boyfriend and recommending the Castle Lager. As the music boomed against the advertisements and the arse colours and lights flickered on and off I lost count of time and simply soaked myself with the stuff.[136]

In *Harvest of Thorns*, advertisements are similarly depicted as a central—and vaguely fraudulent—part of the look and feel of urban life in the 1960s:

At the end of the road, ahead of her, a huge poster she had not seen before showed life-size members of a healthy, happy family all beaming over a loaf of bread and a brand of margarine which was being advertised. The husband in the poster had a corpulent waist and a handsome beard. The two children, a boy and a girl, had shiny skins and looked so alike you could think they were a real family. The mother, who was cutting the margarine, wore an afro wig and a short cotton dress. Her face was the colour of Fanta and her plump shiny legs the colour of Coke. The wife looked as if she had been using lightening creams for a long time.[137]

Since independence, the bitter memory among the educated elite of certain ad campaigns directed at the "African market" has been voiced increasingly in public, particularly through the Consumer Council of Zimbabwe, which now publishes a regular column in the *Herald*. Alec Pongweni's 1983 University of Zimbabwe paper is a typical expression of this wariness and distance. Pongweni speaks of the copywriter's "disorganization" of African culture and notes that "the languages (English, Ndebele and Shona) used for advertising in this country have become a powerful weapon which, if unscrupulously and ruthlessly used, could dislocate the cultural outlook of the people beyond recognition."[138] Another critic, writing in the *Herald* more recently, declared that "Most of the adverts shown in this country borrow very heavily from Western advertising. While this may be very successful with the Western consumers, it may have very little meaning with the generality of the Zimbabwean people."[139]

In the case of demonstration vans, fashion shows, contests, stunts, and point-of-sale promotions, African attitudes toward advertising have been quite different, both among elites and the general population. In interviews, demonstrations and similar public displays from the late 1960s and early 1970s were recalled vividly and with considerable enthusiasm by many informants, particularly those who had been living in rural districts at that time. Demonstrations were praised as entertaining and educational, a great attraction. Many of our informants bemoaned a seeming decline in the frequency and scale of demonstrations since independence.[140] In particular, the regular handing out of free samples, amid growing intimations of scarcity in 1991, was recalled with considerable nostalgia. More generally, contests and give-aways were remembered with similar enthusiasm, though these continue to be common and are thus less imbued with a sense of loss.[141] Some of the other promotions common in the 1970s were regarded at the time and are remembered today with greater skepticism. Fashion shows invoked many of the same emotions connected to cosmetics, while most of those who today recall various public shows and stunts like Lever's "Mr. Power-Foam" campaign for Omo, which included township tours for local boxers dressed as the character, were either mildly bemused at the memory or could not fully disentangle their recollection of these campaigns from recent government educational projects aimed at public health or from the crude "hearts and minds" programs carried out by the Rhodesian armed forces.

Point-of-sale promotions are also recalled today with some ambivalence. The exploitative nature of commercial pricing and retail tactics in the township themselves, which persists today, means that the motives and actions of township storekeepers are still looked on with suspicion, though stores in the center of Harare seem to be regarded today as more equitable in their pricing and range of goods. The specific memory of "truck" trading may have been subsumed into generalized recollections of the pre-1950s colonial era.[142] Moreover, this memory had some unexpected effects: the first generation of African storekeepers, who appeared in different communities at different times, had to defend themselves against presumptions that they too would "do Africans down"—presumptions that were often accurate.[143] The transformation of distribution and the coming of supermarkets in the 1960s and 1970s were largely welcomed at the time for their effects on prices and for freeing Africans from past strictures on what they could see, touch, or buy in stores. The supermarket in the 1970s was a spectacle of abundance and mass consumption that engendered interest and excitement. Elaine S. Abelson has recently described an evocative parallel to this experience in the new sensual spectacle presented by the first urban department stores at the close of the nineteenth century in the United States. These enthralled female customers engendered a new "lust for goods," which occasionally materialized as shoplifting.[144]

The negative side of Africans' historical experiences with stores and commodification in between the first rush of interest in traders at the turn of the century and the 1970s was evident to the retailers who opened supermarkets. The Jarzin brothers, former "truck" merchants, opened their first Harare supermarket in 1974 and found themselves desperately trying to explain that their new store represented an entirely new phase in the relationship between retailers and African customers: "We literally had to meet them at the door, show them where to get a shopping basket, and show them how to shop. . . . We had to persuade them that they wouldn't be arrested for theft if they took goods from the shelves—and then we had to show them that they should only take the amount of goods they could pay for."[145] Some contemporary Zimbabweans continue to regard stores, even supermarkets, with suspicion, which is probably indicative both of the fragmentary realization of access to mass consumption in the postindependence economy and a result of recent scarcity.

That demonstrations and store promotions have attracted much more conscious commentary among Africans than print, radio, and television advertisements is one sign that the real locus of popular African scrutiny of commodification has lain in products themselves. In demonstrations and other consumer theatricals, the products have been the actors and thus the focus of audience attention. In other contests, discourses about the form, packaging, and nature of products have motivated Zimbabweans to express a critical distance from commodification. The most historically potent of these instances were outbreaks of negative rumors in the 1950s about the content and manufacture of goods like tinned food, beer, margarine, and soap. More generally, many Africans have devoted considerable time to the intense inspection of the outward appearance of manufactured goods. The specific legacy of alienation produced by shoddy "truck" goods and by a much more sweeping and generic suspicion of all settlers has meant that even the smallest changes in the packaging or design of goods have been interpreted by consumers to signify mysterious—and sinister—transformations of the essential substance of the products themselves.

Such skeptical regard has touched even the physical heart of commodity relations: the material form of money itself. Particularly in rural districts, many Zimbabweans refused until recent times to fully accept money merely as the physical incarnation of a disembodied transactional principle. This in turn has produced a persistent tendency to avoid banks, a development that frustrated the colonial state. The settler legend that peasants buried untold wealth was partly an attempt to justify the deprivations and confinements of colonial rule, but it also had some basis in fact.[146] This fetishistic examination of the form of money made changes in currency a serious trial for the Rhodesian state. One settler farmer recalled: "Do you know that the sovereign's head faces a different way with each reign [on coins]? Probably not, but the natives spotted it. When King George came to the throne, and the new coins were issued, the natives utterly refused to touch them at first, declaring them spurious. For months I could pay wages only in old coinage."[147] In another instance, at the close of the nineteenth century, Stanley Portal Hyatt reported that one of his customers asked him to exchange "bad shillings," new Edward VII coins, for "good shillings," those that still had Victoria's image imprinted on them.[148] Similarly, when the post-UDI state changed the currency to dollars to stress its independence from Great Britain, many rural people initially refused to acknowledge the change.[149]

This skeptical gaze has been turned on many other goods as well. As noted in chapter 5, Rhodesian executives discovered that minuscule changes in the packaging or presentation of a product could precipitate a massive drop in sales among African consumers, who were interpreting the change in the image of the product as indicative of a concealed transformation in the actual substance of the item. One informant described a change made in the lettering on the package of his favorite cigarette brand; for him, this was a "new" cigarette. When he tried it after this packaging change, he found its flavor "quite strange."[150]

From the perspective of some manufacturers, this scrutiny was not always a bad thing, in that it reinforced the intense brand loyalties that were the bedrock of their activities in the "African market." However, such intense brand identification, even in the economies of the United States and Europe, can sometimes jeopardize the proprietary rights of companies by converting their specific product into a generic type of object. The Xerox Corporation has struggled to prevent photocopying becoming known as "xeroxing," just as the makers of Frisbees have insisted that their product is only one kind of "flying disk." Similarly, in Zimbabwe, toothpaste has threatened to become "colgate." Moreover, this popular tendency to inflexibly insist on a fixed image and form for brand-name products made the "new and improved" campaign, a staple part of bids for greater market share in the West, a much more delicate and complicated prospect for Rhodesian marketers. This in turn endangered the essential core dynamic of modern manufacturing capital, its constant need for a visible public display of movement, change, and expansion.

This latter threat particularly has been demonstrated by the failure of a number of newly introduced commodities to catch on among Africans during the 1970s. As I have noted previously, the 1970s in Zimbabwe in some ways constituted a break in the momentum of commodification. The intensification of the war disrupted advertising campaigns as well as the hegemony of colonial institutions over African bodies and African manners. The cultural power of transnational corporations was to some extent muted by the partial closing of the economy after UDI, and the resources necessary for the production of a wide range of new items were in short supply. Struggles over the meaning and use of certain goods within African material culture became far more fractious and potent. Though goods with previously defined meanings, like soap, lighteners, cosmetics, and a host of others, continued to be important features in the conduct of everyday life for most Africans, new goods introduced in the "African market" during this time had to define or identify needs only weakly referenced to the "prior meanings" established through the firm hegemony of colonial rule and racial capitalism. The popular gaze of skepticism that had long been turned sporadically on the form of commodities was thus newly empowered. Among the products for the body and for cleansing that failed to take hold as fundamental needs during this era were liquid detergents, deodorants, and shampoos. Liquid detergents were conceptually flawed in a straightforward way. In a market where a large proportion of consumers used natural outlets of running water for washing clothes, a liquid simply did not work. Deodorants and shampoos, on the other hand, were used by 21 percent and 35 percent of the wealthiest female consumers respectively, with considerably lower use in other income groups and almost no male use. By comparison, among white women, an average of 90 percent used shampoo and 89 percent used deodorants.[151] Attempts by advertisers to suggest to Africans that a body that was not treated with these goods was irremediably repellent, dirty, or unsuccessful never really penetrated the public imagination. The cleansing of hair had always been an important part of hygienic practice among Africans, but it was something that many felt could be satisfactorily accomplished either with *ruredzo* or with goods with multiple uses like soap, hair straighteners, or even fabric softeners. Deodorants were generally considered a form of purifying the body that most Africans have found excessive and pointless, addressed to needs that were adequately satisfied by soap, smearing, or even perfumes. In interviews, many informants shrugged disinterestedly at the mention of deodorants.

The fact that these commodities have not become fundamental needs provides more evidence about the intertwining of the power of colonial rule with the process of commodification. In those cases where decades of hegemonic labor intensively produced or intruded upon African notions of subjectivity, remaking "common sense," opposition to the lifeworlds conjured around manufactured things by their makers has percolated up through sporadic outbreaks of skepticism and rumor, intensive scrutiny of the form and nature of products, and the quiet maintenance of creative resources for the remaking of the meaning of objects. The changing universe of goods became a vital part of the definition of social difference and identity within African communities and between Africans and white settlers. Things became both receptacles for and active producers of history, memory, and knowledge. Though independence has brought a significant change in the tone and direction of the "public sphere" in Zimbabwe, much of this history of changing relations between African people and industrially produced things continues to weigh upon the present.

Needs, Desire, and Social Justice in Contemporary Zimbabwe

William Reddy has pointed out that "thought about commodities, like thought about persons and institutions, is necessarily encyclopedic in scope."[152] Acquiring knowledge about the history of commodification in Zimbabwe or any other society is therefore largely a question of which volume of the encyclopedia one chooses to pull from the shelf. Any account of the making of commodity culture and the world of consumption in Zimbabwe must be accompanied by much larger details from social and economic history in order to be intelligible.

Each distinctive type of commodity demands a different accounting of history. Alcohol, for example, is connected to precolonial peasant accumulation, to the rich array of customary ceremonies that demand beer, to the building of beerhalls by the colonial state, to shebeens, female brewers, and the underground cultural life of southern African cities. Canned foods were made popular by changing patterns of diet, by migrant labor, by the culture of domestic servility, by the discovery of new physical and sensual tastes, by famine and scarcity. Clothing, automobiles, musical recordings, houses, sculpture, mealie meal, bicycles: they all require "prior meanings" for their particular significance to become historically intelligible.

In this study, I have mapped out the history that informed the consumption of toiletries. I have shown how white attitudes toward black bodies inspired institutions that remade practices of the body, domesticity, and manners. The bodily racism of settler society constantly lurked about the edges of the lives of those Africans whose social aspirations were most identified with "modernity," but the shifting and loosely defined "hygienic ideals" of nineteenth-century Zimbabwean cultures also continuously reinvented and reproduced themselves in everyday life during the twentieth century. Pre-1945 merchant capital brought manufactured goods into the material culture of everyday life for rural peasants, compound workers, and the swelling population of urban townships. By helping to draw all of these Africans into the web of capitalist exchange, "truck" merchants brought cultural possibilities and acquisitive deceptions, possible freedoms and definite restrictions. The development of "new needs" among Africans was perceived by settlers as the essence of "civilization" and as a subtle threat to their power. These attitudes shaped African access to goods and African perceptions of those goods.

Racially coded attitudes toward consumption also loomed large in debates over the building of postwar secondary industry. The inevitable structural realignment of colonial Zimbabwean society that followed shaped the whole of commodity culture for the next three decades; the availability, price, and exchange of soap and other toiletries among Africans in different areas of the country were determined by these transformations. Postwar marketing, a direct by-product of the expansion of secondary industry, took all of these past connotations and distilled them into new and powerful forms that wormed their way into most economic and cultural aspects of African life.

Scholarly studies of the making of the colonial political economy are correct to identify the pivotal role of "new needs" in racial capitalism and settler rule. I have demonstrated the degree to which the production of these needs was not a consequence of monolithic power acting against powerless subjects, but instead grew out of a massively complex intersection of micropowers and macropowers, local desires and collective interests, imagination and restriction. Moreover, capitalism did not invariably act to flatten out and homogenize African lives for the sake of some global "modernity," but instead actively worked to reproduce and redefine numerous forms of social difference. The development of commodity culture oscillated between "common sense" and subtle resistances, but it was also explicitly contentious, public, and political.

To reveal that the development of needs in Zimbabwe was a vital part of the reproduction of colonial social relations may seem to invite a new critique of these needs as a form of false consciousness, burdens on attempts to engineer social justice in postcolonial Zimbabwe. Public discourses about need in modern capitalism relentlessly disguise themselves as concerned with natural and essential desires produced by the autonomous choices of informed consumers. Restoring a sense of history to commodities and consumption refutes these disingenuous claims. However, any contemporary radical social critique of consumption that follows on a scholarly reconstruction of the history of needs and the meanings of goods must not seek to casually dispense with the heterogeneity of the commodity culture that has grown out of colonial rule in Zimbabwe. Consumer needs are no less real for having a history, no less deeply felt for having been part of the world that global capitalism and colonialism have made. Needs, once made, do not casually go away and cannot be legislated or ordered out of existence. Their making and unmaking is beyond the capacity of any one institution or power. The historical failure of state socialism has more than a little to do with a failure to understand these facts. Future attempts to radically reimagine the socioeconomic basis of southern African societies, though unquestionably needed, will do well to remember the resiliency and rootedness of desire.

APPENDIX
BUDGETARY CHARTS,
1957–1970

SALISBURY 1957–1958*

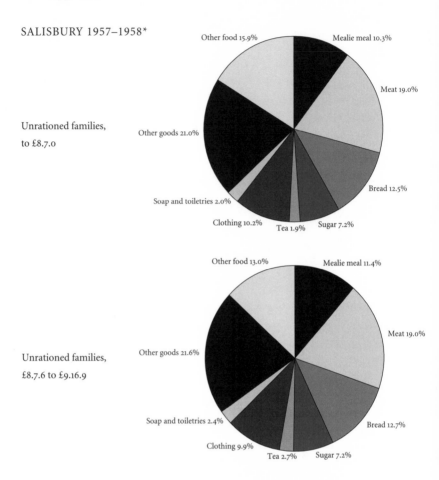

Unrationed families,
to £8.7.0

Other food 15.9%   Mealie meal 10.3%

Meat 19.0%

Other goods 21.0%

Bread 12.5%

Soap and toiletries 2.0%

Clothing 10.2%   Tea 1.9%   Sugar 7.2%

Unrationed families,
£8.7.6 to £9.16.9

Other food 13.0%   Mealie meal 11.4%

Meat 19.0%

Other goods 21.6%

Bread 12.7%

Soap and toiletries 2.4%

Clothing 9.9%   Tea 2.7%   Sugar 7.2%

* The surveys used to construct these charts were explicitly political documents, used both to demonstrate that Africans were better off than anticolonial critics supposed and to encourage strategically limited increases in African wages. The urban African budget surveys were conducted variously by the Central Statistical Offices of Southern Rhodesia, the Federation of Rhodesia and Nyasaland, and post-1965 Rhodesia; the categories used to describe data varied considerably over time. Some of the early budgets in particular included data on the expenditures and income of single men living in hostels, which I have not included here. The surveys concentrated on unrationed families: those families who were not receiving a monthly allotment of food from an employer. Though several of the surveys also tracked the consumption of rationed families, I have chosen not to chart these results here. Each of the surveys was issued in a preliminary and final form, entitled *First Report On . . .* and *Second Report On . . . .* The reports are listed as *Report on the Urban African Budget Survey, Salisbury (Bulawayo, Umtali*, etc.). The Zimbabwe National Archives today, in most but not all cases, holds all versions of each survey. Wherever possible, I have used the final copy of the report. I have not charted the consumption of all the commodities traced by

these reports, but only those few that seem to represent characteristic patterns corresponding to income or to social aspiration. Several of the surveys also estimated aggregate annual consumption by all Africans of certain goods, figures that I have not tried to represent in these charts because of the inconsistency of the data available. These surveys also contain a wealth of other information about incomes, family size, religious affiliations, residency patterns, and other demographic points of interest and are worthy of close examination by scholars studying the postwar era. I have made use here only of three surveys done in Salisbury from 1957 to 1969 and two surveys done in Umtali; other surveys were done in the Midlands, Bulawayo, and in Northern Rhodesia.

There were almost no consistent variations between Bulawayo and Salisbury in these surveys, and the similar colonial cities of Gwelo and Fort Victoria also produced figures that were generally consonant with the Salisbury results. Umtali (later renamed Mutare), located in the far eastern portion of colonial Zimbabwe, was closely linked to trade with the Mozambican coast through the port of Beira, as opposed to the rest of the colonial cities, which tended to receive imported commodities through South Africa. The relatively close overlap between African consumption patterns in Salisbury and Umtali underlines the geographic extent of the economic and social transformation discussed in this book.

The conversion of the monetary values used in these surveys to 1995 Zimbabwean dollars is complicated by a number of issues. The post-UDI transition from the old British monetary system to dollars is one such intricacy, and UDI-era currency controls—kept partially intact by the postcolonial Zimbabwean state—provide another complication. Recent changes in the Zimbabwean economy, including a home-grown version of "structural adjustment," make comparisons even more difficult. Here I am simply at the limits of my own present expertise in economic history and must defer such a comparison to a later study or to someone more skilled than myself.

Other sources for budgetary data which mostly confirm consumption patterns indicated in these charts are:

a) Joint National Council budgets: See ZNA RH 12/2/1/1, Joint National Council, Minutes of the Race Relations Subcommittee. Budgets prepared by the Native Labour Department were discussed at meetings on July 24, 1950, and August 28, 1950. At the pivotal December 6, 1950, meeting, urban budgets calculated by the chief government statistician, J. H. Shaul, were entered into the record. As noted in footnote 18 of chapter 4, Shaul had previously acknowledged in the 1940s that the state's information gathering capacity was weak regarding African communities. The JNC budgets, intended to press representatives of primary industry into acknowledging the insufficiency of wages, deliberately excluded all items that were not classed as absolute "needs."

b) Rhodesia Manufacturing and Milling Budgets. These budgets were commissioned by RMM and collected by T. M. Samkange and the noted Zimbabwean scholar Stanlake Samkange, based on their survey of 294 RMM workers in colonial Harare. There were results for "general workers"; the budgets also included more limited data on "boss boys," "policemen," and "drivers" in the company. Among the interesting notes on these better-paid workers is that they all used a sizably increased amount of toiletries (S. J. T. Samkange and T. M. Samkange, *Survey of Personal, Employment, Expenditure and Accommodation Data of African Employees of the Rhodesian Milling Company* [Salisbury: Rhodesian Milling and Manufacturing Company, 1960]).

c) Howman Commission budgets, ZNA ZBI 1/1/1 to 2/1/2; A. K. H. Weinrich, *Mucheke: Race, Status and Politics in a Rhodesian Community* (Paris: UNESCO), 1976; and R. W. M. Johnson, "An Analysis of African Family Expenditure," *Rhodesian Journal of Economics* 5:1 (1971). The last is particularly interesting because it contains data on rural expenditures.

SALISBURY 1957–1958

Unrationed families,
£10.0.0 to £12.7.0

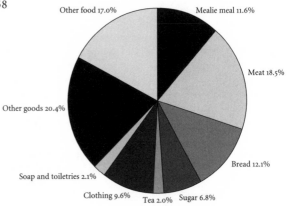

Other food 17.0%   Mealie meal 11.6%

Meat 18.5%

Other goods 20.4%

Bread 12.1%

Soap and toiletries 2.1%

Clothing 9.6%   Tea 2.0%   Sugar 6.8%

Unrationed families,
£12.7.6 to £15.15.0

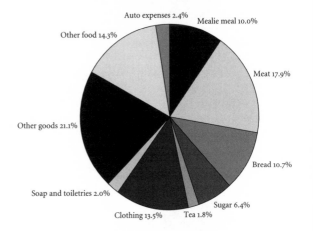

Auto expenses 2.4%   Mealie meal 10.0%

Other food 14.3%

Meat 17.9%

Other goods 21.1%

Bread 10.7%

Soap and toiletries 2.0%

Sugar 6.4%

Clothing 13.5%   Tea 1.8%

Unrationed families,
£15.17.6 and over

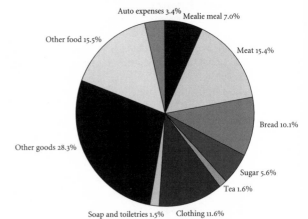

Auto expenses 3.4%   Mealie meal 7.0%

Other food 15.5%

Meat 15.4%

Bread 10.1%

Other goods 28.3%

Sugar 5.6%

Tea 1.6%

Soap and toiletries 1.5%   Clothing 11.6%

UMTALI AND GWELO 1958–1959

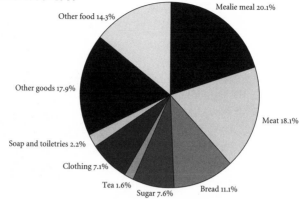

Unrationed families,
to £7.3.7

Mealie meal 20.1%

Other food 14.3%

Other goods 17.9%

Meat 18.1%

Soap and toiletries 2.2%

Clothing 7.1%

Tea 1.6%

Sugar 7.6%

Bread 11.1%

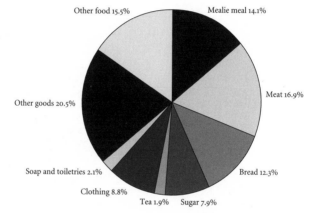

Unrationed families,
£7.50 to £8.15.0

Other food 15.5%

Mealie meal 14.1%

Other goods 20.5%

Meat 16.9%

Soap and toiletries 2.1%

Clothing 8.8%

Tea 1.9%

Sugar 7.9%

Bread 12.3%

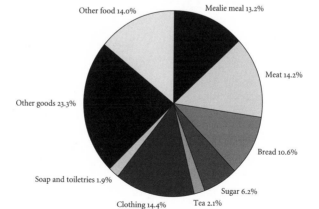

Unrationed families,
£8.15.2 to £10.11.8

Other food 14.0%

Mealie meal 13.2%

Other goods 23.3%

Meat 14.2%

Soap and toiletries 1.9%

Clothing 14.4%

Tea 2.1%

Sugar 6.2%

Bread 10.6%

UMTALI AND GWELO 1958–1959

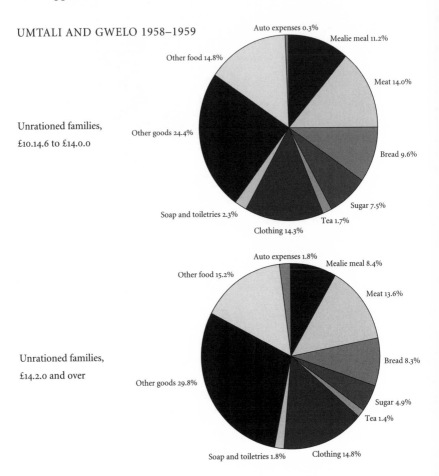

Unrationed families,
£10.14.6 to £14.0.0

Auto expenses 0.3%
Mealie meal 11.2%
Meat 14.0%
Bread 9.6%
Sugar 7.5%
Tea 1.7%
Clothing 14.3%
Soap and toiletries 2.3%
Other goods 24.4%
Other food 14.8%

Unrationed families,
£14.2.0 and over

Auto expenses 1.8%
Mealie meal 8.4%
Meat 13.6%
Bread 8.3%
Sugar 4.9%
Tea 1.4%
Clothing 14.8%
Soap and toiletries 1.8%
Other goods 29.8%
Other food 15.2%

SALISBURY 1963

Unrationed families,
average income £11.15.10

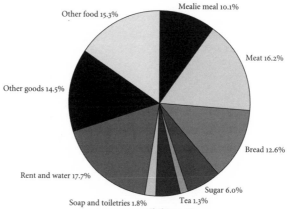

Mealie meal 10.1%
Other food 15.3%
Meat 16.2%
Other goods 14.5%
Bread 12.6%
Rent and water 17.7%
Sugar 6.0%
Soap and toiletries 1.8%   Tea 1.3%
Clothing 4.6%

Unrationed families,
average income £15.7.0

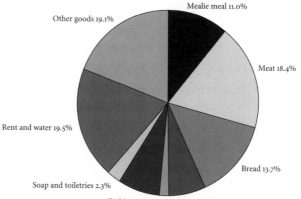

Mealie meal 11.0%
Other goods 19.1%
Meat 18.4%
Rent and water 19.5%
Bread 13.7%
Soap and toiletries 2.3%
Clothing 7.6%   Sugar 6.7%
Tea 1.7%

Unrationed families,
average income £18.8.7

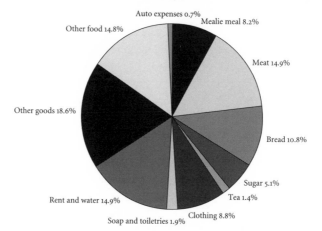

Auto expenses 0.7%
Other food 14.8%   Mealie meal 8.2%
Meat 14.9%
Other goods 18.6%
Bread 10.8%
Sugar 5.1%
Rent and water 14.9%   Tea 1.4%
Soap and toiletries 1.9%   Clothing 8.8%

SALISBURY 1963

Unrationed families,
average income £21.18.8

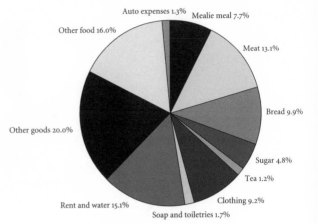

Unrationed families,
average income £28.9.6

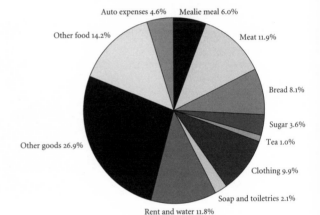

UMTALI 1965

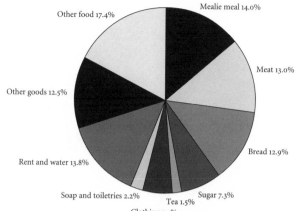

Other food 17.4%
Mealie meal 14.0%
Meat 13.0%
Other goods 12.5%

Unrationed families,
average income £9.16.3

Bread 12.9%
Rent and water 13.8%
Soap and toiletries 2.2%   Tea 1.5%   Sugar 7.3%
Clothing 5.4%

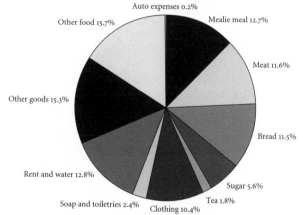

Auto expenses 0.2%
Other food 15.7%
Mealie meal 12.7%
Meat 11.6%
Other goods 15.3%

Unrationed families,
average income £12.7.0

Bread 11.5%
Rent and water 12.8%
Sugar 5.6%
Soap and toiletries 2.4%   Clothing 10.4%   Tea 1.8%

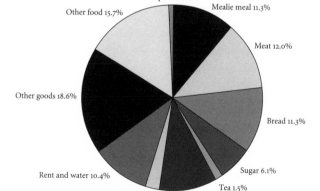

Auto expenses 0.6%
Other food 15.7%
Mealie meal 11.3%
Meat 12.0%
Other goods 18.6%

Unrationed families,
average income £14.14.11

Bread 11.3%
Rent and water 10.4%
Sugar 6.1%
Tea 1.5%
Soap and toiletries 2.3%   Clothing 10.3%

UMTALI 1965

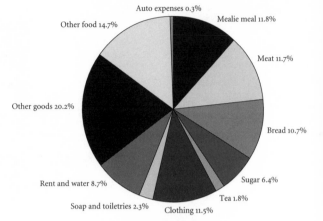

Auto expenses 0.3%
Mealie meal 11.8%
Other food 14.7%
Meat 11.7%
Unrationed families,
average income £17.19.11
Other goods 20.2%
Bread 10.7%
Rent and water 8.7%
Sugar 6.4%
Tea 1.8%
Soap and toiletries 2.3%   Clothing 11.5%

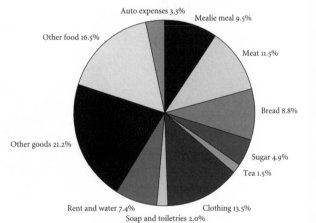

Auto expenses 3.3%
Mealie meal 9.5%
Other food 16.5%
Meat 11.5%
Unrationed families,
average income £22.10.2
Other goods 21.2%
Bread 8.8%
Sugar 4.9%
Tea 1.5%
Rent and water 7.4%   Clothing 13.5%
Soap and toiletries 2.0%

SALISBURY 1969

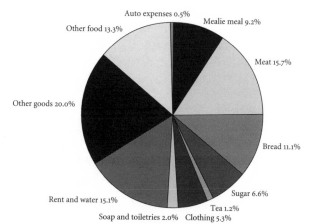

Unrationed families,
average income $26.94

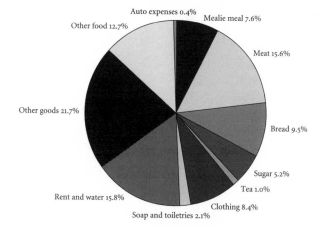

Unrationed families,
average income $38.15

Unrationed families,
average income $46.64

SALISBURY 1969

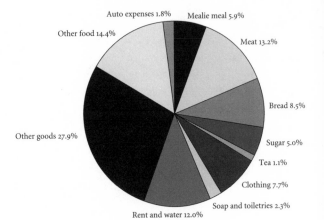

Unrationed families,
average income $58.82

Unrationed families,
average income $99.97

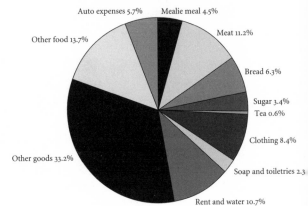

NOTES

Introduction

1 Dambudzo Marechera, "Are There People Living There?" *The House of Hunger* (London: Heinemann, 1978), p. 149.

2 Ibid., p. 151.

3 See, for example, David Eltis, "The Relative Importance of Slaves and Commodities in the Atlantic Trade of Seventeenth-Century Africa," *The Journal of African History* 35:2 (1994); Walter Rodney, *A History of the Upper Guinea Coast* (New York: Monthly Review Press, 1970).

4 Ian Phimister, *An Economic and Social History of Zimbabwe, 1890–1948: Capital Accumulation and Class Struggle* (London: Longman, 1988), p. 69.

5 Giovanni Arrighi, "Labour Supplies in Historical Perspective: A Study of the Proletarianization of the African Peasantry in Rhodesia," in *Essays on the Political Economy of Africa,* ed. Giovanni Arrighi and John Saul (New York: Monthly Review Press, 1973), p. 199.

6 See, for example, Steven Langdon, "Multinational Corporations, Taste Transfer and Underdevelopment: A Case Study from Kenya," *Review of African Political Economy* 2 (1975); Charles Medawar, *Insult or Injury? An Enquiry Into the Marketing and Advertising of British Food and Drug Products in the Third World* (London: Social Audit Ltd., 1979).

7 Judith Williamson, *Decoding Advertisements: Ideology and Meaning in Advertising* (London: Marion Boyars, 1978), p. 99.

8 Arjun Appadurai, "Commodities and the Politics of Value," *The Social Life of Things* (Cambridge: Cambridge University Press, 1986), p. 41.

9 Jan Vansina, *Living with Africa* (Madison: University of Wisconsin Press, 1994). Vansina makes reference to my book in a footnote in which he suggests that a "social history of soap" is emblematic of a turn in Africanist scholarship away from issues of immediate importance to contemporary African audiences. I would not presume to simplistically judge the relevance of my work to hypothetical and generic "Africans," but I hope to make clear throughout the book the

immediacy and importance—for Africans and others—of the history discussed in this book.

10   See Marx's "Preface to the First German Edition" of *Capital*.

11   Nikolas Rose, "Fetishism and Ideology: A Review of Theoretical Problems," *Ideology and Consciousness* 2 (1977), p. 27. Also see John P. Burke, "Reification and Commodity Fetishism Revisited," *Canadian Journal of Political and Social Theory* 3:1 (Winter 1979).

12   Though I was tempted at several points in this work to address the subject of gift exchange in Zimbabwe, I decided not to muddy opaque waters even further, given the already broad and overly ambitious scope of this study.

13   Mary Douglas and Baron Isherwood, *The World of Goods* (New York: Basic Books, 1979), p. 12.

14   Jean Comaroff and John Comaroff, "Goodly Beasts, Beastly Goods: Cattle and Commodities in a South African Context," *American Ethnologist* 17:2 (May 1990), p. 211.

15   Igor Kopytoff, "The Cultural Biography of Things: Commoditization as Process," in *The Social Life of Things*, ed. Arjun Appadurai (London: Cambridge University Press, 1986).

16   Comaroff and Comaroff, "Goodly Beasts," p. 211.

17   Wolfgang Fritz Haug, *Critique of Commodity Aesthetics: Appearance, Sexuality, and Advertising in Capitalist Society* (Minneapolis: University of Minnesota Press, 1986).

18   See, for example, Raymond Williams, "Advertising: The Magic System," *Problems in Materialism and Culture* (London: Verso, 1980); Stuart Ewen, *Captains of Consciousness* (New York: McGraw-Hill Book Company, 1976); Sut Jhally, *The Codes of Advertising: Fetishism and the Political Economy of Meaning in the Consumer Society* (New York: St. Martin's Press, 1987); Sut Jhally, Stephen Kline, and William Leiss, *Social Communication in Advertising: Persons, Products and Images of Well-Being* (New York: Methuen, 1986).

19   Jean Baudrillard, "For a Critique of the Political Economy of the Sign," in *Selected Writings of Jean Baudrillard*, ed. Mark Poster (Stanford: Stanford University Press, 1988).

20   For three excellent explorations of this issue, see Edmond Preteceille and Jean-Pierre Terrail, *Capitalism, Consumption and Needs* (Oxford: Basil Blackwell Publishers, 1985); William Leiss, *The Limits of Satisfaction: An Essay on the Problem of Needs and Commodities* (Toronto: The University of Toronto Press, 1976); Alexis Kontos, "Through a Glass Darkly: Ontology and False Needs," *Canadian Journal of Political and Social Theory* 3:1 (Winter 1979).

21   Rose, "Fetishism and Ideology," p. 47.

22   James Ferguson and Akhil Gupta, "Beyond 'Culture': Space, Identity and the Politics of Difference," *Cultural Anthropology* 7:1 (February 1992), p. 19.

23   Bill Livant, "The Imperial Cannibal," in *Cultural Politics in Contemporary America*, ed. Ian Argus and Sut Jhally (New York: Routledge, 1989), p. 34.

24 Megan Vaughan, *Curing Their Ills: Colonial Power and African Illness* (Stanford: Stanford University Press, 1991).

25 Jean Baudrillard's later work is the most characteristic expression of this kind of fallacious approach. For a perceptive critique of Baudrillard, see Preteceille and Terrail, *Capitalism, Consumption and Needs*.

26 Haug, *Critique of Commodity Aesthetics*, p. 16.

27 E. S. Atieno Odhiambo and David William Cohen, *Siaya* (London: James Currey, 1989).

28 Richard Wilk, "Consumer Goods as Dialogue about Development," *Culture and History* 7 (1992):83.

29 Marshall Berman, *All That Is Solid Melts into Air: The Experience of Modernity* (New York: Penguin Books, 1988), p. 83.

30 See Vaughan, *Curing Their Ills*, introduction.

31 Haug, *Critique of Commodity Aesthetics*, p. 83.

32 Judith Butler, *Gender Trouble: Feminism and the Subversion of Identity* (New York: Routledge, 1990).

33 I am indebted to Professor JoAnne Brown of Johns Hopkins University for forcefully pointing this out to me.

34 David William Cohen, "The Undefining of Oral Tradition," *Ethnohistory* 36:1 (Winter 1989).

1 Cleanliness and "Civilization": Hygiene and Colonialism in Southern Africa

1 Philip Molefe and Phillip Van Niekerk, "Adventures of a Black Man in a White Town," *Weekly Mail*, 6:40, Oct. 19–Oct. 25, 1990, p. 1.

2 Diana Jeater, *Marriage, Perversion and Power: The Construction of Moral Discourse in Southern Rhodesia 1894–1930* (Oxford: Clarendon Press, 1993).

3 See, for further details, Frank Bottomley, *Attitudes to the Body in Western Christendom* (London: Lepus Books, 1979); Lawrence Wright, *Clean and Decent: The Fascinating History of the Bathroom and the Water Closet* (London: Routledge & Kegan Paul, 1960), esp. ch. 3. Also see Michel Feher, ed., *Fragments for a History of the Human Body*, 3 vols. (New York: Zone Books), and George Vigarello, *Concepts of Cleanliness: Changing Attitudes in France Since the Middle Ages*, trans. Jean Birrell (Cambridge: Cambridge University Press, 1988).

4 A number of recent texts support this claim. See, for example, Vigarello, *Concepts of Cleanliness*; Margaret Pelling, "Appearance and Reality: Barber-Surgeons, the Body and Disease," *London 1500–1700: The Making of the Metropolis* (London: Longman, 1986); Alain Corbin, *The Foul and the Fragrant: Odor and the French Social Imagination*, trans. Aubier Montaigne (Cambridge: Cambridge University Press, 1986).

5 For example, Florence Nightingale.

6   Frank Mort, *Dangerous Sexualities: Medico-Moral Politics in England Since 1830* (London: Routledge and Kegan Paul, 1987).

For further details from modern English social history, see Greta Jones, *Social Hygiene in Twentieth Century Britain* (London: Croom Helm, 1986); Judith Walkowitz, *Prostitution and Victorian Society: Women, Class, and the State* (Cambridge: Cambridge University Press, 1980); Gareth Stedman Jones, *Outcast London: A Study of the Relationship Between Classes in Victorian Society* (Oxford: Clarendon Press, 1971); Edward J. Bristow, *Vice and Vigilance: Purity Movements in Britain Since 1700* (London: Gill & Macmillan, 1977); Carol Dyhouse, "Working-Class Mothers and Infant Mortality in England, 1895–1914," *Journal of Social History* 12:2 (1978); Catherine Hall, "The Early Formation of Victorian Domestic Ideology," in *The Victorian Family: Structures and Stresses*, ed. Sandra Burman (London: Croom Helm, 1978); Jane Lewis, *The Politics of Motherhood: Maternal and Child Welfare in England 1900–39* (London: Croom Helm, 1980); Jeanne Peterson, *The Medical Profession in Mid-Victorian England* (Berkeley: University of California Press, 1978).

Literature specifically on cleanliness and hygiene in the United States includes Marilyn Thornton Williams, *Washing 'The Great Unwashed': Public Baths in Urban America, 1840–1920* (Columbus: Ohio University Press, 1991); Susan E. Cayleff, *Wash and Be Healed: The Water Cure Movement and Women's Health* (Philadelphia: Temple University Press, 1987); Suellen M. Hoy, "Municipal Housekeeping: The Role of Women in Improving Urban Sanitation Practices, 1800–1917," in *Pollution and Reform in American Cities*, ed. Martin Melosi (Austin: University of Texas Press, 1980); Jacqueline S. Wilkie, "Submerged Sensuality: Technology and Perception of Bathing," *Journal of Social History* (Summer 1986), pp. 649–64.

7   Anna Davin, "Imperialism and Motherhood," *History Workshop* (Spring 1978), p. 38.

8   See, for example, Nancy Stepan, *The Idea of Race in Science: Great Britain 1800–1960* (Hamden, Conn.: Archon Books, 1982), and George W. Stocking Jr., *Victorian Anthropology* (New York: The Free Press, 1987), for some detailed descriptions of these processes.

9   See Stepan, *Idea of Race*, ch. 4 and 5. Also see G. Searle, *Eugenics and Politics in Britain 1900–1914* (Leyden: Nordhoff International, 1976).

10 Sander Gilman, *Difference and Pathology: Stereotypes of Sexuality, Race and Madness* (Ithaca: Cornell University Press, 1985); and *The Jew's Body* (New York: Routledge, 1991). Also see Catherine Burns, "Reproductive Labors: The Politics of Women's Health in South Africa, 1900 to 1960" (Ph.D. diss., Northwestern University, 1994), and Jan Nederveen Pieterse, *White on Black: Images of Africa and Blacks in Western Popular Culture* (New Haven: Yale University Press, 1992).

11 Philip Setel, "'A Good Moral Tone': Victorian Ideals of Health and the Judgement of Persons in Nineteenth-Century Travel and Mission Accounts from East Africa,"

Working Papers in African Studies, no. 150 (Boston: Boston University African Studies Center, 1991), p. 13. For other examples, see Valerie Pakenham, *The Noonday Sun: Edwardians in the Tropics* (London: Methuen, 1985), p. 99. Also see Kenneth Ballhatchet, *Race, Sex and Class Under the Raj: Imperial Attitudes and Policies and Their Critics, 1793–1905* (London: Weidenfeld and Nicholson, 1980).

12   J. P. R. Wallis, ed., *The Matabele Journals of Robert Moffat*, 2 vols., Oppenheimer Series (London: Chatto & Windus, 1945), from Moffat's Third Journey to Matabeleland, May–November 1854, vol. 1, p. 193.

13   Thomas Morgan Thomas, *Eleven Years in Central South Africa* (London: John Snow and Co., 1872), p. 171.

14   Crook Papers, Pitt Rivers Museum Archives, University of Oxford (1878).

15   Edward Tabler, ed., *To The Victoria Falls Via Matabeleland: The Diary of Major Henry Stabb, 1875* (Cape Town: C. Struik Ltd, 1967).

16   See Jean Comaroff, "The Diseased Heart of Africa: Medicine, Colonialism and the Black Body," in Shirley Lindenbaum and Margaret Lock, eds., *Knowledge, Power, and Practice: The Anthropology of Medicine and Everyday Life* (Berkeley: University of California Press, 1993).

17   Constance Fripp, ed., *Gold and The Gospel in Mashonaland* (London: Chatto & Windus, 1949), p. 54.

18   ZNA N 9/1/1–26, Native Department Annual Reports, Gutu District, 1909. (Native commissioners were asked by the chief native commissioner to explicitly comment on African hygiene in 1909 and again periodically every few years.)

19   ZNA N 9/1/1–26, Native Department Annual Reports, Hartley District, 1909.

20   ZNA N 9/5/8, Native Department, Chief Native Commissioner's Office, Correspondence re: School Inspections, 1922–1923. See Carol Summers, *From Civilization to Segregation: Social Ideals and Social Control in Southern Rhodesia* (Athens: Ohio University Press, 1994) for some especially useful discussions of the term "civilization" in Rhodesian discourse.

21   See ZNA N 9/1/1–26 and S 2076, Native Department Annual Reports, esp. Ndanga District, 1899, for discussion of the infectious nature of hut taxes. Also see ZNA N 4/1/1, 23 July 1903, Chief Native Commissioner Circular, which was accompanied by packets of permanganate of potash "for the purpose of disinfecting all hut tax money collected." Also see Vaughan, *Curing Their Ills*, ch. 6.

22   Nathan M. Shamuyarira, *Crisis in Rhodesia* (London: Andre Deutsch Ltd., 1965), p. 119.

23   C. Frantz and Cyril Rogers, *Racial Themes in Southern Rhodesia: The Attitudes and Behavior of the White Population* (New Haven: Yale University Press, 1962), pp. 235 and 370.

24   "The Truth About the Colour Bar," *The Citizen* (Salisbury), February 24, 1956.

25   Dudley Kidd, *The Essential Kafir* (London: Adam and Charles Black, 1904), p. 34.

26   Jeannie M. Boggie, *First Steps in Civilising Rhodesia*, 4th ed. (Salisbury: Kingstons Ltd, 1966), p. 81.

27   Wilfred Robertson, *Rhodesian Rancher* (London: Blackie & Son Ltd., 1935), p. 36.
28   David Caute, *Under the Skin: The Death of White Rhodesia* (Evanston, Ill.: Northwestern University Press, 1983), p. 108.
29   Jacklyn Cock, *Maids and Madams: Domestic Workers under Apartheid* (London: Women's Press, 1989).
30   Boggie, *A Husband and a Farm in Rhodesia* (n.p.: Salisbury, 1939), p. 106.
31   Boggie, *First Steps*, p. 82.
32   Frantz and Rogers, *Racial Themes*, p. 370.
33   See John Pape, "Black and White: The 'Perils of Sex' in Colonial Zimbabwe," *Journal of Southern African Studies* 16:4 (December 1990); also see Jeater, *Marriage, Perversion and Power*.
34   Margaret Wood, *Pastels under the Southern Cross* (London: Smith & Elder, 1911), p. 171.
35   Helen Caddick, *A White Woman in Central Africa* (London: T. Fisher Unwin, 1900), p. 83.
36   Felix Bryk, *Dark Rapture: The Sex Life of the African Negro* (New York: Walden Publications, 1939), pp. 50–51.
37   Wood, *Pastels*, pp. 125–126.
38   ZNA CA 1/1/1, diary of Algernon Capell, entry for April 1905.
39   Interview, Mr. Muzende, Murewa, May 7, 1991.
40   Representatives of Lever Brothers traveling in Western and Equatorial Africa obviously had a particular interest in local techniques for soap making. In 1921, one representative described the local manufacture of soap in Nigeria:

> Cassava peelings and a Weed, called by the Hausa "Kaimin Kadangere," are sun-dried and then burnt in a wood fire. The ash of peelings, weeds and wood is placed in a calabash, pierced with small holes like a colander, and water poured through. The water "filters" through. The water is then boiled and in about two hours it begins to turn a dull red colour. It is then called "Ruwan Toka."
>
> Aliede (palm kernel oil) is then added in the proportion of six measures of Aliede to ten of the Ruwan Toka. The pot is boiling during this process and is kept boiling for about five hours. At the end of that time it thickens and rises. Another measure of Ruwan Toka is then added in order to thin it slightly. The mixture, still boiling, is then stirred with a stick. The "soap" then begins to appear and is skimmed and placed in another calabash. (H. Lever to W. H. Lever, 1921, Unilever Archives, Blackfriars, London)

41   The Rhodesian ethnographer Michael Gelfand described the making of *mvuka*: "It was made from a plant called *chitura* [this seems to have been either an inaccurate understanding on Gelfand's part or a term particular to a local dialect, as none of the people I talked with had heard of the word] which was dried and then stamped into a round ball convenient for washing" (Michael Gelfand, *Diet*

and Tradition in African Culture [London: E & S Livingstone, 1971], p. 175). Similar balls of pounded soaplike materials appear in a number of nineteenth-century travel narratives of south-central and southern Africa. For example, Emil Holub, traveling in present-day Zambia in 1875, described "little balls of queer-looking soap" (cited in Boggie, *First Steps*, p. 230).

42   I am extremely leery of any claims that see "Africa" as a whole or unitary cultural entity, but I have noted materially similar practices described in scholarship concerning other parts of the continent. Meanings invested in such bodily practices, however, have been clearly local and specific. For two intriguing examples, see Gillian Feeley-Harnik, *A Green Estate: Restoring Independence in Madagascar* (Washington, D.C.: Smithsonian Institution Press, 1991). Feeley-Harnik's work is immensely rich in its rendering of local discourses concerning purity, washing, and the body. Also see Pamela Feldman-Savelsberg, " 'Then We Were Many': Bangangté Women's Conceptions of Health, Fertility and Social Change in a Bamiléké Chiefdom, Cameroon" (Ph.D. diss., Johns Hopkins University, 1989), especially ch. 3.

43   Kidd, *Essential Kafir*, pp. 30–31.

44   Crook Papers, Pitt Rivers Museum Archives, University of Oxford, p. 51.

45   Edwin Smith and Andrew Dale, *The Ila-Speaking Peoples of Northern Rhodesia* (New York: University Books, 1968), vol. 1, p. 92. A similar remark appears in A. Werner, *The Natives of British Central Africa* (New York: Negro Universities Press, 1906, rpt. 1969): "Both children and grown-ups require plenty of oil to keep their skins from cracking and chapping; they neither look nor feel well without it" (p. 102).

46   Philip and Iona Mayer, *Townsmen or Tribesmen: Conservatism and the Process of Urbanization in a South African City* (Cape Town: Oxford University Press, 1961), p. 4.

47   Not all colonial ethnographies concerning Zimbabwe agree on this derivation, but I think it is basically sound, especially in light of the regionally pervasive, long-term tendency to use vocabularies of "dirtiness" and "cleanliness" to describe antagonists. Nathan Shamuyarira described *chiSwina* as meaning more specifically "those people who are supposed to have remained cleaning and eating the intestines of an animal when other tribes were moving on" (Shamuyarira, *Crisis*, p. 118).

48   Tseuneo Yoshikune, "Black Migrants in a White City: A Social History of African Harare, 1890–1925" (Ph.D. diss., University of Zimbabwe, 1989).

49   Such claims came up both in my own casual conversations with a number of acquaintances and in some formal interviews. For example, interviews 1–3, Tashinga squatter camp, Mbare Township, Harare, April 19, 1991.

50   ZNA AOH/58, interview, Mbangwa Ngomambi, 1979.

51   Michael Gelfand and Yvonne Swart, "The Nyora," *NADA* 30 (1953), p. 6.

52   " . . . the Amakalanga and Amaswina ladies dress very differently, and so far from

shaving their heads, usually have long hair, often covered with a quantity of grease and a *bright black substance* [my emphasis], and ornamented with strings of all sorts of beads" (Thomas, *Eleven Years*, p. 172).

53   Caddick, *White Woman*, p. 8.

54   Rose Blennerhassett and Lucy Sleeman, *Adventures in Mashonaland, by Two Hospital Nurses* (London: Macmillan and Co., 1893), p. 251.

55   See Barrie Reynolds, *The Material Culture of the Peoples of the Gwembe Valley* (Manchester: Manchester University Press, 1968), ch. 5.

56   On these points, see for example the accounts of the toilette of an elite Ndebele woman in the Crook Papers, Pitt Rivers Museum Archives, pp. 51–53.

57   Lawrence Vambe, *An Ill-Fated People: Zimbabwe Before and After Rhodes* (Pittsburgh: University of Pittsburgh Press, 1972), p. 179.

58   See, for example, Crook's accounts of his journey north from Bulawayo into Mashonaland: Crook Papers, Pitt Rivers Museum Archives.

59   "Some Notes on the Mrowzi Occupation of the Sebungwe District," *NADA* 21 (1944), p. 30.

60   M. F. C. Bourdillon, *The Shona Peoples: An Ethnography of the Contemporary Shona, with Special Reference to Their Religion* (Gweru: Mambo Press, 1976), pp. 278–79.

61   Guy Taylor, "The Matabele Head Ring," *NADA* 3 (1925), p. 41.

62   Crook Papers, Pitt Rivers Museum, p. 43.

63   Ibid., pp. 106–107.

64   See, for example, photographs in William Edwards, "From Birth to Death," *NADA* 7 (1929).

65   Tsitsi Dangarembga, *Nervous Conditions* (London: Women's Press Ltd., 1988), p. 3.

66   See Gelfand, *Diet and Tradition*, pp. 175–76, for additional descriptions of these practices.

67   Third interview, Tashinga squatter camp, April 19, 1991.

68   Ibid. Some of the most typical discussions occurred at the Tashinga squatter camp, but the issue of *kufadza murume* came up in many other interviews in Mbare during the months of April and May. We generally but not invariably refrained from pursuing this topic when speaking to multigenerational groups, but among members of their own age-set, most people we spoke with were relaxed and amused in discussing this question. In fact, considering *kufadza murume* usually provoked extremely animated debate, drew otherwise silent members of the group into the discussion, and elicited a high degree of (relatively nonacrimonious) passion among the discussants. Other affiliations and ideologies, many of them of recent vintage, like Christianity, or even specifically in the case of the Tashinga camp, the Rhema Church, also clearly played a role in shaping this debate.

On the issue of *rudo* in marriage, see A. K. H. Weinrich, *Women and Racial Discrimination in Rhodesia*, (Paris: UNESCO, 1979).

69   This practice is described in A. Francis Davidson, *South and South Central Africa:*

A Record of Fifteen Years' Missionary Labours Among Primitive Peoples (Elgin, Ill.: Brethren Publishing House, 1915): "The wash basin is the mouth. The mouth is filled with water, which is allowed to run in a thin stream on the hands until they are washed, and then the hands are filled in the same way to wash the face. I was greatly interested once in the operation of bathing twins. She [the mother] spread a blanket on a large rock in the sun. Then she took a gourd of water and filled her mouth. . . . She kept the water in her mouth a short time to take off the chill, then picked up one child, held it out and, with a thin stream of water pouring from her mouth, washed the entire body of the child thoroughly" (p. 197).

70 Discussed in Edwards, "From Birth to Death"; Michael Gelfand, *Growing Up in Shona Society* (Gweru: Mambo Press, 1979), pp. 3–4.

71 Gelfand, *Growing Up*, p. 10. Gelfand wrote this work in the 1970s, but he had been carefully collecting ethnographic material for forty years, and fragmentary nineteenth-century accounts suggest that these teachings were, in a similar form, part of childhood at that time as well. The instruction and socialization of the young were carried out in the nineteenth century by a combination of immediate parents, grandparents, and elder relatives (see below). Gelfand also describes a "cleanliness game" whose historicity is less certain:

> The instructor walks round a circle of seated children, singing "Spot, Spot" and they reply *"Zakariende"* (Keep on going round). He sings in reply *"Ndinotsvaga wangu"* (I am looking for someone for myself). The children answer *"Zakariende"* and the instructor still going round the circle sings *"Musuiki wendiro"* (one who cleans plates). Again the children sing *"Zakari-ende."* Still walking round the circle the instructor sings *"Anondzisuka"* (I am looking for one who cleans) and the children reply *"Zakariende."* *"Dzikati mbeke"* (bright clean) sings the instructor. *"Zakariende"* reply the children. *"Semwedzi wapana"* (like a new moon) continues the instructor. *"Zakariende,"* sing the children and finally, putting his hand on the head of one of the children, the instructor sings, *"Simuka hande"* (rise and follow me). The child he has chosen rises and follows him round the circle. The game is repeated until only one child is left. The others jeer at this one, whom they pretend is the dirty one (Gelfand, *Growing Up*, p. 158).

A similar song is recorded in Colin Style and O-lan Style, eds., *Mambo Book of Zimbabwean Verse in English* (Gweru: Mambo Press, 1986), p. 33.

72 Gelfand, *Diet and Tradition*, p. 175.

73 See, for example, the discussion of washing and smearing in women's initiation in Anita Spring, "Women's Rituals and Natality Among the Luvale of Zambia" (Ph.D. diss., Cornell University, September 1976), p. 126.

74 See ZNA FR 2/1/1, Ivan Fry, Reminiscences, pp. 86–87.

75 Many of those we spoke with insisted that this practice had been highly regional and had died out *kare kare*.

76 See, for example, Paul Chidyausiku, *Broken Roots: A Biographical Narrative on the Culture of the Shona People in Zimbabwe* (Gweru: Mambo Press, 1984), pp. 109–110.

77 Gelfand and Smart, "Nyora," pp. 8–9.

78 See Mary Douglas, *Purity and Danger: An Analysis of Concepts of Pollution and Taboo* (London: Ark Paperbacks, 1984); W. D. Hammond-Tooke, *Patrolling the Herms: Social Structure, Cosmology and Pollution Concepts in Southern Africa,* Raymond Dart Lecture no. 18, Institute for the Study of Man in Africa (Johannesburg: Witwatersrand University Press, 1981). Also see Luise White, "Blood Brotherhood Revisited: Kinship, Relationship, and the Body in East and Central Africa," *Africa* 64:3 (1994).

79 Knight-Bruce, *Memories of Mashonaland* (London: Edward Arnold, 1895), p. 28. Some of the reasoning behind the particular valorization of "Nguni" peoples can be found in John Wright, "Politics, Ideology, and the Invention of the 'Nguni,'" in *Resistance and Ideology in Settler Societies,* ed. Tom Lodge, vol. 4, Southern African Studies Series (Johannesburg: Ravan Press, 1986).

80 ZNA A 3/12/13–18, Administrator's Office, Papers on Health and Medicine, 1910–1923.

81 Stanley Portal Hyatt, *The Old Transport Road* (Bulawayo: Books of Rhodesia, 1914, rpt. 1969), p. 91.

82 As just one example of the omnipresence of such stories, naturalist and humorist Gerald Durrell's account of animal collecting in colonial Cameroons includes this anecdote: "The entire village returned that evening, silently and cautiously, and watched me having my bath. . . . There must have been fifty people watching me as I covered myself with soap and sang lustily. . . . It did not worry me, for I am not unduly modest, and as long as my audience (half of which consisted of women) were silent and made no ribald remarks I was content that they should watch . . . the smaller members of the village . . . had never seen a European . . . [they] wanted to settle various bets among each other as to whether or not I was white all over" (Gerald Durrell, *The Bafut Beagles* [Hammondsworth: Penguin Books, 1954, rpt. 1988], p. 167).

83 Sam Kemp, *Black Frontiers: Pioneer Adventures with Cecil Rhodes' Mounted Police in Africa* (London: George G. Harrup and Co., 1932), pp. 92–93.

84 Crook Papers, Pitt Rivers Museum, p. 42.

85 Tabler, *To the Victoria Falls,* p. 76.

86 Maurice Nyagumbo, *With the People: An Autobiography from the Zimbabwe Struggle* (London: Allison & Busby, 1980), p. 56.

87 H. M. Davidson Moyo, "Letter to the Editor," *Native Mirror* 2:5 (1934), p. 10.

88 Peter Ndzou, "Letters to the Editors: Digests," *Native Mirror* 2:10 (1935), p. 4.

89 M. Wilbert Sigauke, "Do You Shake Hands?" *Bantu Mirror,* September 2, 1939, p. 8.

90 "Chief Mkoba and Headman Madigane (representing Chief Sogwala) . . . very strongly objected to the granting of this application . . . they said the Church of England permitted their converts to eat unclean meat, and drink kaffir beer"

(ZNA S 1542/S2, Chief Native Commissioner, Department of Native Affairs, Correspondence and Papers re: Schools, 1933–1935).

91   J. Bruce-Bays, "The Injurious Effects of Civilization Upon the Physical Condition of the Native Races of South Africa," in *Report of the South African Association for the Advancement of Science*, Sixth Meeting, Grahamstown 1908 (Capetown: The South African Association for the Advancement of Science, 1909), pp. 264 and 266.

92   George McCall Theal, *The Yellow and Dark-Skinned People of Africa South of the Zambesi* (New York: Negro Universities Press, 1910, rpt. 1963), p. 345.

93   "Let's Ask Questions: Good Old Habits of the Bantu," *Bantu Mirror*, April 1, 1939, p. 7.

2   *Education, Domesticity, and Bodily Discipline*

1   Neville Jones, "Community Work of Home Demonstrators During Training," in *Village Education in Africa: Report of the Inter-Territorial Jeanes Conference, Salisbury* (South Africa: Lovedale Press, 1935), p. 79.

2   See, for example, Michael Worboys, "Manson, Ross and Colonial Medical Policy: Tropical Medicine in London and Liverpool, 1899–1914," in *Disease, Medicine and Empire: Perspectives on Western Medicine and the Experience of European Expansion*, ed. Roy MacLeod and Milton Lewis (London: Routledge, 1988), pp. 21–37, esp. comments about the background and interests of Ronald Ross, pp. 22–23, and on the making of colonial policy, pp. 28–33.

3   Donald Denoon, "Temperate Medicine and Settler Capitalism: On the Reception of Western Medical Ideas," in *Disease, Medicine and Empire: Perspectives on Western Medicine and the Experience of European Expansion*, ed. Roy MacLeod and Milton Lewis (London: Routledge, 1988), p. 125. Also see for some examples, Andrew Balfour, *War Against Tropical Disease: Being Seven Sanitary Sermons Addressed to All Interested in Tropical Hygiene and Administration* (London: Wellcome Bureau of Scientific Research, 1920), esp. p. 20.

4   Maynard Swanson, "The Sanitation Syndrome: Bubonic Plague and Urban Native Policy in the Cape Colony, 1900–1909," *Journal of African History* 18:3 (1977), pp. 387–410. Also see Shula Marks and Neil Andersson, "Typhus and Social Control: South Africa, 1917–1950," in *Disease, Medicine and Empire: Perspectives on Western Medicine and the Experience of European Expansion*, ed. Roy MacLeod and Milton Lewis (London: Routledge, 1988), pp. 257–83.

5   One discussion of this can be found in Michael Gelfand, *Tropical Victory: An Account of the Influence of Medicine on Southern Rhodesia, 1890–1923* (Cape Town: Juta & Co. Ltd, 1953), pp. 56–60. Also see his introduction to an archival collection of Rhodesian nursing correspondence, *Mother Patrick and Her Nursing Sisters* (Cape Town: Juta & Co. Ltd., 1964), esp. p. 74.

6   Yoshikune, "Black Migrants."

7   ZNA H 2/9/2, Salisbury Town Clerk's Notes, April 1917.

8   Swanson, "Sanitation Syndrome," p. 408.

9   Hugh Marshall Hole, *Old Rhodesian Days* (London: Frank Cass and Company, 1928), p. 45.

10 Southern Rhodesia Native Affairs Commission, *Report of the Native Affairs Commission, 1910–1911*, p. 18.

11 ZNA E 2/5/6, Correspondence and Curricula, Department of Education, 1890–1923.

12 Ibid. Also see later materials on the visit of the Phelps-Stokes Commission to the Rhodesias, ZNA S 840/2/37, Correspondence, Department of Native Education, 1923–1926.

13 One missionary's typical comment on these objectives: "As the natural inclination of the native is toward laziness and filthiness in personal habits, we were opening the door and taking all who desired to come and giving them a home, our aim being to take them out of their degraded home surroundings." (Davidson, *Fifteen Years' Missionary Labours*, p. 128.)

14 "Report of the Southern Rhodesian Missionary Conference," *Native Mirror* 7, July 1932, p. 15.

15 For a listing of textbooks in use before 1930, consult D. G. Brackett and M. Wrong, "Notes on Hygiene Books Used in Africa," *Africa* 3:4 (October 1930). Also see the supplement to this list by the same authors in the January 1932 issue of *Africa*.

16 See, for one typical account of an inspection, Department of Native Education, *Report of the Director of Native Education*, 1929, p. 56.

17 ZNA N 9/5/8, Department of Native Affairs, Correspondence, 1922–1923.

18 "Catechism of Health," *Rhodesia Native Quarterly* 1:1 (July–September 1926), pp. 5–6. Another early Rhodesian textbook, *The First Book of Health and Hygiene in Chiswina*, which is no longer available in any of the archives I visited, was published by the Society for Promoting of Christian Knowledge in 1927.

19 G. E. P. Broderick, *Personal Hygiene for Elementary Schools in Africa* (London: United Society for Christian Literature, 1939), p. 25.

20 Bishop Gaul, *Mashonaland Quarterly* 34 (November 1900), p. 3.

21 Broderick, *Personal Hygiene*, p. 27.

22 See Norbert Elias, *The History of Manners* (New York: Pantheon Books, orig. German ed. 1939, English ed. 1978), for a stimulating discussion of manners and the "civilizing process." Also see Nancy Rose Hunt, "Colonial Fairy Tales and the Knife and Fork Doctrine in the Heart of Africa," in *African Encounters with Domesticity*, ed. Karen Tranberg Hansen (New Brunswick, N.J.: Rutgers University Press, 1992).

23 "Catechism," p. 5.

24 Broderick, *Personal Hygiene*, pp. 55–56.

25 "Catechism," p. 6.

26   Morgenster Mission, *Hygiene: Standard VI* (Fort Victoria: Morgenster Mission Press, 1950), p. 25.

27   Broderick, *Personal Hygiene*, p. 31.

28   These tensions were persistent throughout colonial pedagogy. See, for example, Michael Adas, "Scientific Standards and Colonial Education in British India and French Senegal," in *Science, Medicine and Cultural Imperialism*, ed. Teresa Meade and Mark Walker (New York: St. Martins Press, 1991), pp. 4–35.

29   See, for example, ZNA S 840/S/41, Syllabi for Native Training Schools, Department of Native Education, 1923. For a contrasting portrayal of how white pupils were regarded as having inborn inclinations toward cleanliness and the capacity to understand "science," see R. J. Challis, "The European Educational System in Southern Rhodesia, 1890–1939," supplement to *Zambezia* (Salisbury: University of Zimbabwe, 1980), p. 52.

30   A. A. Jacques, "Teaching of Hygiene in Native Primary Schools," *Africa* 3:4 (October 1930), pp. 501–502, and W. Millman, "Health Instruction in African Schools: Suggestions for a Curriculum," *Africa* 3:4 (October 1930).

31   See Jean Comaroff and John Comaroff, *Of Revelation and Revolution: Christianity, Colonialism and Consciousness in South Africa*, vol. 1 (Chicago: University of Chicago Press, 1991), p. 209.

32   A. P. Wingate, *Doctor Kalulu* (London: Longman, Green & Co., 1956), pp. 1–2.

33   Nyagumbo, *With the People*, p. 24.

34   Shamuyarira, *Crisis*, p. 115.

35   Ibid., p. 118.

36   Lawrence Vambe, *From Rhodesia to Zimbabwe* (Pittsburgh: University of Pittsburgh, 1976), p. 81.

37   "Mr J. H. Hofmeyr on Native Education," *Bantu Mirror* January 21, 1938, p. 11.

38   *Mahumbwe* has been discussed by most Shona ethnographers; see Bourdillon, *The Shona Peoples*; Gelfand, *Growing Up*; and Chidyausiku, *Broken Roots*, for some of the best accounts.

39   Elizabeth Schmidt, *Peasants, Traders and Wives: Shona Women in the History of Zimbabwe, 1870–1939* (Portsmouth, N.H.: Heinemann Educational Books, 1992), p. 123.

40   Vambe, *From Rhodesia to Zimbabwe*, pp. 13–14.

41   S. N. Sigidi (Tjolotjo Government School), "Letter to the Editor," *Native Mirror* 9 (1932), p. 11.

42   C. C. Freshman Sithole, "Letter to the Editor," *Native Mirror* 2:3 (1934), p. 10.

43   S. Maposa, "Cleanliness Is the Foundation of Christianity," *Bantu Mirror* February 4, 1938.

44   Southern Rhodesia Department of Native Education, *Report of the Director of Native Education*, 1929, p. 57.

45   See Pape, "Black and White," and Schmidt, *Peasants, Traders and Wives*, ch. 6.

46 See Karen Tranberg Hansen, *Distant Companions: Servants and Employers in Zambia, 1900–1985* (Ithaca: Cornell University Press, 1989).

47 Southern Rhodesia Native Affairs Commission, *Report of the Native Affairs Commission 1910–1911*, p. 6.

48 Ibid., p. 18.

49 Teresa A. Barnes, "The Fight for Control of African Women's Mobility in Colonial Zimbabwe, 1900–1939," *Signs* (Spring 1992), pp. 586–608; Schmidt, *Peasants, Traders and Wives*, esp. ch. 4; Summers, "Native Policy"; also Martin Chanock, "Making Customary Law: Men, Women, and Courts in Colonial Northern Rhodesia," in *African Women and the Law: Historical Perspectives*, ed. Margaret Jean Hay and Marcia Wright (Boston: Boston University Press, 1982), pp. 53–67.

50 See, for example, ZNA A 3/18/30/11.

51 ZNA A/3/3/18, Administrator's Office, Correspondence.

52 See, for example, ZNA N 3/7/2 and ZNA NSA 2/3/1.

53 See Schmidt, *Peasants, Traders and Wives*, ch. 5.

54 ZNA S 840/2/16, Department of Native Education, Correspondence re: Laundry, Monte Cassino Mission, 1920–1927.

55 ZNA S 840/2/4, Department of Native Education, Correspondence re: Laundry, St. Monica's Mission Penhalonga, 1913–1927.

56 M. Waters, "Home Economics and Practical Hygiene," Department of Native Education Occasional Paper No. 1 (Salisbury: Government Printers, May 1929), p. 2.

57 ZNA S 840/1/37, Department of Native Education, Correspondence re: Nengubo Mission, 1924.

58 ZNA ZBJ 1/1/3, testimony of Llellwyen Raath, location superintendent, Gwelo.

59 David Flood, "The Jeanes Movement: An Early Experiment," *NADA* 10:3 (1971), p. 16.

60 ZNA N/9/5/8, Department of Native Affairs, Correspondence Re: School Inspections 1922–1923.

61 Flood, "Jeanes Movement," p. 17.

62 See, for example, ZNA S 1561/S6, Department of Native Affairs, Correspondence re: Jeanes Teachers/Home Demonstrators, Joyce Rudd to J. S. March, 1934.

63 In 1934, there were fourteen Jeanes home demonstrators: Priscilla Moyo (Hope Fountain), Bellina Zulu (Matopos), Daisy Dube (Mt. Selinda), Dorothy Mtisi (Ziki School, Bikita), Mary Mhlanga (Lower Gwelo), Rebecca Ncube (Essexvale), Mary Mutukudzi (Epworth), Elizabeth Mkombochota (Mutambara), Susan Kumalo (Gutu), Aletta Mphisa (Morgenster), Nellie Mhandla (Matopos), J. Mujoma (Nyadiri), Maria Gwebu (Enkeldoorn), Rachel Hlazo (Domboshawa). See ZNA S 1561/S6, Correspondence, Department of Native Affairs.

64 Mary Mutukudzi, "My Work at Epworth and Other Stations," in *Village Education in Africa: Report of the Inter-Territorial Jeanes Conference, Salisbury* (South Africa: Lovedale Press, 1935), p. 98.

65 Elizabeth Mhombochota, "Type of Work Done at Mutambara Mission and Re-serve," in *Village Education in Africa: Report of the Inter-Territorial Jeanes Conference, Salisbury* (South Africa: Lovedale Press, 1935), p. 101.

66 Priscilla Moyo, "The Effects of Jeanes Work on the Outlook of Kraal Women and Their Homes," in *Village Education in Africa: Report of the Inter-Territorial Jeanes Conference, Salisbury* (South Africa: Lovedale Press, 1935), p. 107.

67 Rachel Hlazo, "Best Method," *Village Education in Africa: Report of the Inter-Territorial Jeanes Conference, Salisbury* (South Africa: Lovedale Press, 1935), p. 108.

68 James Rutsate, "Reports by Jeanes Visiting Teachers," in *Village Education in Africa: Report of the Inter-Territorial Jeanes Conference, Salisbury* (South Africa: Lovedale Press, 1935), p. 212.

69 ZNA N 9/1/1–26, Native Department Annual Reports, Murewa District, 1911.

70 ZNA N 3/9/1–2, Native Department, Chief Native Commissioner's Correspondence re: Education, 1920–1923.

71 See Carnegie Corporation, *Village Education,* and ZNA files on teachers, especially S 1561 and S 1542/J1.

72 Rebecca Ncube, "What a Home Demonstrator Does," in *Village Education in Africa: Report of the Inter-Territorial Jeanes Conference, Salisbury* (South Africa: Lovedale Press, 1935), p. 105.

73 Moyo, "Effect of Jeanes Work," p. 107.

74 ZNA S 1563, Department of Native Affairs, Annual Reports, 1934.

75 Elizabeth Mhombochota, "Type of Work," p. 102.

76 Peter Fraenkel, *Wayaleshi* (London: Weidenfield and Nicolson, 1959), p. 47.

77 ZNA ZAY 2/2, 1938 Economic Development Committee.

78 James Kundhlande, "Reports by Jeanes Visiting Teachers," in *Village Education in Africa: Report of the Inter-Territorial Jeanes Conference, Salisbury* (South Africa: Lovedale Press, 1935), p. 208.

79 Mutukudzi, "My Work," pp. 98–99.

80 Kundhlande, "Reports," p. 209.

81 ZNA ZBJ 1/1/3, testimony of Kukula et al., Southern Rhodesia Bantu Congress.

82 See Shamuyarira, *Crisis,* and Ngwabi Bhebe, *Benjamin Burombo: African Politics in Zimbabwe* (Harare: College Press, 1989), esp. pp. 107–108.

83 "To Keep the Blanket Full," *Native Mirror* 2 (1931), p. 13.

84 "Mtshabezi Mission," *Native Mirror* 6 (1932).

85 H. H. Morley Wright, "Wake Up Africans! A Call to the Africans," *Native Mirror* 12 (October 1933).

86 For example, authorities sponsored seven "hygiene lectures" in Bulawayo Location in 1944, and estimated that over twenty-four thousand people attended (ZNA ZBJ 1/2/1, testimony, Superintendent, Bulawayo Location).

87 "The Road to Civilisation," *Bantu Mirror,* November 4, 1944. This comparison was a favorite among British colonials, especially in settler colonies. Like many tropes of colonial literatures, it persists in contemporary texts. For example, the

travel guide *Literary Africa* (New York: William Morrow), published in 1988, faithfully reproduces this metaphor in its introduction (p. ix) while gesturing toward anticolonial sentiments.

88   A good deal of the output of the Southern Rhodesia Information Service consisted of such materials, such as the "Fact Paper" series of pamphlets, which included titles like "Hygiene in the Home" and "The Way to Better Dressing."

89   See Vaughan, *Curing Their Ills,* ch. 8, for more details generally on colonial health education films.

90   ZNA S 932/8/1, Southern Rhodesia Ministry of Information, Correspondence re: Educational Films, 1943–1949.

91   ZNA S 932/69, Southern Rhodesia Ministry of Information, Correspondence and Papers re: Broadcasting and the BBC, 1946–1953.

92   "Chiweshe Reserve News," *Bantu Mirror* October 30, 1937.

93   ZNA S 2113/1–2, Southern Rhodesia Ministry of Information, Papers and Drafts re: "Meet the African," 1962.

94   See chapters 4 and 5.

95   "The Home Teacher," *Bantu Mirror* 25:25 (September 10, 1960), p. 4.

96   Nancy Rose Hunt, "Domesticity and Colonialism in Belgian Africa: Usumbura's *Foyer Social,* 1946–1960," *Signs: A Journal of Women in Culture and Society* 15:3 (Spring 1990), pp. 447–74.

97   See Schmidt, *Peasants, Traders and Wives,* pp. 145–49.

98   See "Wayfaring," *Native Mirror* 2 (1931), p. 25.

99   ZNA FO 5/1/1/35, Federation of Women's Institutes, RBC Broadcast, November 28, 1949.

100 Fwisr members did not always respect these implicit boundaries, and sometimes tried to intervene in other political and social arenas after 1945, especially during the Federation period. For more on women's institutes in settler society, and the social roles played by white women, see Karen Tranberg Hansen, "White Women in a Changing World: Employment, Voluntary Work and Sex in Post–World War II Northern Rhodesia," in *Western Women and Imperialism: Complicity and Resistance,* ed. Nupur Chaudhuri and Margaret Strobel (Bloomington: Indiana University Press, 1992).

101 Elsie K. Preston and Dorothy Stebbing, "Life and Work of Miss Catherine Mabel Langham," *NADA* 10:2 (1970), pp. 18–24.

102 ZNA FO 5/1/1/35, Federation of Women's Institutes, Broadcast in "News & Views," October 11, 1950.

103 Fwisr leaders in the 1960s and 1970s represented Helen Mangwende as a compliant and dedicated subject following their commands; other witnesses describe her as the power behind the making of the Murewa African Women's Club and the prime force behind building a national network of such clubs.

104 *Mai* (mother) and *mbuya* (grandmother) are forms of respectful address for adult Shona-speaking women; I have adopted their usage where appropriate in this book.

105 Stephen Peet, personal communication, May 9, 1990.

106 The basic story told in *The Wives of Nendi* was clearly meant by Mai Helen to represent her actual struggle with one or more villages in Murewa District; these struggles are still remembered, though somewhat vaguely, by Murewa residents today, as noted in chapter 1.

107 ZNA MS 418/1/7, manuscripts of Barbara Tredgold, 1963 FAWC pamphlet.

108 Ibid.

109 ZNA ORAL/229, interview with Mary Robinson.

110 *Homecraft*, July 1965, p. 13.

111 ZNA MS 418/1/7, manuscripts of Barbara Tredgold, "African Women Are Winning Their Silent Revolution," draft speech for 1970 FWISR convention.

112 Interview, Mbuya Gwata, Murewa, May 2, 1991. As noted previously, many FWISR publications represented Helen Mangwende as a good but compliant African woman who undertook her ventures under the guidance of white supervisors. There are many reasons to think that Mbuya Gwata's insistence on Helen Mangwende's independence and initiative is an equally potent representation of the genealogy of the clubs. Not only was the FWISR generally obsessed with reinforcing the impression of white supervision in every aspect of the homecraft movement, but much of the testimony before the 1961 Mangwende Commission of Inquiry (following Chief Mangwende's removal from office by the Rhodesian state) makes it clear that authorities within the Native Department and local settlers were profoundly jealous of both Mai Helen and her husband's status, abilities, and influence, especially following their meeting with the Queen of England. Other recollections of Helen Mangwende from this time also suggest that she operated of her own accord—for example, her role in the making of *The Wives of Nendi*.

113 For example, S. Bain, *Homemakers' Training Manual: For Trainers in African Women's Clubs* (Salisbury: The Rhodesia Literature Bureau, 1970).

114 The series began in *Homecraft*, June 1966, pp. 6–7.

115 See Sita Ranchod-Nilsson, " 'Educating Eve': The Women's Club Movement and Political Consciousness Among Rural African Women in Southern Rhodesia, 1950–1980," in *African Encounters with Domesticity,* ed. Karen Tranberg Hansen (New Brunswick, N.J.: Rutgers University Press, 1992).

116 See Ranchod-Nilsson, " 'Educating Eve,' " for details.

117 ZNA MS 418/1/7, manuscripts of Barbara Tredgold, "African Women Are Winning Their Silent Revolution," draft speech for 1970 FWISR convention.

118 Ranchod-Nilsson, " 'Educating Eve,' " p. 197.

119 See, for example, Hansen, "White Women," pp. 260–61.

3 *Buckets, Boxes, and "Bonsella":*
Precolonial Exchange, the "Kaffir Truck" Trade, and African "Needs"

1 D. N. Beach, *The Shona and Zimbabwe, 900–1850* (Gweru: Mambo Press, 1980); H. H. K. Bhila, *Trade and Politics in a Shona Kingdom: The Manyika and Their*

Portuguese and African Neighbours 1571–1902 (Harare: Longman, 1982); S. I. G. Mudenge, *A Political History of Munhumutapa* (Harare: Zimbabwe Publishing House, 1988).

2   S. I. G. Mudenge, "The Role of Foreign Trade in the Rozvi Empire: A Reappraisal," *Journal of African History* 15:3 (1974), pp. 373–91.

3   Mudenge, *Political History,* p. 194.

4   Ngwabi Bhebe, "The Ndebele Trade in the Nineteenth Century," University of Rhodesia Department of History, Henderson Seminar, Paper no. 25, August 1973. Unpublished ms.

5   ZNA BA 14/2/1, Frederick Hugh Barber, trading lists circa 1870s.

6   See Barry A. Kosmin, "The Inyoka Tobacco Industry of the Shangwe People: A Case Study of the Displacement of a Pre-Colonial Economy in Southern Rhodesia, 1898–1938," *African Social Research* 17 (June 1974). Also see Ackson Kanduza, "The Tobacco Industry in Northern Rhodesia, 1912–1938," *International Journal of African Historical Studies* 13:2 (1980).

7   See chapter 4 for more discussion of soap trading and manufacture.

8   ZNA FR 2/2/1, Reminiscences of Ivan Fry, p. 84. Also see C. Martin, "Manyika Beads of the XIX Century," *NADA* 17 (1940).

9   Tabler, *To the Victoria Falls,* p. 68.

10 Bhebe, "The Ndebele Trade," p. 1.

11 See, for example, V. E. M. Machingaidze, "The Development of Settler Capitalist Agriculture in Southern Rhodesia" (D.Phil. diss., University of London, 1980), pp. 491–523.

12 Barry Kosmin, "Ethnic and Commercial Relations in Southern Rhodesia: A Socio-Historical Study of the Asian, Hellene and Jewish Populations 1898–1943" (Ph.D. diss., University of Rhodesia, 1974), p. 300.

13 Ibid., p. 190.

14 I am indebted to Terence Ranger for his comments on the sociological composition of the "truck" traders.

15 Stanley Portal Hyatt, *Off the Main Track* (London: T. Werner Laurie, 1911), p. 6.

16 See Kosmin, "Ethnic and Commercial Relations"; also see Vishnu Padayachee and Robert Morrell, "Indian Merchants and *Dukawallahs* in the Natal Economy, c. 1875–1914," *Journal of Southern African Studies* 17:1 (March 1991); Volker Wild, "Black Competition or White Resentment? African Retailers in Salisbury, 1935–1953," *Journal of Southern African Studies* 17:2 (June 1991).

17 See, for example, the comments of Salisbury retailers Almaleh, Latif, Suleman, and Merdjan in ZNA ZBJ 1/1/4.

18 George Kapnias, interview, Harare, April 16, 1991.

19 Hyatt, *The Old Transport Road,* p. 88; Hyatt, *Off the Main Track,* p. 65.

20 Goromonzi and Chishawasha around colonial Salisbury and other villages near Bulawayo and south toward Beitbridge and Plumtree; areas near the mines of the north and northwest; Murewa and Mtoko along northeastern trade routes with

colonial Malawi and Mozambique; the districts of the Midlands; some villages in the eastern Highlands.

21 Yoshikune, "Black Migrants."

22 Robertson, *Rhodesian Rancher*, p. 39.

23 Nicolas Thomas, *Entangled Objects: Exchange, Material Culture and Colonialism in the Pacific* (Cambridge, Mass.: Harvard University Press, 1991), p. 103.

24 ZNA S 1542/S9, Department of Native Affairs, Correspondence, 1933–1935.

25 ZNA S 1542/T11, Department of Native Affairs, Correspondence, 1936–1947.

26 See, for example, the report of the Native Commissioner, Chilimanzi, 1899. The irritation of merchants at this trade surfaces in the minute books of the Salisbury Chamber of Commerce (ZNA SA 5/1/1–7). See the minutes for March 18, 1902, which called for a special duty on brass wire (ZNA SA 5/1/2).

27 ZNA ZBJ 1/1/4, testimony of G. T. Thornicroft, Mashonaland Kafir Truck Association.

28 ZNA ZBJ 1/1/4, testimony of Phillip Badham.

29 ZNA S1542/T11, C. E. Sparrow to Native Commissioner, Gwelo, April 1936.

30 Vambe, *Ill-Fated People*, pp. 220–23.

31 Ibid., p. 221.

32 Ibid., p. 222.

33 ZNA ZBJ 1/1/2, testimony of C. S. Godfrey.

34 ZNA ZBJ 1/1/4, testimony of Mohammed Hajee Latif.

35 ZNA ZBJ 1/2/1, testimony of Bulawayo Chamber of Commerce.

36 ZNA ZBJ 1/2/2, vol. 2, testimony of Native Commissioner, Mazoe.

37 ZNA S246/727, Native Department correspondence.

38 Vambe, *Ill-Fated People*, p. 221.

39 See chapter 4 for more on Selesnik's case.

40 ZNA ORAL/CH 4, interview, Hulbert Patrick Charles.

41 See T. O. Ranger, "Literature and Political Economy: Arthur Shearly Cripps and the Makoni Labour Crisis of 1911," *Journal of Southern African Studies* 9:1 (October 1982).

42 Vambe, *Ill-Fated People*, p. 222.

43 ZNA ZBJ 1/1/1, vol. 2, testimony of Elias Choto.

44 Hyatt, *Off the Main Track*, pp. 36–37.

45 For more on the Maize Control Acts, see Machingaidze, "Development of Settler Capitalist Agriculture"; Benjamin Davis and Wolfgang Dopcke, "Survival and Accumulation in Gutu: Class Formation and the Rise of the State in Colonial Zimbabwe, 1900–1939," *Journal of Southern African Studies* 14:1 (October 1987); K. P. Vickery, "Saving Settlers: Maize Control in Northern Rhodesia," *Journal of Southern African Studies* 11 (1985).

46 ZNA N/9/1/1–26, Native Department Annual Reports, Native Commissioner, Mtoko, 1914.

47 ZNA ZBJ 1/1/2, testimony of Joyce Rudd.

248 Notes to Chapter Three

48 ZNA ZBJ 1/2/2, vol. 1, testimony of the Native Commissioner, Gutu.

49 ZNA ZBJ 1/2/2, vol. 2, testimony of the Native Commissioner, Mtoko.

50 ZNA ZBJ 1/1/1, vol. 3, testimony of Andrew Midzi.

51 ZNA ZBJ 1/1/1, vol. 2, testimony of P. G. W. Mbofana.

52 ZNA ZBJ 1/1/1, vol. 3, testimony of B. J. Mnyanda.

53 ZNA ZBJ 1/1/1, vol. 2, testimony of Kamdeya (Wedza Native Council).

54 ZNA ZBJ 1/1/4, testimony of Arthur Landau.

55 ZNA ZBJ 1/2/1, testimony of the Bulawayo and Salisbury Chambers of Commerce.

56 Both credit and the general place of the mine store in the life of wage laborers in Zimbabwe are covered in Charles van Onselen, *Chibaro: African Mine Labour in Southern Rhodesia, 1900–1933* (London: Pluto Press, 1976).

57 Southern Rhodesia, *Debates of the Legislative Council*, vol. 2 (1909–1913), Second Session, Fifth Council (May 1912), p. 22.

58 Ibid.

59 Patrick Harries, *Work, Culture and Identity: Migrant Labourers in Mozambique and South Africa, 1860–1910* (Portsmouth, N.H.: Heinemann, 1994), pp. 173–82.

60 Quoted in Kosmin, "Ethnic and Commercial Relations," p. 181.

61 ZNA, ORAL/CH 4, interview with Hulbert Patrick Charles, 1978.

62 Southern Rhodesia, *Debates of the Legislative Council*, vol. 2, Third Session, Fourth Council (1910).

63 See, for example, the comments of the secretary of Native Affairs to the Salisbury Chamber of Commerce in 1938 (ZNA SA 5/1/6).

64 See John Iliffe, *Famine in Zimbabwe* (Gweru: Mambo Press, 1990).

65 ZNA N 3/33/2 Chief Native Commissioner's Correspondence.

66 ZNA ZBJ 1/1/2, testimony of William Gardiner Jamieson.

67 ZNA ZBJ 1/1/1, vol. 1, testimony of Hugh Simmonds.

68 ZNA ZBJ 1/2/2, testimony of B. J. Mnyanda.

69 ZNA ZBJ 1/1/1, vol. 2, testimony of W. H. H. Nicolle.

70 ZNA N 3/24/33, Native Department, Correspondence, 1915–1923.

71 Shamuyarira, *Crisis*, pp. 120–21.

72 Wild, "Black Competition," pp. 179–81, 183.

73 Leslie Bessant and Elvis Muringai, "Peasants, Businessmen and Moral Economy in the Chiweshe Reserve, Colonial Zimbabwe, 1930–1968," *Journal of Southern African Studies* 19:4 (December 1993), p. 573.

74 ZNA N 6/2/1, Native Department Correspondence re: 1915 Conference of Superintendents of Natives.

75 ZNA N 6/2/2, Ass't Native Commissioner, Mazoe to Superintendent of Natives, Salisbury, September 14, 1915.

76 ZNA N 6/2/2, Rhodesian Native Labour Bureau to Salisbury Chamber of Commerce, October 7, 1915.

77 ZNA N 3/7/2, Chief Native Commissioner to General Manager B&M&R Railways, November 1922.

78   ZNA ZBJ 1/1/4, testimony of G. T. Thornicroft, Mashonaland Kafir Truck Association.

79   Indeed, less dramatic and coercive forms of the practice remain in present-day Harare: stores in the Charter Road area still employ young men to try to entice customers inside, and bus companies at the main terminal in Mbare township also employ "touts" to try to recruit customers for their buses.

80   Shamuyarira, *Crisis,* pp. 99–100.

81   Thomas Richards, *The Commodity Culture of Victorian England* (Stanford: Stanford University Press, 1990), pp. 2–3.

82   Grant McCracken, *Culture and Consumption: New Approaches to the Symbolic Character of Consumer Goods and Activities* (Bloomington: Indiana University Press, 1988), p. 28. Also see Appadurai, *The Social Life of Things;* William Reddy, *The Rise of Market Culture: The Textile Trade and French Society, 1750–1900* (Cambridge: Cambridge University Press, 1984); Neil McKendrick, *The Birth of a Consumer Society* (Bloomington: Indiana University Press, 1982); Chandra Mukerji, *From Graven Images* (New York: Columbia University Press, 1983); Rosalind H. Williams, *Dream Worlds: Mass Consumption in Late Nineteenth-Century France* (Los Angeles: University of California Press, 1982).

83   This connection was pointed out to me in a lecture given by Ken Warren of the University of Chicago, on his interesting and original research on historical "misrecognitions" across and through the African diaspora.

84   Issues of the *Rhodesia Herald* in the month of February 1902 dealt in various ways with Earl Grey's speech and the consequent mini-furor it sparked. The speech was first mentioned and summarized in the February 8 issue.

85   *Rhodesia Herald,* February 22, 1902, p. 3.

86   Ibid.

87   ZNA N 9/1/1–26, Native Department Annual Reports, 1902–1903, Charter District.

88   Hole, *Old Rhodesian Days,* p. 47.

89   "Raise Your Own Race," *Native Mirror* 2:3 (1934).

90   ZNA ZBJ 1/1/3, testimony of John Ralstein.

91   ZNA ZBJ 1/2/3, testimony of John Ralstein.

92   ZNA ZBJ 1/1/4, testimony of Jake Hanan.

93   ZNA ZBJ 1/2/1, testimony of the Mashonaland Kafir Truck Association.

94   ZNA ZBJ 1/2/1, testimony of the Salisbury Chamber of Commerce.

95   ZNA ZBJ 1/1/3, testimony of Maurice Lewis.

96   ZNA ZBJ 1/1/2, testimony of Caragianis Angelos.

97   ZNA ZBJ 1/1/3, testimony of Benny Goldstein.

98   Quoted in Jeannie M. Boggie, ed., *Experiences of Rhodesia's Pioneer Women* (Bulawayo: Philpott and Collins, 1938), p. 176. This kind of resigned understanding of the relative balance of power between trader and customer in the initial stages of colonial rule was found all over Africa and formed an ideological sub-

strate in struggles between multinational investors and trader-agents before World War II. For example, a Lever Brothers agent in the Congo wrote to his employers in 1924, "It is of interest to note that here, in the Tropics, it is the Native who is the deciding factor in determining the profit of the Trader and the Trade seems to accept this fact, whilst in Europe, it is the Manufacturers who decided as to their profits and selling prices, and the consumer appears to accept their decision." W. H. Lever acidly replied to the agent, "It is the quality and competition that decides the price of all articles" (Unilever Archives, W. H. Lever to Leverhulme, November 6, 1924).

99   ZNA N 3/33/2, Chief Native Commissioner's Correspondence.
100   ZNA S 1051, Department of Native Affairs Annual Reports, 1943, Wankie District.
101   See, for example, ZNA S2076, Native Affairs Department Annual Reports, 1920.
102   ZNA ZBJ 1/1/2, testimony of John Fotozana.
103   ZNA ZBJ 1/2/3, testimony of representatives of the Umtali Native Welfare Society.
104   ZNA ZBJ 1/1/1, vol. 3, testimony of Aaron Jacha.

4   Manufacturing, the "African Market," and the Postwar Boom

1   Phimister, *Economic and Social History of Zimbabwe*, p. 240.
2   In West Africa, local production of soap posed a considerable challenge to colonial capitalists who tried to manufacture soap there.
3   See chapter 2. Lately, cooperatives in Zimbabwe have also tried to manufacture soap on a small scale, but their products are regarded by most consumers as a poor value. When we discussed "bad soaps" in interviews, cooperative soaps were almost invariably cited.
4   Quoted in D. K. Fieldhouse, *Unilever Overseas: The Anatomy of a Multinational 1895–1965* (Stanford: The Hoover Institution Press, 1978), pp. 380–81.
5   Unilever Archives, Leverhulme to R. Lintermans, 21 November 1924, from S. W. Lusanga.
6   Fieldhouse, *Unilever Overseas*, pp. 100–104.
7   Unilever Archives, Leverhulme to C. Schlesinger, 5 October 1921.
8   Unilever Archives, C. Schlesinger to Leverhulme, June–September 1921.
9   Fieldhouse, *Unilever Overseas*, pp. 112–13.
10   See, for example, ZNA S 1542/S12, Department of Native Affairs, Correspondence, vols. 1 and 2.
11   ZNA S 910/17/14, Department of Customs & Excise, Correspondence, 1929–1948.
12   See Ian Phimister, "Secondary Industrialisation in Southern Africa: The 1948 Customs Agreement Between Southern Rhodesia and South Africa," *Journal of Southern African Studies* 17:3 (September 1991).
13   ZNA S 2094/1/7, Assize Office, Correspondence, 1926–1947.
14   Ibid.

15 ZNA S 246/484, Department of Internal Affairs, Correspondence re: Standardisa-
    tion of Soap Act, 1929–1930.

16 ZNA S 104, Industrial Development Advisory Committee, Minutes, 1940–1944.

17 Though soap makers during this time described these maneuvers as depressing
    the price of soap and thus increasing its circulation among Africans, specific price
    figures kept systematically are hard to come by. The Rhodesian state only spo-
    radically tracked cost-of-living figures before World War II, and given the elas-
    ticity of prices and the institutional nature of the "truck" trade, statistical surveys
    based on "European" retailers are not very reliable indicators for prices of soap or
    other commodities purchased by Africans. (See ZNA ZBJ, testimony of J. H.
    Shaul, and J. H. Shaul, "Distributive Trades of Southern Rhodesia," *South African
    Journal of Economics* 21:2 [June 1953], for further details on the gaps in official
    statistics kept before 1945.) One suggestive document that did track soap prices
    from 1914 to 1928 in European markets shows a slowly falling price for both
    imported and locally produced soaps during that time (*Official Yearbook of the
    Colony of Southern Rhodesia* [Salisbury: Rhodesian Printing and Publishing Com-
    pany, n.d.], Nos. 1–4).

18 ZNA ORAL/HO 6, interview with Joseph Harrison Hodgson.

19 ZNA ZAC 1/1/1, 1912–1913 Cost of Living Commission, testimony of David Mc-
    Cullogh.

20 ZNA ORAL/229, interview with Mary Eileen Robinson, 1979.

21 ZNA N 3/24/33, Native Department, Correspondence re: "Undesirable Euro-
    peans," 1915–1923.

22 ZNA ORAL/BE 5, interview with R. G. Bennett.

23 See Kontos, "Through a Glass Darkly," for some critical reflections on the "needs-
    wants" distinction in Western discourse.

24 Boggie, *First Steps*, p. 132.

25 ZNA ZBI 1/1/1–2, 1943 Howman Commission, testimony of F. L. Hadfield.

26 "Let's Ask Questions," *Native Mirror* January 6, 1940, p. 8.

27 Joyce Rudd, "Make Your Own Things," *Native Mirror* 2:2 (1934).

28 Colonial concerns over the possible satiric intent of Africans adopting European
    fashion have been explored expertly in T. O. Ranger, *Dance and Society in Eastern
    Africa, 1890–1970: The Beni Ngoma* (London: Heinemann, 1975). Also see Chris-
    traud Geary, "Patterns from Without, Meaning from Within: European-Style
    Military Dress and German Colonial Politics in the Bamum Kingdom (Cam-
    eroon)," Discussion Papers in the African Humanities, no. 1 (Boston: Boston
    University African Studies Center, 1989); and Anne Alfhild Hendrickson's upcom-
    ing book on dress and bodily discipline among the Herero of Namibia.

29 ZNA ZBJ 1/1/2, testimony of A.N.C., Bulawayo.

30 ZNA ORAL/BE 5, interview, R. G. Bennett.

31 ZNA N 9/1/1–26, Native Department Annual Reports, Inyanga District, 1914.

32 *Native Mirror* 10: April 1933, p. 4.

33   Morgenster Mission, *Hygiene: Standard VI,* pp. 20–21, 24–25.

34   "Your Clothes! Do Fine Feathers Make Fine Birds?" *Native Mirror* 10 (April 1933), p. 4.

35   "Pathfinders: Manners Maketh the Man," *Bantu Mirror* 2:3 (February 27, 1937), p. 4.

36   Rudd, "Community Work," p. 92. See chapter 2 for more details.

37   For example, in Richard Werbner's "family biography" *Tears of the Dead,* one man recounts how a white policeman beat him for wearing a hat and parting his hair (Richard Werbner, *Tears of the Dead* [Washington, D.C.: Smithsonian Institution Press, 1991], p. 53). Memoirs by prominent Zimbabwean activists and intellectuals have also recounted many such incidents. See Vambe, *From Rhodesia to Zimbabwe* and *An Ill-Fated People,* and Nyagumbo, *With the People,* for especially powerful examples.

38   See chapter 6 for a discussion of white attitudes toward cosmetic use by African women.

39   Boggie, *Husband,* p. 106.

40   "Women's Column," *Bantu Mirror* 2:10 (April 17, 1937), p. 2.

41   See, for example, "The Training of African Mothers," *Native Mirror* 14 (April 1934).

42   Phimister, *Economic and Social History,* p. 182.

43   Adam Ashforth, *The Politics of Official Discourse in South Africa* (Oxford: Clarendon Press, 1990).

44   ZNA ZAY, 1938 Economic Development Commission. Records of the Commission include five volumes of testimony and evidence, ZAY 2/1/1–4 and ZAY 2/2.

45   ZNA ZAY 2/1/1–4, testimony of J. H. Brown, Controller of Customs and Excise.

46   See Machingaidze, "Development of Settler Capitalist Agriculture," for more details.

47   ZNA ZBJ 1/2/1, testimony of Hugh Simmonds.

48   ZNA ZBJ 1/2/1, testimony of M. Stronge, Bikita District.

49   ZNA ZBJ 1/1/4, testimony of Salisbury and Bulawayo Chambers of Commerce.

50   The desperation lying behind the framing of the Godlonton's problematic emerged throughout the hearings in the odd programmatic cul-de-sacs pursued by the commissioners. For example, the commissioners took an avid interest in the formation of African cooperatives, both production cooperatives and cooperative stores that would replace "truck" retail outlets. It seemed appealing, both as a way to break up "truck" trading and as a way to provide a source of money to African communities. The commissioners found that African enthusiasm for cooperatives was strong. The commissioners' ardor ebbed, however, when they found that enthusiasts, African and settler alike, radically diverged in their visions from a whole range of unwritten colonial axioms about the need for the state to maintain tight control over all African projects. The commissioners were gradually forced to recognize that it was impossible to increase the economic viability of African communities without impinging on the foundations of colonial rule.

51 William Margolis, "The Position of the Native Population in the Economic Life of Southern Rhodesia" (thesis, Pretoria: University of South Africa, 1938), pp. 101–104.

52 "Presidential Remarks," *Annual Report of Chambers of Industries of Rhodesia* (Salisbury: n.p., 1945).

53 ZNA ZBJ 1/1/2, testimony of Nigel Phillip.

54 "The Segregation Question, Part IX," *Bantu Mirror* 2:7 (August 14, 1937).

55 ZNA S 915/1–2, Department of Commerce and Industry, Correspondence and Other Material, 1944–1949. Departmental analysis: P. B. Gibbs, "The Economics of Secondary Industry in Southern Rhodesia," 1944.

56 "Presidential Remarks," *Annual Report of Chambers of Industries of Rhodesia* (Salisbury: n.p., 1943 and 1944).

57 ZNA SA 5/1/7, Salisbury Chamber of Commerce Minute Books, 1939–1947.

58 ZNA S 915/1–2, Department of Commerce and Industry, Correspondence and Other Material, 1944–1949. Departmental analysis: P. B. Gibbs, "The Economics of Secondary Industry in Southern Rhodesia," 1944.

59 "Profile of a Commercialist—Benny Goldstein," *Commerce of Rhodesia* 2:7 (July 1951).

60 ZNA ZBJ 1/2/1, testimony of Hugh Simmonds.

61 Ndabaningi Sithole, "African Progress in the New World Order," *Bantu Mirror* March 11, 1944, p. 1.

62 Ndabaningi Sithole, "Letter to the Editor," *Bantu Mirror* September 2, 1944, p. 3.

63 M. G. Manyawu, "The Africans and Comfort," *Bantu Mirror* January 12, 1946, p. 7.

64 ZNA ZBJ 1/1/3, testimony of Dhiliwayo, Jeremiah Sobantu et al.

65 ZNA ZBI 2/1/1–2, testimony of Mazabisa.

66 ZNA S 104, Industrial Development Advisory Committee, Minutes, 1940–1944.

67 Phimister, *Economic and Social History,* p. 258.

68 Phimister, "Secondary Industrialisation," pp. 434–35.

69 ZNA RH 12/2/1/1, Joint National Council, Minutes of the Race Relations Subcommittee, July 24, 1950, p. 7.

70 Ibid., p. 8.

71 ZNA RH 12/2/1/1, Joint National Council, Minutes of the Race Relations Subcommittee, August 28, 1950, p. 10.

72 Ibid., p. 11.

73 ZNA RH 12/2/1/1, Joint National Council, Minutes of the Race Relations Subcommittee, December 6, 1950.

74 Quoted in *Bantu Mirror,* April 15, 1944, p. 1.

75 ZNA RH 12/2/1/1, Joint National Council, Minutes of the Race Relations Subcommittee, December 6, 1950, p. 16.

76 Ibid., p. 18.

77 "Presidential Remarks," *Annual Report of Chambers of Industries of Rhodesia* (Salisbury: n.p., 1945).

78   Southern Rhodesia Department of Trade and Industrial Development, *Secondary Industry in Southern Rhodesia* (Salisbury, 1953), p. 10.

79   Industrial growth and investment in postwar Africa was an explicit topic of concern in the closing years of World War II. See, for example, Theresa Cahan, "Secondary Industries for Tropical Africa," *Africa* 14:4 (October 1943). In the flood of analyses and manuals that followed, one of the most interesting and revealing is T. L. V. Blair, *Africa: A Market Profile* (New York: Frederick Praeger, 1965). A curious example of how such discourses reproduced themselves among African business elites can be found in Ishmael O. Kitinya, *A Guide to Salesmanship for Africa* (Nairobi: Africa Book Services, 1976).

80   *Report of the United Kingdom Trade Mission to the Union of South Africa, Southern Rhodesia and Northern Rhodesia* (London: HMSO, 1931); British Board of Trade, *The African Native Market in the Federation of Rhodesia and Nyasaland* (London: HMSO, 1954). Also see Fergus Chalmers Wright, *African Consumers in Nyasaland and Tanganyika* (London: HMSO, 1955).

81   N. R. Bertram, "An Industrialist's View of the Determinants Required for Growth of Secondary Industry in Rhodesia," *Rhodesian Journal of Economics* 3:2 (June 1969), p. 15.

82   *Central African Market* (Salisbury: George Hindley Ltd., August 1959), p. 2.

83   ZNA RH 12/2/4/1, Garfield Todd, address to annual meeting, Rhodesian Federated Chambers of Commerce, 1957.

84   See chapters 5 and 6 for more details.

85   See the preface in Federation of Rhodesia & Nyasaland Central Statistical Office, *Report on the Urban African Budget Surveys, Northern Rhodesia* (Salisbury, 1960).

86   See Langdon, "Multinational Corporations," for a useful picture of another postwar African soap making industry, in Kenya.

87   These and other details of Lever Brothers' (Zimbabwe) history are mentioned briefly in Fieldhouse, *Unilever Overseas,* but most of the details I discuss above have been gleaned from articles in *The Nutshell,* the house magazine of Lever Brothers (Rhodesia), published from 1962 to 1978, in particular "Our Twenty Years of Progress," 1:13 (Summer 1968). Also useful in this context is Eric Rosenthal, *As Pioneers Still: An Appreciation of Lever Brothers' Contribution in South Africa, 1911–1961* (Cape Town: n.p., 1961).

88   Olivine is still an important competitor of Lever Brothers in Zimbabwe, and was locally owned until its acquisition in the early 1980s by the Heinz Corporation.

89   Southern Rhodesia Department of Trade and Industrial Development, *Secondary Industry,* p. 22.

90   *The Nutshell* 1:14 (Winter 1968).

91   Federation of Rhodesia and Nyasaland Ministry of Commerce and Industry, *Manufacture of Toilet Preparations: Opportunity for Industry No. 11* (Salisbury, 1963).

92   Market Research Africa (Rhodesia), *Profile of Rhodesia: European* (Salisbury,

1970); *Profile of Rhodesia: African* (Salisbury, 1970); Market Research Africa (Rhodesia), *Product Data: Consumer Close-Up* (Salisbury, 1972).

93 George Kapnias, interview, Harare, April 16, 1991.

94 Charles Nyereyegona, "Marketing to the Urban African," *Marketing Rhodesia* 1:4 (May 1973), p. 70.

95 ZNA S 2385, Chief Native Commissioner's Office, Department of Native Affairs, Correspondence re: African Trading Stores, 1947–1950.

96 See Wild, "Black Competition or White Resentment?"

97 George Kapnias, interview, Harare, April 16, 1991; Mr. MacIntosh, interview, Harare, April 22, 1991.

98 George Kapnias, interview, Harare, April 16, 1991.

99 "Kumboyedza Looks for More Outlets," *Rhodesian Retailer* (September 1973), pp. 8–9; "Jazz Takes Its First Step Towards Chain Operation," *Rhodesian Retailer* (May/June 1974), pp. 6–7; "Jarzin Experiments: Shop for African Farmers," *Rhodesian Retailer* (September/October 1974), pp. 16–17.

100 See, for example, M. A. H. Smout, "Shopping Centres and Shopping Patterns in Two African Townships of Greater Salisbury," *Zambezia* 2:1 (December 1971).

101 For example, the 1960 Samkange report on RMM workers noted: "Most employees spend more than they earn each month. They buy on credit from shops. Indian traders actually come to the hostels to collect the money" (S. J. T. Samkange and T. M. Samkange, *Survey of Personal, Employment, Expenditure and Accommodation Data of African Employees of the Rhodesian Milling Company* [Salisbury: Rhodesian Milling and Manufacturing Company, 1960], p. 26).

5   The New Mission: Advertising and Market Research in Zimbabwe, 1945–1979

1 Werner Willi Max Eiselen, "The Elasticity of the Bantu Consumer," *Second Advertising Convention in South Africa* (Durban: Society of Advertisers, 1959), p. 108.

2 Advertisers and their supporters often seized on this parallel themselves. At the 1959 convention, Eiselen's superior, the South African minister of Bantu Education and Development, "likened advertisers to missionaries" (Eiselen, "Elasticity," p. 108). A Colgate-Palmolive executive also used the metaphor in describing the company's attempts to get schoolchildren to brush their teeth regularly (Cornell Butcher, Francis Makosa, and Wellington Chikombero, interview, Harare, November 26, 1990).

3 This recalls the points raised by Megan Vaughan about the colonial interest in African subjectivities (see Vaughan, *Curing Their Ills*). As I noted in my introduction, advertisers got a good deal closer to actualizing such subjectivities, which I believe makes their comparison with missionaries a particularly apt one. Another important connection is the explicitly transnational orientation of both institutions. For advertisers in particular, the complexities of transnational, national, and

local culture and identity since 1945 give their activities a particularly convoluted feel, in common with many other postwar institutions. For some recent interesting meditations on some of the problems involved in the intersection between subjectivity, difference, identity, and transnational institutions, see Arjun Appadurai, "Disjuncture and Difference in the Global Cultural Economy," *Public Culture* 2:2 (Spring 1990); Ferguson and Gupta, "Beyond 'Culture': Space, Identity and the Politics of Difference"; and Michael Kearney, "Borders and Boundaries of State and Self at the End of Empire," *Journal of Historical Sociology* 4:1 (March 1991).

4   W. H. O'Grady, Opening Address, *Second Advertising Convention in South Africa* (Durban: Society of Advertisers, 1959), p. 9.

5   Advertising in a colonial or African context has not been studied a great deal, but I found tremendously suggestive Pierre Thizier Seya's 1981 Ph.D. dissertation on the history and practice of advertising in the Ivory Coast ("Transnational Capitalist Ideology and Dependent Societies: A Case Study of Advertising in the Ivory Coast" [Ph.D. diss., Stanford University, 1981]) and Efurosibinia Emmanuel Adegbija's 1982 Ph.D. dissertation analyzing the discourse of advertising in Nigeria ("A Speech Act Analysis of Consumer Advertisements" [Ph.D. diss., Indiana University, 1982]).

6   See Ewen, *Captains of Consciousness*, esp. pp. 41–48.

7   J. E. Maroun, "Second Address, 'Bantu Market' Session," *Third Advertising Convention in South Africa: The Challenge of a Decade* (Johannesburg: Statistic Holdings Ltd., September 1960), p. 121.

8   "Questions From 'Bantu Market' Session," *Third Advertising Convention in South Africa: The Challenge of a Decade* (Johannesburg: Statistic Holdings Ltd., September 1960), p. 133.

9   These themes have been examined in many texts, among them the aforementioned Ewen, *Captains of Consciousness;* Stuart Ewen and Elizabeth Ewen, *Channels of Desire: Mass Images and the Shaping of American Consciousness* (New York: McGraw Hill, 1982), Kathy Myers, *Understains: The Sense and Seduction of Advertising* (London: Comedia Publishing, 1986); John Sinclair, *Images Incorporated: Advertising as Industry and Ideology* (London: Croom Helm, 1987); and Williams, "Advertising: The Magic System."

10   Maroun, "Second Address," pp. 121–22, 124–25.

11   ZNA RH 12/2/4/1, Annual Congress, Rhodesian Federated Chambers of Commerce, 1957, address of the President of the Associated Chambers of Commerce (ACCOR).

12   E. S. Gargett, "The African in the Market Place," *Marketing Rhodesia* 3:2 (November 1974), p. 60.

13   Joe Van Den Bergh, "Marketing in the Tribal Trust Lands," *Marketing Rhodesia* 2:3 (February 1974), p. 60.

14   See Julie Frederickse, *None But Ourselves: Masses vs. Media in the Making of Zimbabwe* (Harare: Anvil Press, 1982).

15 Vaughan, *Curing Their Ills*, p. 12.

16 Maroun, "Second Address," p. 124.

17 ZNA ZBJ 1/1/4, testimony of Phillip Badham.

18 ZNA ZBJ 1/1/3, testimony of Benny Goldstein.

19 *African Commerce: The Export and Import Journal of Africa*, September 1901, p. 257.

20 *Report of the United Kingdom Trade Mission*, 1931, p. 14.

21 "Natives Have £20,000,000 to Spend," *Commerce of Rhodesia* 4 (1953), pp. 111–12.

22 P. P. Marolen, "Tomorrow Has Already Dawned: The Aspirations of the African Consumer and Businessman," *Marketing Rhodesia* 3:3 (February 1975), pp. 8–9.

23 *The Urban Bantu Market: Understanding Its Complexities and Developing Its Potential*, Seminar Proceedings, Durban (Johannesburg: National Development and Management Foundation, 1969), p. 7.

24 M. M. Dell, "First Address, 'Bantu Market' Session," *Third Advertising Convention in South Africa: The Challenge of a Decade* (Johannesburg: Statistic Holdings Ltd., September 1960).

25 A. P. Van der Reis, "Motivational Factors in Bantu Buying Behaviour," *Research Report no. 15* (Pretoria: Bureau of Market Research, 1966), p. iv.

26 See Vaughan, *Curing Their Ills*, ch. 8, for more on filmmaking. Also see ZNA S 932/8/1, Ministry of Information, Correspondence re: Educational Films, 1943–1949.

27 For one Rhodesian example, see Guy Taylor, "Primitive Colour Vision," *NADA* 1 (1923), pp. 69–72.

28 Eiselen, "Elasticity," p. 114.

29 Van den Bergh, "Marketing," p. 65.

30 Ibid., p. 63.

31 British Board of Trade, *African Native Market*, p. 27.

32 E. G. Tabor, "Is There an Excalibur of the African Market?" *Marketing Rhodesia* 5:3 (March 1977), p. 40.

33 Nimrod Mkele, "Advertising to the Bantu," *Second Advertising Convention in South Africa* (Durban: Society of Advertisers, 1959), p. 131.

34 Andrew Knox, *Coming Clean: A Postscript after Retirement from Unilever* (London: William Heinemann Ltd., 1976), p. 22.

35 British Board of Trade, *African Native Market*, p. 27.

36 F. N. Howland, "Place of Advertising in Native Trade," *Commerce of Rhodesia* (October 1953), p. 55.

37 Nyereyegona, "Marketing," p. 69.

38 Tabor, "Is There an Excalibur," p. 40.

39 Mike Daffy, "Radio: The Mass Medium with Personal Appeal," *Rhodesian Retailer* (July 1973), p. 21.

40 Mkele, "Advertising," p. 131.

41 Howland, "Place of Advertising," p. 55.

42   "Questions From 'Bantu Market' Session," p. 130.

43   Louis Buckle, "Marketing for the African Consumer," *Marketing Rhodesia* 1:1 (1972), p. 17.

44   Mkele, "Advertising," p. 123.

45   *The Rhodesian Grocer* 2:1 (March 1956), p. 26.

46   Federation of Rhodesia and Nyasaland, *Manufacture of Toilet Preparations*, p. 5.

47   Rhodesia, Ministry of Information, *A People's Progress*, 1969.

48   Mkele, "Advertising," p. 129.

49   Nyereyegona, "Marketing," p. 71. It is worth noting that Nyereyegona and other advertisers counseled pitching selected nondurables to male consumers, and further argued that all major durable purchases were the province of African men.

50   Howland, "Native Advertising," p. 56.

51   Mkele, "Advertising," p. 130.

52   "People in Jobs," *The Nutshell* 1:14 (Winter 1968), p. 5.

53   Such giveaways were typical throughout Africa during the colonial era. For example, during Lever's visit to the Congo in 1924–1925, he gave the go-ahead to SAVCO managers to distribute promotional mirrors at a low cost alongside soap purchases (Unilever Archives, Lever to the Special Committee, Kinshasha 1925).

54   ZNA SA 5/1/5, Salisbury Chamber of Commerce Minute Books, 1921–1931.

55   ZNA S 2390/751/39, Department of Native Affairs, Correspondence re: Regulation of Advertisements, 1927–1932. A law was passed in 1929 to regulate such advertising.

56   *Native Mirror* 3 (October 1932).

57   Dallas Smythe, "Communications: Blindspot of Western Marxism," *Canadian Journal of Political and Social Theory* 1:3 (Fall 1977); Jhally, *The Codes of Advertising*. Also see Bill Livant, "The Audience Commodity: On the Blindspot Debate," *Canadian Journal of Political and Social Theory* 3:1 (1979).

58   See Debra Spitulnik, "Radio Culture in Zambia: Audiences, Public Words, and the Nation-State" (Ph.D. diss., University of Chicago, 1994).

59   ZNA ZBI 2/1/1–2, testimony of F. L. Hadfield.

60   British Board of Trade, *African Native Market*, p. 26.

61   Market Research Africa (Rhodesia), *Consumer Contact 1974: Readership Data* (Salisbury, 1975).

62   One piece of Zimbabwean market research on hoardings in 1976 suggested that the "audience of Hoardings is constant," "some 56,000 per hoarding site" (Dillon Ad Signs Ltd., "A Study into Hoarding Audiences Amongst the African" [October 1976]).

63   The characteristic language of all such press and poster advertising in Zimbabwe has been perceptively analyzed in Alec J. Pongweni, "The Language of Advertising and Its Sociocultural Implications for the Consumer," Centre for Applied Social Sciences, University of Zimbabwe, October 1983, unpublished paper.

64   Van der Reis, "Motivational Factors."

65 ZNA S 932/69, Ministry of Information, Correspondence and Papers re: Broadcasting and the BBC, 1946–1953. For details on the setup and functioning of CABS, also see Fraenkel, *Wayaleshi*.

66 ZNA FO 5/1/1/35, Federation of Women's Institutes, Correspondence re: Broadcasting, 1939–1959.

67 Market Research Africa (Rhodesia), *National Listenership Survey: RBC African Service* (Salisbury, 1971).

68 Charles Wilson, *The History of Unilever*, vol. 2 (London: Cassell & Company Ltd., 1954), p. 364.

69 Market Research Africa, *National Listenership Survey*, pp. 10, 24.

70 Dillon Enterprises, *Rhodesian African Cinema Survey*, 1972. Also, Roger Dillon, interview, Harare, April 3, 1991. Again, see Vaughan, *Curing Their Ills*, ch. 8, for another discussion of these issues from a wider perspective.

71 This overall description of the work of the OAP is taken from *The Nutshell*, especially "People in Jobs" 1:14 (Winter 1968).

72 Interviewees in the advertising business suggested to me that this was an uncommon practice, as it was held to be unpredictable, but that it had been periodically employed, especially with regard to tobacco.

73 J. G. Hillis, "An Account of a Market Research in a Tribal Trust Area," *NADA* 10:3 (1971), p. 30.

74 ZNA S 2113/1–2, Ministry of Information, 1962.

75 Tabor, "Is There an Excalibur," p. 40.

76 *The Nutshell* 1:16 (Winter 1969), p. 15. Once again, the link between advertising tactics and military propaganda during the liberation war is emphasized by this example.

77 Cornell Butcher, Francis Makosa, and Wellington Chikombero, interview, Harare, November 26, 1990.

78 Maroun, "Second Address," p. 121.

79 "Questions From 'Bantu Market' Session," p. 130.

80 Ibid., p. 133.

81 Douglas Kadenhe, Harare, interview, May 23, 1991.

82 These categories are drawn from P. J. du Plessis, ed., *Consumer Behaviour: A South African Perspective* (Pretoria: Southern Book Publishers, 1990), and the "Black Sociometer" used by the South African firm Market Research: Africa, which I examined during a visit to their Johannesburg offices in late May 1991.

83 Nick Green and Reg Lascaris, *Communication in the Third World: Seizing Advertising Opportunities in the 1990s* (Cape Town: Tafelberg Publishers/Human & Rousseau Ltd., 1990).

84 Janice C. Simpson, "Buying Black," *Time* 140:9 (August 31, 1992), pp. 52–53.

85 Williamson, *Decoding Advertisements*, p. 167.

86 This received further impetus from the commitments of some multinationals to a policy of "-ization," as Lever Brothers has referred to it, in which the management

of foreign subsidiaries are culled from local employees wherever possible. Lever Brothers in particular pursued this strategy in the 1960s and 1970s.

87 Douglas Kadenhe, interview, Harare, May 23, 1991.

88 Marolen, "Tomorrow," p. 3.

89 "Interview with Ben Mucheche," *Commerce Rhodesia* (June 1974), p. 15.

90 Nyereyegona, "Marketing," pp. 64, 68.

91 Tabor, "Excalibur," pp. 36–37.

92 Williams, "Advertising: The Magic System," p. 189.

93 Maurice Mathewman, interview, Harare, May 13, 1991.

94 See, for example, Alec J. Pongweni, "The Language of Advertising and Its Sociocultural Implications for the Consumer," Centre for Applied Social Sciences, University of Zimbabwe, October 1983, unpublished ms, pp. 15–16.

95 See Tabor, "Excalibur," p. 38, for some of these examples. I was also told similar stories in a number of interviews, among them with Maurice Mathewman, Harare, May 13, 1991; Roger Dillon, Harare, April 3, 1991; Cornell Butcher et al., Harare, November 26, 1990; Jack Wazara, Harare, May 15, 1991; and Douglas Kadenhe, Harare, May 23, 1991.

96 Advertising remains an industry, and its fundamental place in the architecture of capitalism in Zimbabwe has not particularly changed. Nor is it entirely colorblind; some of the current advertisements I saw in 1990–1991 still disproportionately feature white models and actors. One measurement of the impact of the total transformation of official attitudes toward race in Zimbabwe, however, cropped up in my interview with the head of a South African marketing firm, who not only trotted out the old canard about color perception, but also claimed that Africans react differently to vivid images because they "spend most of their lives in the dark" (Clive Corder, interview, Johannesburg, May 30, 1991).

97 The Bata advertisement featured a young warrior searching for *lobola* and finding a Bata store inside of a baobab tree. He brings the magical shoes back to the headman after a series of misadventures and wins his bride.

98 Mkele, "Advertising," p. 127.

99 ZNA S 2113/1–2, Ministry of Information, 1962.

100 "Hygiene," *Homecraft Magazine* (May 1970), p. 6.

101 Hillis, "An Account," p. 30.

102 Rian Malan, *My Traitor's Heart* (London: Vintage, 1990), p. 60. Africans in Zimbabwe today also link Lifebuoy to health, to men, to the achievement of total cleanliness. See chapter 6 for more details.

103 *Bantu Mirror*, August 24, 1946, p. 8.

104 *Bantu Mirror*, January 19, 1946, p. 7.

105 *Bantu Mirror*, October 2, 1948, p. 9.

106 *Bantu Mirror*, January 15, 1944, p. 5.

107 Baudrillard, "For a Critique of the Political Economy of the Sign."

108 See, for example, *Bantu Mirror*, May 6 and May 20, 1950.

109 Mkele, "Advertising," p. 130.
110 *Bantu Mirror* 1:47 (January 1937), p. 6.
111 *African Parade* 8:5 (March 1961), p. 74.
112 *Bantu Mirror,* June 22, 1940, p. 6.
113 *Bantu Mirror,* October 22, 1960, p. 4.
114 *Bantu Mirror,* November 9, 1960, p. 13.
115 *Bantu Mirror,* April 29, 1950.
116 *Radio Post,* September 1959, p. 20.
117 Pongweni, "The Language of Advertising."
118 *Parade and Foto-Action,* March 1969.
119 Mkele, "Advertising," p. 128.
120 For example, Lever Brothers "re-launched" most of its products in the 1960s and 1970s, trying to make over their image with the familiar tag lines "new" and "improved."
121 *Bantu Mirror* 1:48 (January 9, 1937), p. 3.
122 *Bantu Mirror* 25:25 (September 10, 1960), p. 3.
123 See Pieterse, *White on Black,* pp. 195–98.
124 *Parade and Foto-Action,* January 1969, front cover.
125 *Parade and Foto-Action,* January 1969, inside cover.
126 *Parade and Foto-Action,* December 1973, p. 5. It is worth remembering that readers in the early 1970s would have known that any such "fellow guests" would have been exclusively white.
127 These examples are taken from foto-plays running from 1969 to 1973.
128 I am indebted to Ivan Karp, Luise White, and Nancy Rose Hunt for telling me about a number of these types of incidents. For an account of most of the types of rumors mentioned in this paragraph, also see Fraenkel, *Wayaleshi,* ch. 12.
129 Fraenkel, *Wayaleshi,* p. 201.
130 For some general examples of rumors about commodities in southern Africa, see Arthur Goldstuck, *The Rabbit in the Thorn Tree* (Johannesburg: Penguin Books, 1990).
131 See Harry F. Wolcott, *The African Beer Gardens of Bulawayo* (New Brunswick, N.J.: Rutgers Center of Alcohol Studies, 1974). I was told a similar story about the introduction of Chibuku Beer during an interview with business executives (Cornell Butcher, Francis Makosa, Wellington Chikombero, interview, Harare, November 26, 1990).
132 Maurice Mathewman, interview, Harare, May 13, 1991; Douglas Kadenhe, interview, Harare, May 23, 1991.
133 See chapter 6.
134 Roger Dillon, interview, Harare, April 3, 1991.
135 See Appadurai, "Disjuncture and Difference," for illuminating comments on the functioning of "transnational flows" in the modern global economy.
136 Douglas Kadenhe, interview, Harare, May 23, 1991.

137   This section is based extensively on comments by Douglas Kadenhe, who now heads an independent marketing firm and worked for many years for Lever Brothers in British Central Africa.

138   Douglas Kadenhe, interview, Harare, May 23, 1991.

6   Bodies and Things: Toiletries and Commodity Culture in Postwar Zimbabwe

1   Shimmer Chinodya, *Harvest of Thorns* (Harare: Baobab Books, 1989), pp. 79–81.

2   For example, Robert I. Rotberg, citing the London Missionary Society (LMS) archives, notes that Kawimbe Mission in colonial Zambia had by 1902 1,850 pounds of soap among its inventory of trade goods and supplies for converts (Robert I. Rotberg, "Rural Rhodesian Markets," in *Markets in Africa*, ed. Paul Bohannen and George Dalton [Evanston, Ill.: Northwestern University Press, 1962], p. 583).

   Also, as noted in chapter 2, prior to local soap manufacture commencing on a significant scale, a standard lesson in mission classes on domesticity involved soap making. In the last few years, development agencies and their allies have again begun to stress soap making in the home as a useful activity for rural peasants.

3   Ellen Hammon and Ken Rickard, "How to Develop Sex Appeal," *Parade and Foto-Action*, May 1969, p. 16.

4   Chinodya, *Harvest of Thorns*, pp. 143, 229, 243.

5   In a recent volume of oral histories about the experience of women during the war, this demand for soap from ZANLA and ZIPRA forces comes up again and again. See Irene Staunton, ed., *Mothers of the Revolution* (Harare: Baobab Books, 1990).

6   For example, Tapera household, interview, Kambuzuma Township, Harare, May 18, 1991.

7   Richard and Innocent Katema, interview, "Joburg" section of Mbare Township, Harare, April 24, 1991.

8   Chinodya, *Harvest of Thorns*, p. 30.

9   See Fraenkel, *Wayaleshi*, p. 51.

10   For example, Richard and Innocent Katema, interview, "Joburg" section of Mbare Township, Harare, April 24, 1991.

11   Other soaps mentioned with frequency during interviews—Jade and Geisha—are, like cooperative soap, recent introductions to the market, and were strongly praised, especially by those of low income, as being "big," a good value for the price.

12   These practices were discussed in numerous interviews with consumers and executives. The particular appeal of "glycerine" was mentioned in an interview with executives (Cornell Butcher, Francis Makosa, and Wellington Chikombero, interview, Harare, November 26, 1990).

13   Tashinga squatter camp, interviews, Harare, Mbare Township, April 19, 1991.

14   Staunton, *Mothers of the Revolution*, p. 217.

15   Mayer and Mayer, *Townsmen or Tribesmen*, p. 21.

16   For a further analysis of these genres and their historical evolution, see the work of George P. Kahari, especially his recent study *The Rise of the Shona Novel: A Study in Development, 1890–1984* (Gweru: Mambo Press, 1990).

17   J. W. Marangwanda, *Kumazivandadzoka* (Harare: Longman Zimbabwe, 1959, rpt. 1988); Aaron C. Moyo, *Ziva Kwawakabva* (Harare: Longman Zimbabwe, 1977).

18   M. M. Hove, *Confessions of a Wizard* (Gweru: Mambo Press, 1985), p. 60.

19   Ben Sibenke, *My Uncle Grey Bhonzo* (Harare: Longman Zimbabwe, 1982), p. 23.

20   Ibid., p. 24.

21   Ibid., p. 41.

22   *Parade,* April 1991, p. 29.

23   Chinjerai Hove, *Bones* (Harare: Baobab Books, 1988), p. 97.

24   Dangarembga, *Nervous Conditions,* p. 2.

25   Ibid., pp. 58–59.

26   Ibid., p. 90.

27   Hove, *Bones,* p. 84.

28   Dambudzo Marechera, "Black Skin, What Mask," in *The House of Hunger* (London: Heinemann International, 1978), p. 93.

29   Kristina Rungano, *A Storm Is Brewing* (Harare: Zimbabwe Publishing House, 1984), p. 74.

30   Charles Mungoshi, "The Day the Bread Van Didn't Come," in *The Setting Sun and the Rolling World* (London: Heinemann International, 1972, rpt. 1989).

31   Dangarembga, *Nervous Conditions,* pp. 70–71.

32   Ibid., p. 59.

33   Judith Moyo, "Nyamakondo—Haunt of the Kingfishers," in *Mambo Book of Zimbabwean Verse in English,* ed. Colin and O-lan Style (Gweru: Mambo Press, 1986), p. 303.

34   For example, this is noted in David Lan, *Guns & Rain: Guerrillas and Spirit Mediums in Zimbabwe* (London: James Currey, 1985).

35   Mr. Muzembe, interview, Murewa, May 7, 1991. Also, Magomo household, interview, Murewa, May 7, 1991.

36   Chinodya, *Dew in the Morning,* p. 65.

37   I am indebted to my assistant, Tuso Tapera, for some particularly vivid and detailed accounts of how these processes work.

38   Ambi was licensed in 1963; some creams had been advertised as "bleaching" agents before, but had not achieved notable success as lightening products. See *Central African Merchant* (Salisbury), March 1963 issue.

39   James Ferguson, "The Cultural Topography of Wealth: Commodity Paths and the Structure of Property in Rural Lesotho," *American Anthropologist* 94:1 (March 1992).

40 See Phimister, *Economic and Social History,* p. 73.

41 Vambe, *From Rhodesia to Zimbabwe,* p. 57.

42 Michael Gelfand, "Who Is Rich and Poor in Traditional Shona Society," *NADA* 10:4 (1971), p. 50. Gelfand, like other Rhodesian ethnographic writers, had ulterior motives for depicting Shona culture in this way. For Gelfand, this alleged rejection of material wealth was a sign of the laudable communalism and antimaterialism of Shona Society. State officials, on the other hand, used this image as a pretext to lampoon and criticize Africans who crossed the boundaries of consumer propriety.

43 Bourdillon, *Shona Peoples,* p. 115.

44 Vambe, *From Rhodesia To Zimbabwe,* p. 12.

45 ZNA N/9/1/1–26, Native Department Annual Reports, Ndanga-Bikita District, 1910.

46 Nyagumbo, *With the People.*

47 See any of the African budget surveys conducted by the Federation of Rhodesia and Nyasaland or the government of Southern Rhodesia/Rhodesia between 1957 and 1970.

48 See, for example, Yoshikune, "Black Migrants"; Clive Kileff, "Black Suburbanites: An African Elite in Salisbury, Rhodesia," in *Urban Man in Southern Africa,* ed. Clive Kileff and Wade C. Pendleton (Gweru: Mambo Press, 1975); Michael O. West, "African Middle-Class Formations in Colonial Zimbabwe, 1890–1965" (Ph.D. diss., Harvard University, 1990); Steven Thornton, "The Struggle for Profit and Participation by an Emerging African Petty-Bourgeoisie in Bulawayo, 1893–1933," in *Societies of Southern Africa,* vol. 9 (unpublished ms, University of London, 1980); William R. Duggan, "The Rural African Middle Class of Southern Rhodesia," *African Affairs* 79:315 (April 1980); and David B. Moore, "The Ideological Formation of the Zimbabwean Ruling Class," *Journal of Southern African Studies* 17:3 (September 1991).

49 Vambe, *From Rhodesia to Zimbabwe,* pp. 211–12.

50 Williamson, *Decoding Advertisements,* p. 13.

51 For example, in contemporary St. Petersburg [Leningrad], in the aftermath of the end of the Soviet Union, those with access to *valyuta* (hard currency) have embraced the public consumption of expensive foreign goods, once consumed only by communist functionaries in private. These new conspicuous consumers, beneficiaries of what is called "dollar apartheid," have no consistent niche in the structure of the economy, but occupy a heterogeneous array of positions in St. Petersburg: prostitutes, former party officials, the intelligentsia, and black marketeers (Steven Erlanger, " 'Dollar Apartheid' Makes a Few Russians Rich but Resented," *New York Times,* August 23, 1992).

52 Vambe, *From Rhodesia to Zimbabwe,* pp. 192, 195–96.

53 Shamuyarira, *Crisis,* pp. 62–68.

54 Michael Gelfand, *Shona Religion* (Cape Town: Juta & Company, 1962), p. 90.

55   Alfred Gell, "Newcomers to the World of Goods: Consumption among the Muria Gonds," in *The Social Life of Things,* ed. Arjun Appadurai (London: Cambridge University Press, 1986), pp. 110–11.

56   Kileff, "Black Suburbanites," p. 90.

57   Several articles in early issues of *NADA* describe spirits similar to *zvidhoma: makondionerakupi* ("where did you see me?", the taunt uttered by the spirit) or *zituhwani.* The authors claimed that Shona-speaking informants regarded these spirits as "new" (A. Burbridge, "In Spirit-Bound Rhodesia," *NADA* 2 [1924]; H. C. Hugo, "The Spirit-World of the Mashona," *NADA* 3 [1925]). Given the evident comparisons of *zvidhoma* with the cultural image of San-speaking hunters in local Bantu-speaking societies in some descriptions (they are sometimes said to carry bows and are always dimunitive, and in some cases are said to have grotesque faces or huge penises), and also given that *doma* or *chidhoma* is sometimes used as a term for a San-speaker, the "newness" of *zvidhoma* at any one moment may therefore be drawn from a considerably longer history of metaphors generated by the interaction of Bantu-speakers with the Khoisan peoples of southwest Africa. On the other hand, one informant insisted that *zvidhoma* are the spirits of dead people, generally children, captured by witches (Tapera household, interview, Kambuzuma Township, Harare, May 20, 1991).

58   *Sunday Mail* (Harare), May 5, 1991, p. 1.

59   Tuso Tapera, personal communication.

60   Werbner, *Tears of the Dead,* p. 137.

61   Bessant and Muringai, "Peasants, Businessmen and Moral Economy," p. 581.

62   W. L. Chigidi, *Kwaingovawo Kuedza Mhanza* (Harare: The College Press, 1989), p. 6.

63   Mai Mbiriri, interview, "Joberg" section of Mbare Township, Harare, April 25, 1991.

64   This term appears to have been used in West Africa as well to refer to the users of lighteners during the same era. Obioma Nnaemeka, personal communication, June 1992.

65   Musaemura Zimunya, *Country Dawns and City Lights* (Harare: Longman, 1985), p. 37.

66   "Thorough Revolt," *Zimbabwe News* 3:14 (July 20, 1968), pp. 1–2.

67   Pongweni, *Language,* p. 18.

68   Sekai Nzenza, *Zimbabwean Woman: My Own Story* (London: Karia Press, 1988), p. 135.

69   Ibid., p. 82.

70   Chinodya, *Harvest of Thorns,* p. 80.

71   Nzenza, *Zimbabwean Woman,* p. 28.

72   Chinodya, *Dew in the Morning,* p. 112.

73   Kapura household, interview, "Nationals" section of Mbare Township, Harare, May 9, 1991.

74 Richard and Innocent Katema, interview, "Joburg" section of Mbare Township, Harare, April 24, 1991.

75 Douglas Kadenhe, interview, Harare, May 23, 1991.

76 Mkele, "Advertising," p. 135.

77 See Schmidt, *Peasants, Traders and Wives,* chs. 2 and 3, for a richly detailed account of these developments in Goromonzi District.

78 ZNA S 2113/1–2, Ministry of Information, Papers and Drafts, "Meet the African," 1962.

79 Mkele, "Advertising," p. 129.

80 See Schmidt, *Peasants, Traders and Wives,* and Barnes, "The Fight for Control of Women's Mobility."

81 See, for example, references to the shifting culture and observations of white travelers to Zimbabwe in David Caute, "Marechera in Black and White," in *Cultural Struggle and Development in Southern Africa,* ed. Preben Kaarsholm (London: James Currey Ltd., 1991).

82 ZNA ZBJ 1/2/3, testimony of Location Superintendent, Umtali.

83 ZNA ZBJ 1/2/3, testimony of Location Superintendent, Gwelo.

84 "On the Frontiers of Trade: Cosmetics for the African Woman," *Commerce of Rhodesia,* July 1955.

85 Hammon and Rickard, "How to Develop Sex Appeal."

86 Mai Mbiriri, interview, "Joberg" section of Mbare Township, Harare, April 25, 1991.

87 Rudo Gaidzanwa, *Images of Women in Zimbabwean Literature* (Harare: The College Press, 1985), p. 98.

88 Vambe, *From Rhodesia to Zimbabwe,* p. 20.

89 Vambe, *Ill-Fated People,* p. 200.

90 Zimunya, "Be Warned," *Country Dawns and City Lights,* p. 56.

91 Tapera household, interview, Kambuzuma Township, Harare, May 18, 1991.

92 Charira family, first interview, Murewa, May 2, 1992.

93 Richard and Innocent Katema, interview, "Joburg" section of Mbare Township, Harare, April 24, 1991.

94 Tapera household, interview, Kambuzuma Township, Harare, May 18, 1991.

95 Another context in which this comparison has taken place is with regard to those women who fought in the *chimurenga,* who were regarded by rural villagers and many urban men as promiscuous. See Staunton, *Mothers of the Revolution,* and Frederickse, *None But Ourselves,* for some further discussions of this matter.

96 It is fairly reasonable to suggest that one of the unacknowledged reasons that white officials were reluctant to expel all known African prostitutes from townships in the 1930s and 1940s is that some of these same officials, along with many other white men, were customers of these women. Lawrence Vambe, for example, mentions a number of these cases. (More fragmentary evidence also suggests that brothels operating in the early days of the colony had multiracial workers and

clientele. See, for example, ZNA ORAL/BE 5, Interview, R. G. Bennett.) Also see Pape, "Black and White," for more details.

97   Mr. Muzembe, interview, Murewa, May 7, 1991.

98   Doreen (last name unknown), interview, "Nationals" section of Mbare Township, Harare, May 9, 1992.

99   Mai Mbiriri, interview, "Joberg" section of Mbare Township, Harare, April 25, 1991.

100   A. K. H. Weinrich, *Mucheke: Race, Status and Politics in a Rhodesian Community* (Paris: UNESCO, 1976), pp. 108–9.

101   Israel Ochino, Tauisai Tapera, and others, interview, Kambuzuma Township, Harare, May 3, 1991.

102   Weinrich, *Mucheke*, p. 109.

103   Tashinga squatter camp, second interview, Mbare Township, Harare, April 19, 1991.

104   Chizemba household, interview, Mbare Township, Harare, April 25, 1991.

105   Weinrich, *Mucheke*, pp. 131–32.

106   Ibid., p. 114.

107   James Ferguson, "Mobile Workers, Modernist Narratives: A Critique of the Historiography of Transition on the Zambian Copperbelt," part 1, *Journal of Southern African Studies* 16:4 (December 1990).

108   James Ferguson, "The Country and the City on the Copperbelt," *Cultural Anthropology* 7:1 (1992).

109   Zimunya, "City Lights," in *Country Dawns and City Lights*, p. 46.

110   Zimunya, "O Harare!" in *Country Dawns and City Lights*, p. 74.

111   Musaemura Zimunya, "Kimiso (A Version of Christmas)," in *Patterns of Poetry in Zimbabwe*, ed. Flora Veit-Wild (Gweru: Mambo Press, 1988), p. 71.

112   Chinodya, *Dew in the Morning*, p. 45.

113   Charira family, second interview, Murewa, May 2, 1991.

114   David Coplan, *In Township Tonight* (London: Longman, 1985).

115   Jean Comaroff, *Body of Power, Spirit of Resistance: The Culture and History of a South African People* (Chicago: University of Chicago Press, 1985).

116   See Ranger, *Dance and Society*. Also see Geary, "Patterns from Without," and Hendrickson, forthcoming.

117   See, for example, Walter Rodney's discussion of consumption and the slave trade, in which he writes, "Africans displayed a weakness for and indeed an obsession with European commodities, and, given this fatal flaw, the tragedy unfolded inexorably" (Rodney, *History of the Upper Guinea Coast*, p. 253).

118   Chinodya, *Harvest of Thorns*, p. 181.

119   Fraenkel, *Wayaleshi*, p. 36.

120   Kosmin, "Ethnic and Commercial Relations," p. 179.

121   Vambe, *From Rhodesia to Zimbabwe*, p. 14.

122   Hove, *Confessions of a Wizard*, p. 76.

123   Ranger, *Dance and Society,* p. 164.

124   ZNA NSA 2/3/1, Superintendant of Native Commissioners, Correspondence re: Complaints.

125   Chinodya, *Harvest of Thorns,* p. 3.

126   Owners of sewing machines also had an edge in *Radio Homecraft* and FAWC contests. Some of the prizes had considerable value.

127   See, for example Malan, *My Traitor's Heart,* p. 112.

128   Cornell Butcher, Francis Makosa, and Wellington Chikombero, interview, Harare, November 26, 1990.

129   I was told this in numerous casual conversations with Mbare men who fished regularly in Lake McIlwaine near Harare.

130   Cornell Butcher, Francis Makosa, and Wellington Chikombero, interview, Harare, November 26, 1990.

131   Beyond the forms of popular "common sense" that disparage advertising in general, there have been many instances of such skeptical regard toward commodity culture as a whole in modern American society—fed, in part, by the persistence of early industrial and preindustrial ideals of "moral economy" that are critical of "materialism." The translation of certain advertising symbols and spokespeople (say, "Mrs. Olson" from the coffee advertisements) into comedic icons is one example of this distancing effect. Another comes from the regular outbreak of "urban legends" and rumors about product messages and packaging—as in the "discovery" of hidden symbols and images (for example, in the persistent belief that Procter and Gamble's old corporate logo was a Satanist symbol). The history of advertisers' interest in subliminal and subtextual communication, as well as their more contemporary incorporation of postmodernist technique, suggests that such concern has real foundation. Witness, for example, the almost mischievously obvious penile nose of "Joe Camel" in cigarette ads. This kind of scrutiny in the industrialized West is in many ways closely akin to the history of close observation of product packaging and form found in Zimbabwe's recent history.

132   Mr. Muzembe, interview, Murewa, May 7, 1991.

133   Israel Ochino, Tauisai Tapera, and others, interview, Kambuzuma Township, Harare, May 3, 1992.

134   See Vaughan, *Curing Their Ills,* pp. 191–93, for more details.

135   Tapera household, interview, Kambuzuma Township, Harare, May 18, 1991.

136   Marechera, *The House of Hunger,* p. 24.

137   Chinodya, *Harvest of Thorns,* pp. 19–20.

138   Pongweni, "Language of Advertising," pp. 7–8.

139   Lovemore Makunike, "Advertisers Taking Us for Gullible Idiots," *Herald* November 10, 1990, p. 2.

140   The *chimurenga* interrupted all traveling demonstrations during its most severe phase from 1975 to 1979. Corporations always perceived demonstration vans and

the like as an initial phase in their marketing, as a way to introduce African consumers, especially in rural markets, to new goods. As a consequence, few companies have resumed demonstrations with the intensity or frequency of pre-war operations, though many companies still employ at least some demonstrators in their marketing departments.

141 The general popularity of contests has carried far beyond commercial promotions. Most former FAWC members recall *Radio Homecraft* and related competitions with great fondness. In one recent volume of oral testimonies, one woman remembered, "I was very industrious and did a lot of sewing. I made a lot of nice things which always won prizes at shows. The first time I won an iron and the following year I won a portable radio. . . . When the war came . . . I told my children to take my radio. They still take turns to have it and call it 'our mother's power' because I won it with energy and hard work" (Staunton, *Mothers of the Revolution*, p. 38).

142 I simply cannot say with any confidence whether this is the case based on my own interviews and conversations. No one I spoke to—with the exception of the son of a "truck" trader—recalls any of the institutions characteristic of commercial activity in the 1940s and earlier. I simply have not spoken to enough people or to people resident in some of the more notorious districts at the appropriate time. Changes in distribution and later in the regulation of peasant cultivation may also have obscured the specificity of these earlier experiences.

143 See, for example, Bessant and Muringai, "Peasants, Businessmen and Moral Economy."

144 Elaine S. Abelson, *When Ladies Go A-Thieving: Middle-Class Shoplifters in the Victorian Department Store* (Oxford: Oxford University Press, 1989).

145 "Jarzin Experiments," *Rhodesian Retailer*, p. 17.

146 Annual reports by native commissioners between 1900 and 1940 persistently hinted at evidence that money was being kept hidden in most peasant households, despite programs to encourage banking.

147 Robertson, *Rhodesian Rancher*, p. 46.

148 Hyatt, *Off the Main Track*, p. 42.

149 See, for example, the conversation on page 96 of writer Steven Mpofu's collection of short stories entitled *Shadows on the Horizon* (Harare: Zimbabwe Publishing House, 1984).

150 Richard and Innocent Katema, interview, "Joburg" section of Mbare Township, Harare, April 24, 1991.

151 Market Research Africa (Rhodesia), *Profile of Rhodesia: European* (Salisbury: n.p., 1972).

152 William Reddy, "The Structure of a Cultural Crisis: Thinking About Cloth in France Before and After the Revolution," in *The Social Life of Things*, ed. Arjun Appadurai (London: Cambridge University Press, 1986), p. 261.

BIBLIOGRAPHY

Published Books

Abelson, Elaine S. *When Ladies Go A-Thieving: Middle-Class Shoplifters in the Victorian Department Store.* Oxford: Oxford University Press, 1989.

Appadurai, Arjun, ed. *The Social Life of Things.* Cambridge: Cambridge University Press, 1986.

Ashforth, Adam. *The Politics of Official Discourse in Twentieth-Century South Africa.* Oxford: Clarendon Press, 1992.

Atieno Odhiambo, E. S., and David William Cohen. Siaya. London: James Currey, 1989.

Bain, S. *Homemakers' Training Manual: For Trainers in African Women's Clubs.* Salisbury: The Rhodesia Literature Bureau, 1970.

Balfour, Andrew. *War Against Tropical Disease: Being Seven Sanitary Sermons Addressed to All Interested in Tropical Hygiene and Administration.* London: Wellcome Bureau of Scientific Research, 1920.

Ballhatchet, Kenneth. *Race, Sex and Class Under the Raj: Imperial Attitudes and Policies and Their Critics, 1793–1905.* London: Weidenfeld and Nicholson, 1980.

Beach, D. N. *The Shona and Zimbabwe, 900–1850.* Gweru: Mambo Press, 1980.

Bean, Robert. *The Races of Man: Differentiation and Dispersal of Men.* New York: The University Society, 1935.

Berman, Marshall. *All That Is Solid Melts into Air: The Experience of Modernity.* New York: Penguin Books, 1988.

Bhebe, Ngwabi. *Benjamin Burombo: African Politics in Zimbabwe.* Harare: College Press, 1989.

Bhila, H. H. K. *Trade and Politics in a Shona Kingdom: The Manyika and Their Portuguese and African Neighbours 1571–1902.* Harare: Longman, 1982.

Bigland, Eileen. *Pattern in Black and White.* London: Lindsay Drummond, 1940.

Blair, T. L. V. *Africa: A Market Profile.* New York: Frederick Praeger, 1965.

Blennerhassett, Rose, and Lucy Sleeman. *Adventures in Mashonaland, by Two Hospital Nurses.* London: Macmillan and Co., 1893.

Boggie, Jeannie M., *First Steps in Civilising Rhodesia,* 4th ed. Salisbury: Kingstons Ltd, 1966.

———. *A Husband and a Farm in Rhodesia.* Salisbury: 1939.

———, ed. *Experiences of Rhodesia's Pioneer Women.* Bulawayo: Philpott & Collins, 1938.

Bohannen, Paul, and George Dalton, eds. *Markets in Africa.* Northwestern University African Studies, no. 9. Evanston, Ill.: Northwestern University Press, 1962.

Bottomley, Frank. *Attitudes to the Body in Western Christendom.* London: Lepus Books, 1979.

Bourdieu, Pierre. *Distinction: A Social Critique of the Judgement of Taste.* Trans. Richard Nice. Cambridge: Harvard University Press, 1984.

Bourdillon, M. F. C. *The Shona Peoples: An Ethnography of the Contemporary Shona, with Special Reference to Their Religion.* Gweru: Mambo Press, 1976.

Bristow, Edward J. *Vice and Vigilance: Purity Movements in Britain Since 1700.* London: Gill & Macmillan, 1977.

British Board of Trade. *The African Native Market in the Federation of Rhodesia and Nyasaland.* London: HMSO, 1954.

Broderick, G. E. P. *Personal Hygiene for Elementary Schools in Africa.* London: United Society for Christian Literature, 1939.

Brown, William Harvey. *On the South African Frontier: The Adventures and Observations of an American in Mashonaland and Matabeleland.* New York: Charles Scribner's Sons, 1899.

Bryk, Felix. *Dark Rapture: The Sex Life of the African Negro.* New York: Walden Publications, 1939.

Bullock, Charles. *The Mashona.* Westport, Conn.: Negro Universities Press, 1928, rpt. 1970.

Butler, Judith. *Gender Trouble: Feminism and the Subversion of Identity.* New York: Routledge, 1990.

Caddick, Helen. *A White Woman in Central Africa.* London: T. Fisher Unwin, 1900.

Carnegie Corporation. *Village Education in Africa: Report of the Inter-Territorial Jeanes Conference, Salisbury.* South Africa: Lovedale Press, 1935.

Cartwright, Doris. *How to Wash and Iron Things for Your Family.* Cape Town: Longman, Green and Company, 1951.

Caute, David. *Under the Skin: The Death of White Rhodesia.* Evanston, Ill.: Northwestern University Press, 1983.

Cayleff, Susan E. *Wash and Be Healed: The Water Cure Movement and Women's Health* (Philadelphia: Temple University Press, 1987).

Chapman, James. *Travels in the Interior of South Africa.* 2 vols. London: Bell & Daldy, 1868.

Chidyausiku, Paul. *Broken Roots: A Biographical Narrative on the Culture of the Shona People in Zimbabwe.* Gweru: Mambo Press, 1984.

Chigidi, W. L. *Kwaingovawo Kuedza Mhanza.* Harare: The College Press, 1989.

Chinodya, Shimmer. *Dew in the Morning.* Gweru: Mambo Press, 1982.

———. *Harvest of Thorns.* Harare: Baobob Books, 1989.

Cock, Jacklyn. *Maids and Madams: Domestic Workers under Apartheid.* London: Women's Press, 1989.

Comaroff, Jean. *Body of Power, Spirit of Resistance: The Culture and History of a South African People.* Chicago: University of Chicago Press, 1985.

——, and John Comaroff. *Of Revelation and Revolution: Christianity, Colonialism and Consciousness in South Africa.* Vol. 1. Chicago: University of Chicago Press, 1991.

Coplan, David. *In Township Tonight.* London: Longman, 1985.

Corbin, Alain. *The Foul and the Fragrant: Odor and the French Social Imagination.* Trans. Aubier Montaigne. Cambridge: Cambridge University Press, 1986.

Dangarembga, Tsitsi. *Nervous Conditions.* London: Women's Press Ltd., 1988.

Davidson, A. Francis. *South and South Central Africa: A Record of Fifteen Years' Missionary Labours Among Primitive Peoples.* Elgin, Ill: Brethren Publishing House, 1915.

Douglas, Mary. *Purity and Danger: An Analysis of the Concepts of Pollution and Taboo.* London: Ark Paperbacks, 1966, rpt. 1984.

——, and Baron Isherwood. *The World of Goods.* n.p., 1979.

Du Plessis, P. J., ed. *Consumer Behaviour: A South African Perspective.* Pretoria: Southern Book Publishers, 1990.

Durrell, Gerald. *The Bafut Beagles.* Great Britain: Penguin Books, 1954, reprint 1988.

Elias, Norbert. *The History of Manners.* New York: Pantheon Books, orig. German ed. 1939, English ed. 1978.

Ewen, Stuart. *Captains of Consciousness.* New York: McGraw-Hill Book Company, 1976.

——, and Elizabeth Ewen. *Channels of Desire: Mass Images and the Shaping of American Consciousness.* New York: McGraw Hill, 1982.

Feeley-Harnik, Gillian. *A Green Estate: Restoring Independence in Madagascar.* Washington, D.C.: Smithsonian Institution Press, 1991.

Feher, Michel, ed. *Fragments for a History of the Human Body.* 3 vols. New York: Zone Books, 1989.

Fieldhouse, D. K. *Unilever Overseas: The Anatomy of a Multinational 1895–1965.* Stanford: The Hoover Institution Press, 1978.

Finlason, C. E. *A Nobody in Mashonaland.* London: George Vickers, 1893.

Fraenkel, Peter. *Wayaleshi.* London: Weidenfield and Nicolson, 1959.

Frantz, C., and Cyril Rogers. *Racial Themes in Southern Rhodesia: The Attitudes and Behavior of the White Population.* New Haven: Yale University Press, 1962.

Frederickse, Julie. *None But Ourselves: Masses vs. Media in the Making of Zimbabwe.* Harare: Anvil Press, 1982.

Fripp, Constance, ed. *Gold and The Gospel in Mashonaland.* London: Chatto & Windus, 1949.

Gaidzanwa, Rudo. *Images of Women in Zimbabwean Literature.* Harare: The College Press, 1985.

Gelfand, Michael. *Diet and Tradition in African Culture.* London: E & S Livingstone, 1971.

——. *Growing Up in Shona Society.* Gweru: Mambo Press, 1979.

——. *Mother Patrick and Her Nursing Sisters.* Cape Town: Juta & Co. Ltd., 1964.

———. *Tropical Victory: An Account of the Influence of Medicine on Southern Rhodesia, 1890–1923*. Cape Town: Juta & Co. Ltd., 1953.

Gilman, Sander. *Difference and Pathology: Stereotypes of Sexuality, Race and Madness*. Ithaca: Cornell University Press, 1985.

———. *The Jew's Body*. New York: Routledge, 1991.

Goldstuck, Arthur. *The Rabbit in the Thorn Tree*. Johannesburg: Penguin Books, 1990.

Green, Nick, and Reg Lascaris. *Communication in the Third World: Seizing Advertising Opportunities in the 1990s*. Cape Town: Tafelberg Publishers/Human & Rousseau Ltd., 1990.

Gussman, B. W. *African Life in an Urban Area: A Study of the African Population of Bulawayo*. Vol. 2. Bulawayo: The Federation Office, 1953.

Hammond-Tooke, W. D. *Patrolling the Herms: Social Structure, Cosmology and Pollution Concepts in Southern Africa*. Raymond Dart Lecture no. 18. Institute for the Study of Man in Africa. Johannesburg: Witwatersrand University Press, 1981.

Hanna, A. J. *The Story of the Rhodesias and Nyasaland*. London: Faber and Faber, 1960.

Hansen, Karen Tranberg. *Distant Companions: Servants and Employers in Zambia, 1900–1985*. Ithaca: Cornell University Press, 1989.

———, ed., *African Encounters with Domesticity*. New Brunswick, N.J.: Rutgers University Press, 1992.

Harries, Patrick. *Work, Culture and Identity: Migrant Labourers in Mozambique and South Africa, 1860–1910*. Portsmouth, N.H.: Heinemann, 1994.

Haug, Wolfgang Fritz. *Critique of Commodity Aesthetics: Appearance, Sexuality, and Advertising in Capitalist Society*. Minneapolis: University of Minnesota Press, 1986.

Hole, Hugh Marshall. *Old Rhodesian Days*. London: Frank Cass and Company, 1928.

Hove, Chinjerai. *Bones*. Harare: Baobob Books, 1988.

Hove, M. M. *Confessions of a Wizard*. Gweru: Mambo Press, 1985.

Hyatt, Stanley Portal. *Off the Main Track*. London: T. Werner Laurie, 1911.

———. *The Old Transport Road*. Bulawayo: Books of Rhodesia, 1914, rpt. 1969.

Iliffe, John. *Famine in Zimbabwe*. Gweru: Mambo Press, 1990.

Jeater, Diana. *Marriage, Perversion and Power: The Construction of Moral Discourse in Southern Rhodesia 1894–1930*. Oxford: Clarendon Press, 1993.

Jhally, Sut. *The Codes of Advertising: Fetishism and the Political Economy of Meaning in the Consumer Society*. New York: St. Martin's Press, 1987.

———, Stephen Kline, and William Leiss. *Social Communication in Advertising: Persons, Products and Images of Well-Being*. New York: Methuen, 1986.

Jones, Gareth Stedman. *Outcast London: A Study of the Relationship Between Classes in Victorian Society*. Oxford: Clarendon Press, 1971.

Jones, Greta. *Social Hygiene in Twentieth Century Britain*. London: Croom Helm, 1986.

Kahari, George P. *The Rise of the Shona Novel: A Study in Development, 1890–1984*. Gweru: Mambo Press, 1990.

Kallaway, Peter, and Patrick Pearson. *Johannesburg: Images and Continuities: A History of Working Class Life Through Pictures, 1885–1935*. Johannesburg: Ravan Press, 1986.

Karp, Ivan, and Michael Jackson, eds. *Personhood and Agency: The Experience of Self and Other in African Cultures.* Uppsala Studies in Cultural Anthropology No. 14. Washington, D.C.: Smithsonian Institution Press, 1990.

Kitinya, Ishmael O. *A Guide to Salesmanship for Africa.* Nairobi: Africa Book Services, 1976.

Kemp, Sam. *Black Frontiers: Pioneer Adventures with Cecil Rhodes' Mounted Police in Africa.* London: George G. Harrup and Co., 1932.

Kidd, Dudley. *The Essential Kafir.* London: Adam and Charles Black, 1904.

Knight-Bruce, G. W. H. *Memories of Mashonaland.* London: Edward Arnold, 1895.

Knox, Andrew. *Coming Clean: A Postscript after Retirement from Unilever.* London: William Heinemann Ltd., 1976.

Lan, David. *Guns & Rain: Guerrillas and Spirit Mediums in Zimbabwe.* London: James Currey, 1985.

Leiss, William. *The Limits of Satisfaction: An Essay on the Problem of Needs and Commodities.* Toronto: The University of Toronto Press, 1976.

Lindenbaum, Shirley, and Margaret Lock, eds. *Knowledge, Power, and Practice: The Anthropology of Medicine and Everyday Life.* Berkeley: University of California Press, 1993.

Malan, Rian. *My Traitor's Heart.* London: Vintage, 1990.

Marangwanda, J. W. *Kumazivandadzoka.* Harare: Longman Zimbabwe, 1959, rpt. 1988.

Marechera, Dambudzo. *The House of Hunger.* London: Heinemann International, 1978.

Mayer, Phillip, and Iona Mayer. *Townsmen or Tribesmen: Conservatism and the Process of Urbanization in a South African City.* Cape Town: Oxford University Press, 1961.

McCracken, Grant. *Culture and Consumption: New Approaches to the Symbolic Character of Consumer Goods and Activities.* Bloomington: Indiana University Press, 1988.

McKendrick, Neil. *The Birth of a Consumer Society.* Bloomington: Indiana University Press, 1982.

McNeil, R. T., and M. E. Anderson. *Health Education for Tropical Schools.* London: Collins, 1965.

Medawar, Charles. *Insult or Injury? An Enquiry into the Marketing and Advertising of British Food and Drug Products in the Third World.* London: Social Audit Ltd., 1979.

Morgenster Mission. *Hygiene: Standard VI.* Fort Victoria: Morgenster Mission Press, 1950.

Mort, Frank. *Dangerous Sexualities: Medico-Moral Politics in England Since 1830.* London: Routledge and Kegan Paul, 1987.

Moyo, Aaron C. *Ziva Kwawakabva.* Harare: Longman Zimbabwe, 1977.

Mpofu, Steven. *Shadows on the Horizon.* Harare: Zimbabwe Publishing House, 1984.

Mudenge, S. I. G. *A Political History of Munhumutapa.* Harare: Zimbabwe Publishing House, 1988.

Mudimbe, V. Y. *The Invention of Africa.* Bloomington: Indiana University Press, 1988.

Mukerji, Chandra. *From Graven Images.* New York: Columbia University Press, 1983.

Mungoshi, Charles. *The Setting Sun and the Rolling World.* London: Heinemann International, 1972, rpt. 1989.

Myers, Kathy. *Understains: The Sense and Seduction of Advertising.* London: Comedia Publishing, 1986.

Nyagumbo, Maurice. *With the People: An Autobiography from the Zimbabwe Struggle.* London: Allison & Busby, 1980.

Nzenza, Sekai. *Zimbabwean Woman: My Own Story.* London: Karia Press, 1988.

Official Yearbook of the Colony of Southern Rhodesia. Salisbury: Rhodesian Printing and Publishing Company, Nos. 1–4, n.d.

Pakenham, Valerie. *The Noonday Sun: Edwardians in the Tropics.* London: Methuen, 1985.

Pearson, A., and R. Mouchet. *The Practical Hygiene of Native Compounds in Tropical Africa.* London: Bailliere, Tindall and Cox, 1923.

Peterson, Jeanne. *The Medical Profession in Mid-Victorian England.* Berkeley: University of California Press, 1978.

Phimister, Ian. *An Economic and Social History of Zimbabwe, 1890–1948: Capital Accumulation and Class Struggle.* London: Longman, 1988.

Pieterse, Jan Nederveen. *White on Black: Images of Africa and Blacks in Western Popular Culture.* New Haven: Yale University Press, 1992.

Pratt, Mary Louise. *Imperial Eyes: Travel Writing and Transculturation.* London: Routledge and Kegan Paul, 1992.

Preteceille, Edmond, and Jean-Pierre Terrail. *Capitalism, Consumption and Needs.* Oxford: Basil Blackwell Publishers, 1985.

Rabinbach, Anson. *The Human Motor: Energy, Fatigue and the Origins of Modernity.* Los Angeles: University of California Press, 1990.

Ranger, Terence. *Dance and Society in Eastern Africa, 1890–1970: The Beni Ngoma.* London: Heinemann, 1975.

Reddy, William. *The Rise of Market Culture: The Textile Trade and French Society, 1750–1900.* Cambridge: Cambridge University Press, 1984.

Report of the United Kingdom Trade Mission to the Union of South Africa, Southern Rhodesia and Northern Rhodesia. London: HMSO, 1931.

Reynolds, Barrie. *The Material Culture of the Peoples of the Gwembe Valley.* Manchester: Manchester University Press, 1968.

Richards, Ellen H. *The Cost of Cleanness.* New York: John Wiley & Sons, 1908.

Richards, Thomas. *The Commodity Culture of Victorian England.* Stanford: Stanford University Press, 1990.

Robertson, Wilfred. *Rhodesian Rancher.* London: Blackie & Son Ltd., 1935.

Rodney, Walter. *A History of the Upper Guinea Coast.* New York: Monthly Review Press, 1970.

Rosenthal, Eric. *As Pioneers Still: An Appreciation of Lever Brothers' Contribution to South Africa.* Cape Town, 1961.

Rungano, Kristina. *A Storm Is Brewing.* Harare: Zimbabwe Publishing House, 1984.

Schmidt, Elizabeth. *Peasants, Traders and Wives: Shona Women in the History of Zimbabwe, 1870–1939.* Portsmouth, N.H.: Heinemann Educational Books, 1992.

Searle, G. *Eugenics and Politics in Britain 1900–1914*. Leyden: Nordhoff International, 1976.

Shamuyarira, Nathan M. *Crisis in Rhodesia*. London: Andre Deutsch Ltd., 1965.

Sibenke, Ben. *My Uncle Grey Bhonzo*. Harare: Longman Zimbabwe, 1982.

Sinclair, John. *Images Incorporated: Advertising as Industry and Ideology*. London: Croom Helm, 1987.

Smith, Edwin, and Andrew Dale. *The Ila-Speaking Peoples of Northern Rhodesia*. New York: University Books.

Staunton, Irene, ed. *Mothers of the Revolution*. Harare: Baobab Books, 1990.

Stepan, Nancy. *The Idea of Race in Science: Great Britain 1800–1960*. Hamden, Conn.: Archon Books, 1982.

Stocking, George W. Jr. *Victorian Anthropology*. New York: The Free Press, 1987.

Style, Colin, and O-lan Style, eds. *Mambo Book of Zimbabwean Verse in English*. Gweru: Mambo Press, 1986.

Summers, Carol. *From Civilization to Segregation: Social Ideals and Social Control in Southern Rhodesia*. Athens: Ohio University Press, 1994.

Tabler, Edward, ed. *To The Victoria Falls Via Matabeleland: The Diary of Major Henry Stabb, 1875*. Cape Town: C. Struik Ltd., 1967.

Taussig, Michael. *The Devil and Commodity Fetishism in South America*. Chapel Hill, N.C.: University of North Carolina, 1980.

Theal, George McCall. *The Yellow and Dark-Skinned People of Africa South of the Zambesi*. New York: Negro Universities Press, 1910, rpt. 1963.

Thomas, Nicolas. *Entangled Objects: Exchange, Material Culture and Colonialism in the Pacific*. Cambridge, Mass.: Harvard University Press, 1991.

Thomas, Thomas Morgan. *Eleven Years in Central South Africa*. London: John Snow and Co., 1872.

The Urban Bantu Market: Understanding Its Complexities and Developing Its Potential. Seminar Proceedings, Durban. Johannesburg: National Development and Management Foundation, 1969.

Vail, Leroy, ed. *The Creation of Tribalism in Southern Africa*. Los Angeles: University of California Press, 1989.

Vambe, Lawrence. *From Rhodesia to Zimbabwe*. Pittsburgh: University of Pittsburgh Press, 1976.

———. *An Ill-Fated People: Zimbabwe Before and After Rhodes*. Pittsburgh: University of Pittsburgh Press, 1972.

Van Nitsen, R. *L'Hygiène des Travailleurs Noirs dans les Camps Industriels du Haut-Katanga*. Bruxelles: Marcel Hayez pour L'Academie Royale de Belgique, 1933.

Van Onselen, Charles. *Chibaro: African Mine Labour in Southern Rhodesia, 1900–1933*. London: Pluto Press, 1976.

Vansina, Jan. *Living with Africa*. Madison: University of Wisconsin Press, 1994.

Vaughan, Megan. *Curing Their Ills: Colonial Power and African Illness*. Stanford: Stanford University Press, 1991.

Veit-Wild, Flora. *Patterns of Poetry in Zimbabwe*. Gweru: Mambo Press, 1988.

Vigarello, George. *Concepts of Cleanliness: Changing Attitudes in France Since the Middle Ages*. Trans. Jean Birrell. Cambridge: Cambridge University Press, 1988.

Walkowitz, Judith. *Prostitution and Victorian Society: Women, Class, and the State*. Cambridge: Cambridge University Press, 1980.

Wallis, J. P. R., ed. *The Matabele Journals of Robert Moffat*. 2 vols. No. 1: Oppenheimer Series. London: Chatto & Windus, 1945.

——, ed., *The Matabele Mission: A Selection from the Correspondence of John and Emily Moffat, David Livingstone and Others, 1858–1878*. No. 2: Oppenheimer Series. London: Chatto & Windus, 1945.

Weinrich, A. K. H. *Mucheke: Race, Status and Politics in a Rhodesian Community*. Paris: UNESCO, 1976.

——. *Women and Racial Discrimination in Rhodesia*. Paris: UNESCO, 1979.

Werbner, Richard. *Tears of the Dead*. Washington, D.C.: Smithsonian Institution Press, 1991.

Werner, A. *The Natives of British Central Africa*. New York: Negro Universities Press, 1906, rpt. 1969.

White, Landeg. *Magomero: Portrait of an African Village*. Cambridge: Cambridge University Press, 1987.

Williams, Marilyn Thornton. *Washing 'The Great Unwashed': Public Baths in Urban America, 1840–1920*. Columbus: Ohio University Press, 1991.

Williams, Myfawny. *A Child's History of Rhodesia*. London: Simpkin, Marshall, Hamilton, Kent & Co. Ltd., 1925.

Williams, Rosalind H. *Dream Worlds: Mass Consumption in Late Nineteenth-Century France*. Los Angeles: University of California Press, 1982.

Williamson, Judith. *Decoding Advertisements: Ideology and Meaning in Advertising*. London: Marion Boyars, 1978.

Wilson, Charles. *The History of Unilever*. Vols 1 and 2. London: Cassell & Company, 1954.

Wingate, A. P. *Doctor Kalulu*. London: Longman, Green & Co., 1956.

Wolcott, Harry F. *The African Beer Gardens of Bulawayo*. New Brunswick, N.J.: Rutgers Center of Alcohol Studies, 1974.

Wood, Margaret. *Pastels under the Southern Cross*. London: Smith & Elder, 1911.

Wright, Fergus Chalmers. *African Consumers in Nyasaland and Tanganyika*. London: HMSO, 1955.

Wright, Lawrence. *Clean and Decent: The Fascinating History of the Bathroom and the Water Closet*. London: Routledge & Kegan Paul, 1960.

Zimunya, Musaemura. *Country Dawns and City Lights*. Harare: Longman, 1985.

Journal Articles and Chapters in Books

Adas, Michael. "Scientific Standards and Colonial Education in British India and French Senegal." In *Science, Medicine and Cultural Imperialism*, ed. Teresa Meade and Mark Walker. n.p., 1991.

Appadurai, Arjun. "Disjuncture and Difference in the Global Cultural Economy." *Public Culture* 2:2 (Spring 1990).

Arrighi, Giovanni. "Labour Supplies in Historical Perspective: A Study of the Proletarianization of the African Peasantry in Rhodesia." In *Essays on the Political Economy of Africa,* ed. Giovanni Arrighi and John Saul. New York: Monthly Review Press, 1973.

Barnes, Teresa A. "The Fight for Control of African Women's Mobility in Colonial Zimbabwe, 1900–1939." *Signs* (Spring 1992).

Baudrillard, Jean. "For a Critique of the Political Economy of the Sign." In *Selected Writings of Jean Baudrillard,* ed. Mark Poster. Stanford: Stanford University Press, 1988.

Bertram, N. R. "An Industrialist's View of the Determinants Required for Growth of Secondary Industry in Rhodesia." *Rhodesian Journal of Economics* 3:2 (June 1969).

Bessant, Leslie, and Elvis Muringai. "Peasants, Businessmen and Moral Economy in the Chiweshe Reserve, Colonial Zimbabwe, 1930–1968." *Journal of Southern African Studies* 19:4 (December 1993).

Brackett, D. G., and M. Wrong. "Notes on Hygiene Books Used in Africa." *Africa* 3:4 (October 1930).

Bruce-Bays, J. "The Injurious Effects of Civilization Upon the Physical Condition of the Native Races of South Africa." In *Report of the South African Association for the Advancement of Science.* Sixth Meeting, Grahamstown 1908. Capetown: The South African Association for the Advancement of Science, 1909.

Buckle, Louis. "Marketing for the African Consumer." *Marketing Rhodesia* 1:1 (1972).

Burbridge, A. "In Spirit-Bound Rhodesia." *NADA* 2 (1924).

Burke, John P. "Reification and Commodity Fetishism Revisited." *Canadian Journal of Political and Social Theory* 3:1 (Winter 1979).

Cahan, Theresa. "Secondary Industries for Tropical Africa." *Africa* 14:4 (October 1943).

"Catechism of Health." *Rhodesia Native Quarterly* 1:1 (July–September 1926).

Caute, David. "Marechera in Black and White." In *Cultural Struggle and Development in Southern Africa,* ed. Preben Kaarsholm. London: James Currey Ltd., 1991.

Challis, R. J. "The European Educational System in Southern Rhodesia, 1890–1939." Supplement to *Zambezia.* Salisbury: University of Zimbabwe, 1980.

Chanock, Martin. "Making Customary Law: Men, Women, and Courts in Colonial Northern Rhodesia." In *African Women and the Law: Historical Perspectives,* ed. Margaret Jean Hay and Marcia Wright. Boston: Boston University Press.

Cohen, David William. "The Undefining of Oral Tradition." *Ethnohistory* 36:1 (Winter 1989).

Comaroff, Jean. "The Diseased Heart of Africa: Medicine, Colonialism and the Black Body." In *Knowledge, Power and Practice,* ed. Shirley Lindenbaum and Margaret Lock. Berkeley: University of California Press, 1993.

———, and John Comaroff. "Christianity and Colonialism in South Africa." *American Ethnologist* 13 (1986).

——. "Goodly Beasts, Beastly Goods: Cattle and Commodities in a South African Context." *American Ethnologist* 17:2 (May 1990).

Cowen, Michael, and Robert Shenton. "The Origin and Course of Fabian Socialism in Africa." *Journal of Historical Sociology* 4:2 (June 1991).

Daffy, Mike. "Radio: The Mass Medium with Personal Appeal." *Rhodesian Retailer* (July 1973).

Davin, Anna. "Imperialism and Motherhood." *History Workshop* (Spring 1978).

Davis, Benjamin, and Wolfgang Dopcke. "Survival and Accumulation in Gutu: Class Formation and the Rise of the State in Colonial Zimbabwe, 1900–1939." *Journal of Southern African Studies* 14:1 (October 1987).

Dell, M. M. "First Address, 'Bantu Market' Session." *Third Advertising Convention in South Africa: The Challenge of a Decade.* Johannesburg: Statistic Holdings Ltd., September 1960.

Denoon, Donald. "Temperate Medicine and Settler Capitalism: On the Reception of Western Medical Ideas." In *Disease, Medicine and Empire: Perspectives on Western Medicine and the Experience of European Expansion,* ed. Roy MacLeod and Milton Lewis. London: Routledge, 1988.

Du Bow, Saul. "Mental Testing and the Understanding of Race in Twentieth-Century South Africa." In *Science, Medicine and Cultural Imperialism,* ed. Teresa Meade and Mark Walker. New York: St. Martin's Press, 1991.

Duggan, William R. "The Rural African Middle Class of Southern Rhodesia." *African Affairs* 79:315 (April 1980).

Dyhouse, Carol. "Working-Class Mothers and Infant Mortality in England, 1895–1914," *Journal of Social History* 12:2 (1978).

Edwards, William. "From Birth to Death." *NADA* 7 (1929).

Eiselen, Werner Willi Max. "The Elasticity of the Bantu Consumer." *Second Advertising Convention in South Africa.* Durban: Society of Advertisers, 1959.

Eltis, David. "The Relative Importance of Slaves and Commodities in the Atlantic Trade of Seventeenth-Century Africa." *The Journal of African History,* 35:2 (1994).

Erlanger, Steven. "'Dollar Apartheid' Makes a Few Russians Rich but Resented." *New York Times,* August 23, 1992.

Ferguson, James. "The Country and the City on the Copperbelt." *Cultural Anthropology* 7:1 (1992).

——. "The Cultural Topography of Wealth: Commodity Paths and the Structure of Property in Rural Lesotho." *American Anthropologist* 94:1 (March 1992).

——. "Mobile Workers, Modernist Narratives: A Critique of the Historiography of Transition on the Zambian Copperbelt." Part 1. *Journal of Southern African Studies* 16:4 (December 1990).

——, and Akhil Gupta. "Beyond 'Culture': Space, Identity and the Politics of Difference." *Cultural Anthropology* 7:1 (February 1992).

Flood, David. "The Jeanes Movement: An Early Experiment." *NADA* 10:3 (1971).

Gargett, E. S. "The African in the Market Place." *Marketing Rhodesia* 3:2 (November 1974).

Gelfand, Michael. "Who Is Rich and Poor in Traditional Shona Society." *NADA* 10:4 (1971).

——, and Yvonne Swart. "The Nyora." *NADA* 30 (1953).

Gell, Alfred. "Newcomers to the World of Goods: Consumption among the Muria Gonds." In *The Social Life of Things*, ed. Arjun Appadurai. London: Cambridge University Press, 1986.

Geras, Norman. "Essence and Appearance: Aspects of Fetishism in Marx's *Capital*." *New Left Review* 65 (1971).

Hall, Catherine. "The Early Formation of Victorian Domestic Ideology." In *The Victorian Family: Structures and Stresses*, ed. Sandra Burman. London: Croom Helm, 1978.

Hamilton, C. A., and John Wright. "The Making of the amaLala: Ethnicity, Ideology and Relations of Subordination in a Precolonial Context." *South African Historical Journal* 22 (November 1990).

Hansen, Karen Tranberg. "White Women in a Changing World: Employment, Voluntary Work and Sex in Post–World War II Northern Rhodesia." In *Western Women and Imperialism: Complicity and Resistance*, ed. Nupur Chaudhuri and Margaret Strobel. Bloomington: Indiana University Press, 1992.

Heath, Deborah. "Fashion, Anti-Fashion and Heteroglossia in Urban Senegal." *American Ethnologist* 19:1 (February 1992).

Hillis, J. G. "An Account of a Market Research in a Tribal Trust Area." *NADA* 10:3 (1971).

Howland, F. N. "Place of Advertising in Native Trade." *Commerce of Rhodesia* (October 1953).

Hoy, Suellen M. "Municipal Housekeeping: The Role of Women in Improving Urban Sanitation Practices, 1800–1917." In *Pollution and Reform in American Cities*, ed. Martin Melosi. Austin: University of Texas Press, 1980.

Hugo, H. C. "The Spirit-World of the Mashona." *NADA* 3 (1925).

Hunt, Nancy Rose. "Colonial Fairy Tales and the Knife and Fork Doctrine in the Heart of Africa." In *African Encounters with Domesticity*, ed. Karen Tranberg Hansen. New Brunswick, N.J.: Rutgers University Press, 1992.

——. "Domesticity and Colonialism in Belgian Africa: Usumbura's *Foyer Social*, 1946–1960." *Signs: A Journal of Women in Culture and Society* 15:3 (Spring 1990).

"Interview with Ben Mucheche." *Commerce Rhodesia* (June 1974).

Jacques, A. A. "Teaching of Hygiene in Native Primary Schools." *Africa* 3:4 (October 1930).

"Jarzin Experiments: Shop for African Farmers." *Rhodesian Retailer* (September/October 1974).

"Jazz Takes Its First Step Towards Chain Operation." *Rhodesian Retailer* (May/June 1974).

Johnson, R. W. M. "An Analysis of African Family Expenditure." *Rhodesian Journal of Economics* 5:1 (1971).

Kanduza, Ackson. "The Tobacco Industry in Northern Rhodesia, 1912–1938." *International Journal of African Historical Studies* 13:2 (1980).

Kearney, Michael. "Borders and Boundaries of State and Self at the End of Empire." *Journal of Historical Sociology* 4:1 (March 1991).

Kileff, Clive. "Black Suburbanites: An African Elite in Salisbury, Rhodesia." In *Urban Man in Southern Africa*, ed. Clive Kileff and Wade C. Pendleton. Gweru: Mambo Press, 1975.

Kontos, Alexis. "Through a Glass Darkly: Ontology and False Needs." *Canadian Journal of Political and Social Theory* 3:1 (Winter 1979).

Kopytoff, Igor. "The Cultural Biography of Things: Commoditization as Process." In *The Social Life of Things*, ed. Arjun Appadurai. London: Cambridge University Press, 1986.

Kosmin, Barry A. "The Inyoka Tobacco Industry of the Shangwe People: A Case Study of the Displacement of a Pre-Colonial Economy in Southern Rhodesia, 1898–1938." *African Social Research* 17 (June 1974).

"Kumboyedza Looks for More Outlets." *Rhodesian Retailer* (September 1973).

Langdon, Steven. "Multinational Corporations, Taste Transfer and Underdevelopment: A Case Study from Kenya." *Review of African Political Economy* 2 (1975).

Lewis, Jane. *The Politics of Motherhood: Maternal and Child Welfare in England 1900–39* (London: Croom Helm, 1980).

Livant, Bill. "The Audience Commodity: On the Blindspot Debate." *Canadian Journal of Political and Social Theory* 3:1 (1979).

———. "The Imperial Cannibal." In *Cultural Politics in Contemporary America*, ed. Ian Argus and Sut Jhally. New York: Routledge, 1989.

MacVicar, Neil. "Health Teaching in Schools, With Special Reference to Native Schools." In *Report of the South African Association for the Advancement of Science*. Sixth Meeting, Grahamstown 1908. Capetown: The South African Association for the Advancement of Science, 1909.

Marks, Shula, and Neil Andersson. "Typhyus and Social Control: South Africa, 1917–1950." In *Disease, Medicine and Empire: Perspectives on Western Medicine and the Experience of European Expansion*, ed. Roy MacLeod and Milton Lewis. London: Routledge, 1988.

Martin, C. "Manyika Beads of the XIX Century." *NADA* 17 (1940).

Marolen, P. P. "Tomorrow Has Already Dawned: The Aspirations of the African Consumer and Businessman." *Marketing Rhodesia* 3:3 (February 1975).

Maroun, J. E. "Second Address, 'Bantu Market' Session." *Third Advertising Convention in South Africa: The Challenge of a Decade.* Johannesburg: Statistic Holdings Ltd., September 1960.

McGee, T. G. "Mass Markets—Little Markets: Some Preliminary Thoughts on the Growth of Consumption and Its Relationship to Urbanization: A Case Study of Malaysia." In *Markets and Marketing*, ed. Stuart Plattner. Lantham, Mass: University Press of America, 1985.

Mkele, Nimrod. "Advertising to the Bantu." *Second Advertising Convention in South Africa.* Durban: Society of Advertisers, 1959.

Millman, W. "Health Instruction in African Schools: Suggestions for a Curriculum." *Africa* 3:4 (October 1930).

Molefe, Philip, and Phillip Van Niekerk. "Adventures of a Black Man in a White Town." *Weekly Mail* 6:40 (October 19–25, 1990).

Moore, David B. "The Ideological Formation of the Zimbabwean Ruling Class." *Journal of Southern African Studies* 17:3 (September 1991).

Moyo, Judith. "Nyamakondo—Haunt of the Kingfishers." In *Mambo Book of Zimbabwean Verse in English,* ed. Colin Style and O-lan Style. Gweru: Mambo Press, 1986.

Mudenge, S. I. G. "The Role of Foreign Trade in the Rozvi Empire: A Reappraisal." *Journal of African History* 15:3 (1974).

"Natives Have £20,000,000 to Spend." *Commerce of Rhodesia* 4 (1953).

Nyereyegona, Charles. "Marketing to the Urban African." *Marketing Rhodesia* 1:4 (May 1973).

O'Grady, W. H. Opening Address. *Second Advertising Convention in South Africa.* Durban: Society of Advertisers, 1959.

"On the Frontiers of Trade: Cosmetics for the African Woman." *Commerce of Rhodesia,* July 1955.

"Our Twenty Years of Progress." *The Nutshell* (Lever Brothers) 1:13 (Summer 1968).

Padayachee, Vishnu, and Robert Morrell. "Indian Merchants and *Dukawallahs* in the Natal Economy, c. 1875–1914." *Journal of Southern African Studies* 17:1 (March 1991).

Pape, John. "Black and White: The 'Perils of Sex' in Colonial Zimbabwe." *Journal of Southern African Studies* 16:4 (December 1990).

Pelling, Margaret. "Appearance and Reality: Barber-Surgeons, the Body and Disease." In *London 1500–1700: The Making of the Metropolis.* London: Longman, 1986.

"People in Jobs." *The Nutshell* 1:14 (Winter 1968).

Phimister, Ian. "Secondary Industrialisation in Southern Africa: The 1948 Customs Agreement Between Southern Rhodesia and South Africa." *Journal of Southern African Studies* 17:3 (September 1991).

Preston, Elsie K., and Dorothy Stebbing. "Life and Work of Miss Catherine Mabel Langham." *NADA* 10:2 (1970).

"Profile of a Commercialist—Benny Goldstein." *Commerce of Rhodesia* 2:7 (July 1951).

"Questions From 'Bantu Market' Session." *Third Advertising Convention in South Africa: The Challenge of a Decade.* Johannesburg: Statistic Holdings Ltd., September 1960.

Ranchod-Nilsson, Sita. "'Educating Eve': The Women's Club Movement and Political Consciousness Among Rural African Women in Southern Rhodesia, 1950–1980." In *African Encounters with Domesticity,* ed. Karen Tranberg Hansen. New Brunswick, N.J.: Rutgers University Press, 1992.

Ranger, T. O. "Literature and Political Economy: Arthur Shearly Cripps and the Makoni Labour Crisis of 1911," *Journal of Southern African Studies* 9:1 (October 1982).

Reddy, William. "The Structure of a Cultural Crisis: Thinking About Cloth in France Before and After the Revolution." In *The Social Life of Things,* ed. Arjun Appadurai. London: Cambridge University Press, 1986.

Rose, Nikolas. "Fetishism and Ideology: A Review of Theoretical Problems." *Ideology and Consciousness* 2 (1977).

Shaul, J. H. "Distributive Trades of Southern Rhodesia." *South African Journal of Economics* 21:2 (June 1953).

Simpson, Janice C. "Buying Black." *Time* 140:9 (August 31, 1992).

Smout, M. A. H. "Shopping Centres and Shopping Patterns in Two African Townships of Greater Salisbury." *Zambezia* 2:1 (December 1971).

Smythe, Dallas. "Communications: Blindspot of Western Marxism." *Canadian Journal of Political and Social Theory* 1:3 (Fall 1977).

"Some Notes on the Mrowzi Occupation of the Sebungwe District." *NADA* 21 (1944).

Swanson, Maynard. "The Sanitation Syndrome: Bubonic Plague and Urban Native Policy in the Cape Colony, 1900–1909." *Journal of African History* 18:3 (1977).

Tabor, E. G. "Is There an Excalibur of the African Market?" *Marketing Rhodesia* 5:3 (March 1977).

Taylor, Guy. "The Matabele Head Ring." *NADA* 3 (1925).

———. "Primitive Colour Vision." *NADA* 1 (1923).

Thornton, Steven. "The Struggle for Profit and Participation by an Emerging African Petty-Bourgeoisie in Bulawayo, 1893–1933." In *Societies of Southern Africa*, vol. 9. University of London, 1980.

Van den Bergh, Joe. "Marketing in the Tribal Trust Lands." *Marketing Rhodesia* 2:3 (February 1974).

Van der Reis, A. J. "Motivational Factors in Bantu Buying Behaviour." *Research Report no. 15*. Pretoria: Bureau of Market Research, 1966.

Vickery, K. P. "Saving Settlers: Maize Control in Northern Rhodesia." *Journal of Southern African Studies* 11 (1985).

White, Luise. "Blood Brotherhood Revisited: Kinship, Relationship, and the Body in East and Central Africa." *Africa* 64:3 (1994).

Wild, Volker. "Black Competition or White Resentment? African Retailers in Salisbury, 1935–1953." *Journal of Southern African Studies* 17:2 (June 1991).

Wilk, Richard. "Consumer Goods as Dialogue About Development." *Culture and History* 7 (1992).

Wilkie, Jacqueline S. "Submerged Sensuality: Technology and Perception of Bathing." *Journal of Social History* (Summer 1986).

Williams, Raymond. "Advertising: The Magic System." In *Problems in Materialism and Culture*, London: Verso, 1980.

Worboys, Michael. "Manson, Ross and Colonial Medical Policy: Tropical Medicine in London and Liverpool, 1899–1914." In *Disease, Medicine and Empire: Perspectives on Western Medicine and the Experience of European Expansion*, ed. Roy MacLeod and Milton Lewis. London: Routledge, 1988.

Wright, John. "Politics, Ideology, and the Invention of the 'Nguni.'" In *Resistance and Ideology in Settler Societies*, ed. Tom Lodge. Vol. 4. Southern African Studies Series. Johannesburg: Ravan Press, 1986.

Zimunya, Musaemura. "Kimiso (A Version of Christmas)." In *Patterns of Poetry in Zimbabwe,* ed. Flora Veit-Wild. Gweru: Mambo Press, 1988.

Newspapers

Articles from Zimbabwean newspapers and other newspapers used generally in this study are listed below.

Bantu Mirror (Native Mirror)
"Let's Ask Questions," *Native Mirror* February 18, 1939.
"Let's Ask Questions," *Native Mirror* May 6, 1939.
"Let's Ask Questions." *Native Mirror* January 6, 1940.
"Mtshabezi Mission." *Native Mirror* 6 (1932).
"Raise Your Own Race." *Native Mirror* 2:3 (1934).
"Report of the Southern Rhodesian Missionary Conference." *Native Mirror* 7 (July 1932).
"To Keep the Blanket Full." *Native Mirror* 2 (1931).
"The Training of African Mothers." *Native Mirror* 14 (April 1934).
"Wayfaring." *Native Mirror* 2 (1931).
"Your Clothes! Do Fine Feathers Make Fine Birds?" *Native Mirror* 10 (April 1933).
Moyo, H. M. Davidson. "Letter to the Editor." *Native Mirror* 2:5 (1934).
Ndzou, Peter. "Letters to the Editors: Digests." *Native Mirror* 2:10 (1935).
Rudd, Joyce. "Make Your Own Things." *Native Mirror* 2:2 (1934).
Sigidi, S. N. (Tjolotjo Government School). "Letter to the Editor." *Native Mirror* 9 (1932).
Sithole, C. C. Freshman. "Letter to the Editor." *Native Mirror* 2:3 (1934).
Sobantu, J. H. "Letter to the Editor." *Native Mirror* 10 (April 1933).
Wright, H. H. Morley. "Wake Up Africans! A Call to the Africans." *Native Mirror* 12 (October 1933).
"Chiweshe Reserve News." *Bantu Mirror* October 30, 1937.
"The Home Teacher." *Bantu Mirror* September 10, 1960.
"Mr J. H. Hofmeyr on Native Education." *Bantu Mirror* January 21, 1938.
"Let's Ask Questions: Good Old Habits of the Bantu." *Bantu Mirror* April 1, 1939.
"Pathfinders: Manners Maketh the Man." *Bantu Mirror* February 27, 1937.
"The Road to Civilisation." *Bantu Mirror* November 4, 1944.
"The Segregation Question, Part IX." *Bantu Mirror* August 14, 1937.
"Women's Column." *Bantu Mirror* April 29, 1939.
Manyawu, M. G. "The Africans and Comfort." *Bantu Mirror* January 12, 1946.
Maposa, S. "Cleanliness Is the Foundation of Christianity." *Bantu Mirror* February 4, 1938.
Sigauke, M. Wilbert. "Do You Shake Hands?" *Bantu Mirror,* September 2, 1939.

Sithole, Ndabaningi. "African Progress in the New World Order." *Bantu Mirror* March 11, 1944.
——. "Letter to the Editor." *Bantu Mirror* September 2, 1944.

African Parade (Parade and Foto-Action)
Hammon, Ellen, and Ken Rickard. "How to Develop Sex Appeal." *Parade and Foto-Action* May 1969.

The Citizen
"The Truth About the Colour Bar," *The Citizen* (Salisbury), February 24, 1956.

Rhodesia Herald (Harare Herald)
Makunike, Lovemore. "Advertisers Taking Us for Gullible Idiots." *Herald* November 10, 1990.

Others
African Daily News
Bulawayo Chronicle
Central African Market
Central African Merchant
Moto
The Rhodesian Grocer

Manuscripts, Dissertations, and Unpublished Materials

Adegbija, Efurosibinia Emmanuel. "A Speech Act Analysis of Consumer Advertisements." Ph.D. diss., Indiana University, 1982.
Bhebe, Ngwabi. "The Ndebele Trade in the Nineteenth Century." University of Rhodesia, Department of History. Henderson Seminar, paper no. 25, August 1973. Unpublished ms.
British Red Cross Society. *How to Keep Healthy.* Red Cross Booklet no. 1. Lusaka: Publications Bureau, circa 1940s [?].
Burns, Catherine. "Reproductive Labors: The Politics of Women's Health in South Africa, 1900 to 1960." Ph.D. diss., Northwestern University, 1994.
Dillon Ad Signs Ltd. "A Study into Hoarding Audiences Amongst the African." Salisbury, October 1976.
Dillon Enterprises. *Rhodesian African Cinema Survey.* Salisbury, 1972.
Feldman-Savelsberg, Pamela. "'Then We Were Many': Bangangté Women's Conceptions of Health, Fertility and Social Change in a Bamiléké Chiefdom, Cameroon." Ph.D. diss., Johns Hopkins University, 1989.
Geary, Christraud. "Patterns from Without, Meaning from Within: European-Style Military Dress and German Colonial Politics in the Bamum Kingdom (Camer-

oon)." Discussion Papers in the African Humanities, no. 1. Boston: Boston University African Studies Center, 1989.

Kosmin, Barry. "Ethnic and Commercial Relations in Southern Rhodesia: A Socio-Historical Study of the Asian, Hellene and Jewish Populations 1898–1943." Ph.D. diss., University of Rhodesia, 1974.

Machingaidze, V. E. M. "The Development of Settler Capitalist Agricultue in Southern Rhodesia." D.Phil. diss., University of London, 1980.

Margolis, William. "The Position of the Native Population in the Economic Life of Southern Rhodesia." Thesis, Pretoria: University of South Africa, 1938.

Market Research Africa (Rhodesia). *Consumer Contact 1974: Readership Data.* Salisbury, 1975.

Market Research Africa (Rhodesia). *National Listenership Survey: RBC African Service.* Salisbury, 1971.

Market Research Africa (Rhodesia). *Product Data: Consumer Close-Up.* Salisbury, 1972.

Market Research Africa (Rhodesia). *Profile of Rhodesia: African.* Salisbury, 1970.

Market Research Africa (Rhodesia). *Profile of Rhodesia: European.* Salisbury, 1970.

Pongweni, Alec J. "The Language of Advertising and Its Sociocultural Implications for the Consumer." Centre for Applied Social Sciences, University of Zimbabwe, October 1983. Unpublished ms.

Rhodesian Milling and Manufacturing Company. "The Story of Atlas and Atlas Products." Salisbury: Rhodesian Milling and Manufacturing Company, 1934.

Rowlands, Michael. "The Consumption of an African Modernity." University College, London. Unpublished ms.

Samkange, S. J. T., and T. M. Samkange. *Survey of Personal, Employment, Expenditure and Accomodation Data of African Employees of the Rhodesian Milling Company.* Salisbury: Rhodesian Milling and Manufacturing Company, 1960.

Setel, Philip. "'A Good Moral Tone': Victorian Ideals of Health and the Judgement of Persons in Nineteenth-Century Travel and Mission Accounts from East Africa." *Working Papers in African Studies, no. 150.* Boston: Boston University African Studies Center, n.d.

Seya, Pierre Thizier. "Transnational Capitalist Ideology and Dependent Societies: A Case Study of Advertising in the Ivory Coast." Ph.D. diss., Stanford University, 1981.

Spitulnik, Debra. "Radio Culture in Zambia: Audiences, Public Words, and the Nation-State." Ph.D. dissertation, University of Chicago, 1994.

Spring, Anita. "Women's Rituals and Natality Among the Luvale of Zambia." Ph.D. diss., Cornell University, September 1976.

Waters, M. "Home Economics and Practical Hygiene." Department of Native Education, occasional paper no. 1. Salisbury: Government Printers, May 1929.

West, Michael O. "African Middle-Class Formations in Colonial Zimbabwe, 1890–1965." Ph.D. diss., Harvard University, 1990.

Yoshikune, Tseuneo, "Black Migrants in a White City: A Social History of African Harare, 1890–1925." Ph.D. diss., University of Zimbabwe, 1989.

Archival Records

Zimbabwe National Archives (ZNA) Harare, Zimbabwe

Department of Native Affairs

N 9/1/1–26	Annual Reports.
S 1051	Annual Reports.
S 2076	Annual Reports.
S 1563	Annual Reports.
N 4/1/1	Chief Native Commissioner, Circular, July 23, 1903.
N 9/5/8	Chief Native Commissioner, Correspondence re: School Inspections, 1922–1923.
N 6/2/1–3	Superintendents of Natives Conference, 1909.
N 9/5/8	Correspondence re: School Inspections 1922–1923.
N 3/7/2	Chief Native Commissioner, Correspondence.
N 3/9/1–2	Chief Native Commissioner, Correspondence re: Education, 1920–1923.
N 3/33/2	Chief Native Commissioner, Correspondence.
N 3/24/33	Correspondence re: "Undesirable Europeans," 1915–1923.
NSA 2/3/1	Superintendent of Native Commissioners, Correspondence re: Complaints.
S 1542/L11	Correspondence re: Newspapers and Literature, 1933–1940.
S 2390/751/39	Correspondence re: Regulation of Advertisements, 1927–1932.
S 246/727	Correspondence.
S 2385	Chief Native Commissioner's Office, Correspondence re: African Trading Stores, 1947–1950.
S 1542/S9	Correspondence, 1933–1935.
S 1542/T11	Correspondence, 1936–1947.
S 1542/S12	Correspondence.
S 1542/S2	Chief Native Commissioner, Correspondence and Papers re: Schools, 1933–1935.
S 1561/S6	Correspondence re: Jeanes Teachers, 1930–1938.
S 1542/J1	Correspondence re: Jeanes Teachers, 1932–1943.

Department of Native Education/Department of Education
Report of the Director of Native Education.

E 2/5/6	Correspondence and Curricula, 1890–1923.
E 5/2/1/1–2	Correspondence re: School Inspection, 1890–1923.
S 840/2/37	Correspondence, 1923–1926.
S 2076	Correspondence and Reports, 1928–1950.
S 840/S/41	Syllabi for Native Training Schools, 1923.
S 840/2/16	Correspondence re: Laundry, Monte Cassino Mission, 1920–1927.
S 840/2/4	Correspondence re: Laundry, St. Monica's Mission Penhalonga, 1913–1927.

S 840/1/21   Correspondence re: Laundry Instruction, Hope Fountain Mission, 1921–1927.
S 840/1/37   Correspondence re: Nengubo Mission, 1924.
840/1/37   Correspondence re: Empadeni Mission, 1922.

Department of Health
H 2/9/2   Salisbury Town Clerk's Notes, April 1917.
H 2/1/1–2   Correspondence.
H 6/2/1   Correspondence.

Department of Customs & Excise/Assize Office
S 910/17/14   Department of Customs & Excise, Correspondence, 1929–1948.
S 2094/1/7   Assize Office, Correspondence, 1926–1947.

Department of Internal Affairs
S 246/484   Correspondence re: Standardisation of Soap Act, 1929–1930.

Department of Commerce and Industry; Associated Committees and Commissions
S 915/1–2   Correspondence and Other Material, 1944–1949.
S 104   Industrial Development Advisory Committee, Minutes, 1940–1944.
S 109/1   Development Coordinating Commission, Minutes, 1948–1949.
                    Manufacture of Toilet Preparations: Opportunity for Industry No. 11. Salisbury, 1963.

British South Africa Company Administrator's Office
A 3/12/13–18   Papers on Health and Medicine, 1910–1923.
A 3/3/28–31   Correspondence, 1890–1923.
A 3/18/30/11   Correspondence re: Employment of Girls, 1909–1911.
A/3/3/18   Correspondence.

Ministry of Information
S 932/8/1   Correspondence re: Educational Films, 1943–1949.
S 932/69   Correspondence and Papers re: Broadcasting and the BBC, 1946–1953.
S 2113/1–2   Papers and Drafts re: "Meet the African," 1962.
Pamphlet   *The Man and His Ways,* 1969.

1912–1913 Cost of Living Commission
ZAC 1/1/1   Testimony.

1938 Economic Development Committee
ZAY 2/1/1–4   Testimony.
ZAY 2/2   Testimony.

1943 Howman Commission on Economic, Social and Health Conditions of Africans Employed in Urban Areas
ZBI 1/1/1–2   Testimony.
ZBI 2/1/2   Testimony.

1944 Godlonton Commission on Native Production and Trade
ZBJ 1/1/1–4, oral testimonies.
ZBJ 1/2/1–3, written testimonies.

Miscellaneous Records
RH 12/2/1/1   Joint National Council, Minutes of the Race Relations Subcommittee.
RH 12/2/4/1   Rhodesian Federated Chambers of Commerce, Papers.
GEN-P/MAS   Mashonaland Native Welfare Society, Agenda, May 5, 1930.
                        Southern Rhodesia. *Debates of the Legislative Council.*
                        Southern Rhodesia Native Affairs Commission. *Report of the Native Affairs Commission, 1910–1911.*
                        Mangwende Commission of Inquiry. *Records of the 1961 Mangwende Reserve Commission of Inquiry.*
                        Chambers of Industries. *Annual Report of Chambers of Industries of Rhodesia.*
                        Southern Rhodesia Department of Trade and Industrial Development. *Secondary Industry in Southern Rhodesia.* Salisbury, 1953.
                        Federation of Rhodesia and Nyasaland Central Statistical Office.
                        Federation of Rhodesia and Nyasaland. *Manufacture of Toilet Preparations,* n.d.
                        Southern Rhodesia Central Statistical Office.
                        Rhodesia Central Statistical Office. *Reports on the Urban African Budget Survey, 1957–1970.*
                        Photographic Archives.
                        Pamphlet *A People's Progress,* 1968

Federation of Women's Institutes
FO 5/1/1/35   Federation of Women's Institutes, Correspondence and Papers.
MS 418/1/7   Manuscripts of Barbara Tredgold, 1963 FAWC pamphlet.
Miscellaneous papers.

Pitt Rivers Museum Archives University of Oxford
Crook Papers (1878)

Salisbury Chamber of Commerce
SA 5/1/1–7   Minutes.

Manuscripts
BA 14/2/1   Frederick Hugh Barber, trading lists circa 1870s.
CA 1/1/1   Diary of Algernon Capell.
FR 2/1/1   Reminiscences of Ivan Fry.

Interviews
ORAL/HO 3   Interview, Henry George Howman.
ORAL/FI 5   Interview, Harold Carsdale Finkle.

ORAL/229	Interview, Mary Robinson.
ORAL/CH 4	Interview, Hulbert Patrick Charles.
ORAL/BE 5	Interview, R. G. Bennett.
AOH/58	Interview, Mbangwa Ngomambi.
ORAL/HO 6	Interview, Joseph Harrison Hodgson.

Unilever Archives Blackfriars, London.
Correspondence of W. H. Lever.
Diaries of W. H. Lever.

Interviews

Many of our interviews in Harare and Murewa were conducted with entire households, families, or other large groups. Interviews with individuals are distinguished by the use of a single person's name.

Interviews by Timothy Burke
Biegle, Richard. Harare, May 10, 1991.
Consumer Council of Zimbabwe, May 8, 1991.
Butcher, Cornell, Francis Makosa, and Wellington Chikombero, Harare, November 26, 1990.
Corder, Clive. Johannesburg, May 30, 1991.
Dillon, Roger. Harare, April 3, 1991.
Kadenhe, Douglas. Harare, May 23, 1991.
Kapnias, George. Harare, April 16, 1991.
MacIntosh, Mr. Harare, April 22, 1991.
Mathewman, Maurice. Harare, May 13, 1991.
Wazara, Jack. Harare, May 15, 1991.

Interviews by Timothy Burke and Tuso Tapera
Chikote, Sekai, Charles Chikote, M. Kokora, E. Shamba, Beulah Hodze, and others. "Joberg" section of Mbare Township, Harare, April 24, 1991.
Chitundundu household. Mbare Township, Harare, May 18, 1991.
Chizemba household. Mbare Township, Harare, April 25, 1991.
Doreen (last name unknown). "Nationals" section of Mbare Township, Harare, May 9, 1992.
Kapura household. "Nationals" section of Mbare Township, Harare, May 9, 1991.
Katema, Richard and Innocent. "Joburg" section of Mbare Township, Harare, April 24, 1991.
Manyanga, Alice, and Amelia Zdongo. "Joberg" section of Mbare Township, Harare, April 25, 1991.
Mbiriri, Mai. "Joberg" section of Mbare Township, Harare, April 25, 1991.

Moyo household. "Joberg" section of Mbare Township, Harare, April 24, 1991.
Mukuru household. Mbare Township, Harare, May 18, 1991.
Musa household. "Nationals" section of Mbare Township, Harare, May 9, 1991.
Tashinga squatter camp (3 interviews). Mbare Township, Harare, April 19, 1991.
Charira family. Murewa (2 interviews), May 2, 1992.
Gomo family. Murewa, May 7, 1991.
Gwata, Mbuya. Murewa, May 2, 1991.
Tayerera household. May 7, 1991.
Muzende, Mr. Murewa, May 7, 1991.

Interviews by Tuso Tapera
Mutegani, Priscilla. Kambuzuma Township, Harare, May 9, 1992.
Ochino, Israel, Tauisai Tapera, and others. Kambuzuma Township, Harare, May 3, 1991.
Sakarombe, Sophia. Kambuzuma Township, Harare, May 16, 1991.
Tapera household. Kambuzuma Township, Harare, May 18 and 20, 1991.

INDEX

Timothy Burke is Assistant Professor of History at
Swarthmore College.

Library of Congress Cataloging-in-Publication Data
Burke, Timothy.
Lifebuoy men, lux women : commodification,
consumption, and cleanliness in modern Zimbabwe /
Timothy Burke.
p. cm. — (Body, commodity, text : studies of objectifying
practice)
Includes bibliographical references and index.
ISBN 0-8223-1753-2 (cl : alk. paper). — ISBN 0-8223-1762-1
(pa : alk. paper)
1. Soap trade—Zimbabwe—History.   2. Hygiene
products—Zimbabwe—Marketing—History.   3. Rural
health—Zimbabwe—History.   I. Title.   II. Series: Body,
commodity, text.
HD9999.S73Z553   1996   95-44291 CIP